T0251489

Laboratory Techniques
in Plant Bacteriology

Laboratory Techniques in Plant Bacteriology

Prof. S. G. Borkar

CRC Press
Taylor & Francis Group
Boca Raton London New York

CRC Press is an imprint of the
Taylor & Francis Group, an **informa** business

CRC Press
Taylor & Francis Group
6000 Broken Sound Parkway NW, Suite 300
Boca Raton, FL 33487-2742

First issued in paperback 2021

© 2018 by Taylor & Francis Group, LLC
CRC Press is an imprint of Taylor & Francis Group, an Informa business

No claim to original U.S. Government works

ISBN 13: 978-1-03-209600-1 (pbk)
ISBN 13: 978-1-138-63405-3 (hbk)

This book contains information obtained from authentic and highly regarded sources. Reasonable efforts have been made to publish reliable data and information, but the author and publisher cannot assume responsibility for the validity of all materials or the consequences of their use. The authors and publishers have attempted to trace the copyright holders of all material reproduced in this publication and apologize to copyright holders if permission to publish in this form has not been obtained. If any copyright material has not been acknowledged please write and let us know so we may rectify in any future reprint.

Except as permitted under U.S. Copyright Law, no part of this book may be reprinted, reproduced, transmitted, or utilized in any form by any electronic, mechanical, or other means, now known or hereafter invented, including photocopying, microfilming, and recording, or in any information storage or retrieval system, without written permission from the publishers.

For permission to photocopy or use material electronically from this work, please access www.copyright.com (http://www.copyright .com/) or contact the Copyright Clearance Center, Inc. (CCC), 222 Rosewood Drive, Danvers, MA 01923, 978-750-8400. CCC is a not-for-profit organization that provides licenses and registration for a variety of users. For organizations that have been granted a photocopy license by the CCC, a separate system of payment has been arranged.

Trademark Notice: Product or corporate names may be trademarks or registered trademarks, and are used only for identification and explanation without intent to infringe.

Publisher's Note
The publisher has gone to great lengths to ensure the quality of this reprint but points out that some imperfections in the original copies may be apparent.

Library of Congress Cataloging-in-Publication Data
Names: Borkar, S. G. (Suresh Govindrao), 1956- author.
Title: Laboratory techniques in plant bacteriology / author: Suresh G. Borkar.
Description: Boca Raton : Taylor & Francis, 2017.
Identifiers: LCCN 2017013476 | ISBN 9781138634053 (hardback : alk. paper)
Subjects: LCSH: Bacterial diseases of plants–Laboratory manuals. | Bacteriology, Agricultural–Laboratory manuals.
Classification: LCC SB374 .B67 2017 | DDC 632/.32–dc23
LC record available at https://lccn.loc.gov/2017013476

Visit the Taylor & Francis Web site at
http://www.taylorandfrancis.com

and the CRC Press Web site at
http://www.crcpress.com

Dedicated to my teachers, who inspired and trained me in plant bacteriology:

Dr. J. P. Verma
Plant Bacteriologist
Indian Agriculture Research Institute
New Delhi, India

and

Dr. L. Garden
Plant Bacteriologist
Plant Bacteriology Laboratory
Institute National de la Recherché Agronomique
Beaucouzé, Angers, France

Contents

Preface

Plant pathogenic bacteria are known to cause serious plant diseases and economic losses in several important crop plants. Several epidemics of some of the bacterial disease in Western countries and in Asia are well documented, studied, and to a more or less extent, controlled or eradicated. The number of plant pathologists working on plant bacterial diseases or plant pathogenic bacteria are fewer than those working on plant viruses or fungal pathogens of plants. The reason is that fewer scientists are trained in plant bacteriology in most countries. Similarly, only a few countries have specialized plant bacteriology laboratories. Further, the lack of books on laboratory technique in plant bacteriology, discourages young professionals from working in this field.

If this trend continues, there will be few plant bacteriologists available in most countries to deal with future challenges in this branch of plant pathology. Though plant bacteriology is part of the curriculum of master's and doctorate degrees in plant pathology in several countries, to conduct the practicals of the student, there are few technique books on plant bacteriology available.

It is an implicit duty of a plant bacteriologist, trained in plant bacteriology at IARI, New Delhi, and INRA, Beaucouze, Angers, during my master's, doctoral, and postdoctorate program, to write a book on laboratory techniques in plant bacteriology so that those teaching this subject will take maximum practicals to train students technically and prepare them to deal with the future challenges in bacterial diseases of crop plants.

The book *Laboratory Techniques in Plant Bacteriology* covers all the techniques essential to the study of plant pathogenic bacteria. The book consists of 41 chapters wherein chapter-specific techniques are illustrated and cover all aspects in plant bacteriology to train a person as plant bacteriologist.

Professor S. G. Borkar

Acknowledgments

I sincerely acknowledge the inspiration and help received during my career development as a plant bacteriologist from my teachers and colleagues in India and in France who are internationally well known plant bacteriologists. Professor J. P. Verma, Professor P. N. Patel, Professor R. Ride, Professor L. Garden, Professor Lussetti, Professor Samson, and Professor Barzaic, with whom I had a chance to work, and Professor Ponagopoulos, Professor Klement, Professor Lilliot, Professor Braudberry, and Professor Rudolph, with whom I met in my professional career and was inspired by and from whom I learned several techniques of plant bacteriology.

I am thankful to all my scientific colleagues around the world whose techniques were consulted while writing this book. My thanks are also to my parents, the late Shri Govindrao and Smt. Sumitra, whom I consider a legend to bring me to this level. My wife, Dr. Sandhya, and son, Antriksh, also spared me to take on the Herculean task of writing this book.

All my students, who worked with me on plant bacteriology problems for their master's and doctorate degrees, made me feel the necessity to write such a book for them; my colleagues in the department who teach plant bacteriology practical also wanted this type of book to be made available to them. Therefore, I thank them all for validating the need for such a book. I also acknowledge the help received during the preparation of this manuscript from Dr. V. Chimote, associate professor of biotechnology, throughout the chapter molecular techniques. I am thankful to all my master's and doctoral students of plant bacteriology whose work and photographs are included in the following chapters.

I put on record the help of Kavita Sandip Dhawade in the preparation of this manuscript.

Professor S. G. Borkar

Author

Professor S. G. Borkar has been head of the Department of Plant Pathology and Agricultural Microbiologyat Mahatma Phule Agriculture University since 2005. He graduated from Dr. Punjabrao Deshmukh Krishi Vidyapeeth, Akola, in 1977 and obtained his MSc and PhD from IARI, New Delhi, in 1979 and 1983, respectively. Dr. Borkar completed his postdoctorate at INRA, Angers, France, in 1984 and earned his DSc from the International University, Washington, DC, in 1999. He is a fellow of the Indian Phytopathological Society and Eurasian Academy of Environmental Sciences.

After returning from France, he was appointed assistant professor of plant pathology at Jawaharlal Nehru Krishi Vishwa Vidyalaya, Jabalpur (MP), and served the university from January 1985 to December 1989. He was named associate professor of plant pathology at Mahatma Phule Krishi Vidyapeeth, Rahuri, in December 1989, and in May 1994, he was selected as a professor of plant pathology by the Maharashtra Council of Agriculture Education and Research for Mahatma Phule Krishi Vidyapeeth, Rahuri. From 2012 to 2013, also served as dean of the Post Graduate Institute of Mahatma Phule Krishi Vidyapeeth, Rahuri.

Dr. Borkar has published more than 100 research papers in approximately 25 national and 7 foreign journals. He has guided around 30 students to their MSc and PhD degree research. Most of his students serve as ICAR scientists and in agricultural universities. He has received 13 awards from scientific societies and social organizations within the country and abroad. He has developed 4 wheat varieties, 6 patents, several technologies and recommendations, and several new strains of beneficial microbes. Dr. Borkar has published six books and has chaired several sessions in ICAR workshops and national seminars. He has visited several universities abroad in France, Greece, the United Kingdom, and Nepal. He is on the selection committee of different agricultural universities in the country, and is a well-known teacher and scientist at the national and international levels. A strain of *Klebsiella pneumoniae* that infects plants is named after him (*Klebsiella pneumoniae* strain Borkar) by NCBI. His biographical note is included in Asia/Pacific Who's Who of 1998 and in Twentieth Century Admirable Achievers: Distinguished Who's Who of 1999.

1.1.4 STORAGE CABINETS

1. Store your chemicals, working glassware, such as Petri plates, test tubes, beakers, flasks, and pipettes, separately in an individual cabinet allocated for these. (See Figure 1.1.)
2. Store your sample temporary (not more than 48 hours) in well-sealed packets in the cabinet allocated for this purpose.
3. Store all markers, inoculating needles, forceps, scissors, rubber bands, threads, and other miscellaneous items in one place.
4. Store all papers, such as brown paper, butter paper, water-absorbent paper, and tissue papers in one place.
5. Store all the filter papers in one place.

1.1.5 MAINTENANCE OF LABORATORY HYGIENE

Maintain hygienic condition in the laboratory.

1. Do not keep open the microbial culture plates.
2. Put the waste material in a disposable bag at a proper place; the disposable bag should not be kept open. Autoclave the disposable material before final disposal.
3. Clean the laboratory floor with Lysol or any other detergent daily. Sweep the working tables with proper antimicrobial agents or with methylated spirit. Dusting of the instruments at regular intervals is a must.

1.2 LABORATORY INSTRUMENTS

1.2.1 MAINTENANCE OF INSTRUMENTS

Make annual maintenance contracts for the instruments so as to maintain them by a trained technician. Check the proper working of the instruments before putting to final use for experimental

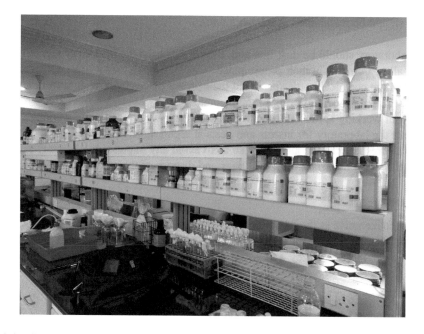

FIGURE 1.1 Chemical storage cabinet in bacteriology laboratory. (Courtesy of Dr. S. G. Borkar, Department of Plant Pathology, Mahatma Phule Krishi Vidyapeeth, Rahuri.)

Author

Professor S. G. Borkar has been head of the Department of Plant Pathology and Agricultural Microbiologyat Mahatma Phule Agriculture University since 2005. He graduated from Dr. Punjabrao Deshmukh Krishi Vidyapeeth, Akola, in 1977 and obtained his MSc and PhD from IARI, New Delhi, in 1979 and 1983, respectively. Dr. Borkar completed his postdoctorate at INRA, Angers, France, in 1984 and earned his DSc from the International University, Washington, DC, in 1999. He is a fellow of the Indian Phytopathological Society and Eurasian Academy of Environmental Sciences.

After returning from France, he was appointed assistant professor of plant pathology at Jawaharlal Nehru Krishi Vishwa Vidyalaya, Jabalpur (MP), and served the university from January 1985 to December 1989. He was named associate professor of plant pathology at Mahatma Phule Krishi Vidyapeeth, Rahuri, in December 1989, and in May 1994, he was selected as a professor of plant pathology by the Maharashtra Council of Agriculture Education and Research for Mahatma Phule Krishi Vidyapeeth, Rahuri. From 2012 to 2013, also served as dean of the Post Graduate Institute of Mahatma Phule Krishi Vidyapeeth, Rahuri.

Dr. Borkar has published more than 100 research papers in approximately 25 national and 7 foreign journals. He has guided around 30 students to their MSc and PhD degree research. Most of his students serve as ICAR scientists and in agricultural universities. He has received 13 awards from scientific societies and social organizations within the country and abroad. He has developed 4 wheat varieties, 6 patents, several technologies and recommendations, and several new strains of beneficial microbes. Dr. Borkar has published six books and has chaired several sessions in ICAR workshops and national seminars. He has visited several universities abroad in France, Greece, the United Kingdom, and Nepal. He is on the selection committee of different agricultural universities in the country, and is a well-known teacher and scientist at the national and international levels. A strain of *Klebsiella pneumoniae* that infects plants is named after him (*Klebsiella pneumoniae* strain Borkar) by NCBI. His biographical note is included in Asia/Pacific Who's Who of 1998 and in Twentieth Century Admirable Achievers: Distinguished Who's Who of 1999.

1 Laboratory Ethics in Plant Bacteriology Laboratory
Instruments and General Guidelines

To carry out scientific experiments, a research laboratory is required. A research laboratory is a well-equipped room with closed cabinets of glassware and chemicals and working platforms with hygienic conditions. The plant bacteriology laboratory necessarily requires aseptic conditions to work with plant pathogenic bacteria. To maintain laboratory standards, the bacteriology laboratory requires certain rules and ethics to follow while working.

1.1 ETHICS IN THE LABORATORY

Put an important notice at the laboratory entrance depicting the following:

1. Sanitize yourself before entering laboratory.
2. No footwear or animals allowed.
3. No food or drink allowed.
4. This is a zone of silence.

1.1.1 CLEANLINESS

Cleanliness should be maintained in the laboratory.

1. The laboratory should be dust-free and, therefore, sweep the laboratory regularly.
2. Keep all instruments in their proper place.
3. Keep all the chemical bottles/packet at their proper places by arranging them alphabetically.
4. Avoid bringing food and drink in the laboratory.
5. Hang out working aprons at their proper places and wear them while working.
6. Sanitize the laboratory for insects and pests at regular intervals.
7. Keep air freshener in the laboratory so as to avoid any unpleasant laboratory smells.
8. Keep the waste material in disposal bags and dispose of them daily.

1.1.2 SILENCE

1. Do not make noise in the laboratory. Strive for silence in the laboratory.
2. Use the laboratory only for experimentation and research, not for general purpose.
3. Avoid discussions, lectures, and gossip in the laboratory.

1.1.3 DRESS CODE

1. Dress properly while working in laboratory.
2. Put on your apron, gloves, mask, and head cover while working in the laboratory.
3. Maintain dress material in separate storage cabinets.

1.1.4 STORAGE CABINETS

1. Store your chemicals, working glassware, such as Petri plates, test tubes, beakers, flasks, and pipettes, separately in an individual cabinet allocated for these. (See Figure 1.1.)
2. Store your sample temporary (not more than 48 hours) in well-sealed packets in the cabinet allocated for this purpose.
3. Store all markers, inoculating needles, forceps, scissors, rubber bands, threads, and other miscellaneous items in one place.
4. Store all papers, such as brown paper, butter paper, water-absorbent paper, and tissue papers in one place.
5. Store all the filter papers in one place.

1.1.5 MAINTENANCE OF LABORATORY HYGIENE

Maintain hygienic condition in the laboratory.

1. Do not keep open the microbial culture plates.
2. Put the waste material in a disposable bag at a proper place; the disposable bag should not be kept open. Autoclave the disposable material before final disposal.
3. Clean the laboratory floor with Lysol or any other detergent daily. Sweep the working tables with proper antimicrobial agents or with methylated spirit. Dusting of the instruments at regular intervals is a must.

1.2 LABORATORY INSTRUMENTS

1.2.1 MAINTENANCE OF INSTRUMENTS

Make annual maintenance contracts for the instruments so as to maintain them by a trained technician. Check the proper working of the instruments before putting to final use for experimental

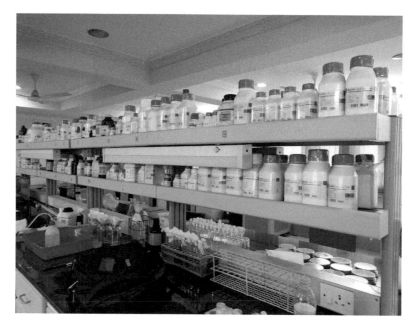

FIGURE 1.1 Chemical storage cabinet in bacteriology laboratory. (Courtesy of Dr. S. G. Borkar, Department of Plant Pathology, Mahatma Phule Krishi Vidyapeeth, Rahuri.)

purposes so as to avoid the loss of important research material. Maintain a log book for each instrument and enter its working hours. Refer to the "directions for use" manual if how to work a particular instrument is not known. Clean and dry the instrument after use and switch off the buttons.

1.2.2   BASIC INSTRUMENTS

1. Laminar air flow cabinet
     Uses: It is the most needed and important instrument in the laboratory. It is required in the working of microbes under aseptic condition. The laminar air flow cabinet is used for isolation of microbes, bacterial plant pathogens, their transfers, preparation of media plates, and for all other purposes where aseptic conditions are required in microbial work. (See Figure 1.2.)
2. BOD incubator
     Uses: BOD incubators are important instruments used for the growth and multiplication of microorganisms. The microbial culture tubes/plates are incubated at required temperatures in BOD incubators to facilitate the growth of microbes. (See Figure 1.3.)
3. Refrigerator
     Uses: The refrigerator is an important instrument in research laboratory to store the microbial cultures, heat-sensitive chemicals, drugs, antigens, antibodies, and so on. These can also be used in exceptional cases to temporarily store unutilized working plates and tubes of solidified media. (See Figure 1.4.)
4. Oven
     Uses: These are basically used for sterilization of glassware at higher temperatures and drying of samples at appropriate temperatures. (See Figure 1.5.)
5. Ultrapure water machine
     Uses: Ultrapure water or distilled sterile water is needed in all microbiological work, in the preparation of cultural media, preparation of chemical solutions, water blanks, and so on. (See Figure 1.6.)

FIGURE 1.2   Laminar air flow cabinet. (Courtesy of Dr. S. G. Borkar, Department of Plant Pathology, Mahatma Phule Krishi Vidyapeeth, Rahuri.)

FIGURE 1.3 BOD incubator. (Courtesy of Dr. S. G. Borkar, Department of Plant Pathology, Mahatma Phule Krishi Vidyapeeth, Rahuri.)

6. Autoclaves
    Uses: To sterilize the working media, water blanks, glassware, and in certain cases, soil material, disposable microbial material, and all other materials that are not degraded/spoiled during steam sterilizations. (See Figure 1.7.)
7. Compound microscopes
    Uses: To identify microbes, their cell structures, and sizes, and to view the internal cellular material. (See Figure 1.8.)

1.2.3 APPLIED INSTRUMENTS

1. Spectrophotometer, UV-spectrophotometer, or spectronic
    Uses: To measure the quantity of various biochemical materials of plants and microbes based on the absorbance at particular wavelength. (See Figure 1.9.)

FIGURE 1.4   Fridge for storage of bacterial cultures. (Courtesy of Dr. S. G. Borkar, Department of Plant Pathology, Mahatma Phule Krishi Vidyapeeth, Rahuri.)

2. Ultracentrifuge

   Uses: To separate the biological component of bacterial cells, separation of plant tissues, and so on. (See Figure 1.10.)

3. Gas chromatography mass spectroscopy

   Uses: To identify the biochemical material and its organic form to name it based on the search library.

4. Electrophoresis apparatus

   Uses: To separate macromolecules like DNA, RNA, and proteins. DNA fragments are separated according to their size. (See Figure 1.11.)

5. Eliza unit

   Uses: This is a diagonostic tool to detect pathogens in disease samples by using solid phase enzyme immunoassay. (See Figure 1.12.)

FIGURE 1.5   Hot air oven for sterilization. (Courtesy of Dr. S. G. Borkar, Department of Plant Pathology, Mahatma Phule Krishi Vidyapeeth, Rahuri.)

FIGURE 1.6   Ultrapure water machine. (Courtesy of Dr. S. G. Borkar, Department of Plant Pathology, Mahatma Phule Krishi Vidyapeeth, Rahuri.)

FIGURE 1.7 Autoclave for sterilization. (Courtesy of Dr. S. G. Borkar, Department of Plant Pathology, Mahatma Phule Krishi Vidyapeeth, Rahuri.)

6. PCR machine

Uses: The thermal cycler (also known as thermocycler, PCR machine, or DNA amplifier) is most commonly used to amplify segments of DNA via the polymerase chain reaction. (See Figure 1.13.)

7. UV transilluminator

Uses: To identify the fluorescence illuminated by bacterial cultures. To identify the *Pseudomonas* fluorescence bacterial cultures. (See Figure 1.14.)

8. Lyophilizer

Uses: To obtain the bacterial culture in lyophilized (dehydrated powder form) state for long-term preservation and storage. (See Figure 1.15.)

9. Plant growth chambers

Uses: To incubate the seedling for its growth; to incubate the pathogen inoculated seedling or plants at appropriate temperature, humidity, and light intensities for symptoms development.

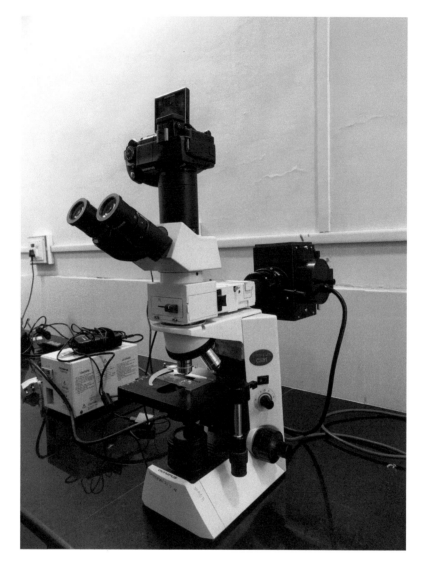

FIGURE 1.8   Compound microscope. (Courtesy of Dr. S. G. Borkar, Department of Plant Pathology, Mahatma Phule Krishi Vidyapeeth, Rahuri.)

1.3   NECESSARY MISCELLANEOUS INSTRUMENTS

1. Hot plate cum magnetic stirrer
    Uses: To heat the liquid material at particular temperatures while stirring.
2. Vortex stirrer
    Uses: To mix the medium/broth suspension in a test tube.
3. Electric mixture
    Uses: To obtain a liquid paste of plant samples or mix the experimental material.
4. Hot water bath
    Uses: To heat the liquid material at particular temperature or to evaporate the biological material.

FIGURE 1.9   Spectrophotometer unit. (Courtesy of Dr. S. G. Borkar, Department of Plant Pathology, Mahatma Phule Krishi Vidyapeeth, Rahuri.)

  5. Vacuum evaporator
       Uses: To evaporate the biological material to obtain it in residual form or powder form.
  6. pH meter
       Uses: To measure and adjust the pH of mediums, buffers, and biological materials.
  7. Shaker
       Uses: To agitate the inoculated mediums for microbial growth.
  8. Pipette washer
       Uses: To suspend the pipette in cleaning solutions before washing.
  9. G-type filters
       Uses: To filter the solutions to make them fungal- and bacteria-free, to sterilize amino acids and sugar solutions, and to obtain bacterial viruses (bacteriophages).

1.4   GENERAL GUIDELINES

  1. Corrosive chemicals and solvents should be stored in separate places. The storage place should be moisture-free and dry. A solvent that emits fumes should be worked with in a solvent cabinet/fume hood.
  2. Close all water taps when not in use. Check all electrical buttons and shut them off when leaving the laboratory. Keep the electrical buttons of all instruments except BOD incubators and the freezer off while leaving your laboratory.
  3. Clean the laboratory and working tables before leaving.
  4. Keep a first aid kit in the laboratory and use in case of minor accidents.
  5. The laboratory should have a fire extinguisher to be used in case of minor fire emergencies.
  6. Depict telephone numbers in a prominent place of the laboratory for the following services: in charge of laboratory, medical doctor, fire emergency, and police.

FIGURE 1.10 High speed centrifuge. (Courtesy of Dr. S. G. Borkar, Department of Plant Pathology, Mahatma Phule Krishi Vidyapeeth, Rahuri.)

7. Any damage to instruments, solvent pills, and breakages should be brought to the notice of the laboratory in-charge.
8. Do not overrun the autoclave and ovens over prescribed times. Handle them as per the prescriptions and directions for use.
9. Prepare and keep a "directions for use" booklet in the laboratory where directions for use of all the available instruments in your laboratory are summarized.
10. Do not smoke or chew tobacco or pan masala in the laboratory.
11. Use only ultrapure water or distilled sterile water in microbiological work.
12. Record your observations in the research observation notebook.
13. Plan your experiment well in advance to have time for the procurement of materials required in the experiment.

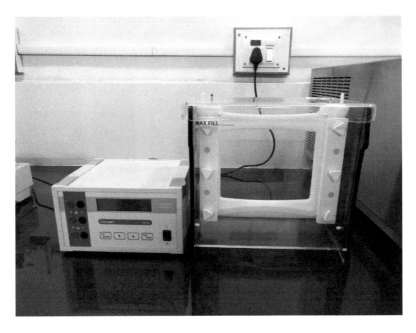

FIGURE 1.11   Gel electrophoresis unit. (Courtesy of Dr. S. G. Borkar, Department of Plant Pathology, Mahatma Phule Krishi Vidyapeeth, Rahuri.)

FIGURE 1.12   ELIZA unit. (Courtesy of Dr. S. G. Borkar, Department of Plant Pathology, MPKV Mahatma Phule Krishi Vidyapeeth, Rahuri.)

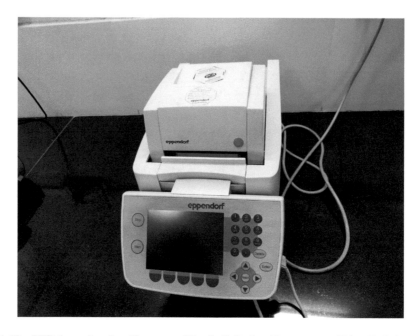

FIGURE 1.13   PCR thermal cycler. (Courtesy of Dr. S. G. Borkar, Department of Plant Pathology, Mahatma Phule Krishi Vidyapeeth, Rahuri.)

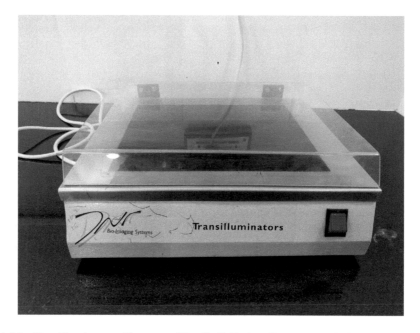

FIGURE 1.14   Transilluminators. (Courtesy of Dr. S. G. Borkar, Department of Plant Pathology, Mahatma Phule Krishi Vidyapeeth, Rahuri.)

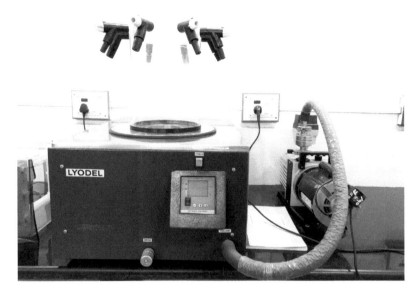

FIGURE 1.15   Lyophilizer. (Courtesy of Dr. S. G. Borkar, Department of Plant Pathology, Mahatma Phule Krishi Vidyapeeth, Rahuri.)

14. Prepare the medium and slants in required quantities to avoid the waste and drying of these materials. Use the chemicals judiciously and in required quantities only.
15. Keep the lids of chemical bottles and solvent bottles tightly closed.
16. Keep soap, liquid soap, and towels or tissue paper near the wash basin.
17. Wash your glassware and used Petri plates at separate places and not in the research laboratory.
18. Make separate arrangements for the washing room and sterilization room; these should be interconnected to the research laboratory.
19. Do not jump to conclusions unless you get the same results repeatedly.

2 Confirmation of Bacterial Pathogen in Diseased Samples

Before proceeding to isolate the bacterium from a diseased plant, it is desirable to confirm the presence of the bacterium in the affected tissues. This can be done through the following tests.

2.1 BY OOZE TEST

This test is more appropriate to detect the localized bacterial infection of leaves, stem, buds, flower, and fruit.

Material Required

Disease sample, glass slide, cover slip, distilled sterile water, microscope, razor blade, tissue paper, and so on.

Procedure

Wash an infected diseased portion in running tap water to remove the dirt and external microbes. Dry the sample with blotter paper/tissue paper. Cut a piece of the infected tissue portion with the razor blade on a sterilized glass slide. Put a few drops of water on the cut portion and put on the cover slip. Keep the slide at room temperature for 5 minutes to allow the bacterial ooze to come out. Examine under low power objective.

Observation

If the infection is due to bacteria, a cloudy mass of bacterial cells will be seen oozing through the cut ends of the plant tissue pieces.

Important Guidelines

Some idea is also obtained as to whether the infection is parenchymatous or vascular. In vascular infections, bacteria oozes out forcefully at distinct points corresponding to the vascular strands; in parenchymatous infections, the oozing of the bacteria is slow, diffused, and throughout the cut ends.

This test works well in a majority of the bacterial diseases. Difficulty may be encountered in diseases like crown gall, hairy root, and leafy gall, where all the affected tissue may not contain the bacterial cells. In tissues that contain large quantities of starch, dispersal of starch grains in water may mask the bacterial ooze and make it difficult to recognize. However, with some experience it is possible to distinguish bacterial ooze from starch grains. (See Figure 2.1.)

2.2 BY MILKY WATER TEST

This test is more appropriate to detect the systemic and vascular infections in leaves, stem, and roots. This test is generally performed for detection of bacterial blight of rice bacterium.

Material Required

Disease sample, distilled sterile water, test tube, razor blade, tissue paper, and so on.

FIGURE 2.1 Bacteria oozing from the cut end of the sample. (Courtesy of Drs. S. G. Borkar and K. S. Birajdar, Department of Plant Pathology, MPKV, Rahuri.)

Procedure

Take an infected disease sample and wash in running tap water to remove the dirt and external microbes. Dry the sample with blotter paper/tissue paper. Take a sterilized test tube with 10 ml of distilled water in it. Cut a piece of infected leaf, stem, or root sample. Immerse a cut portion of the infected sample in test tube water and allow it to stand for 3 to 5 minutes.

Observation

Observe a strand of bacteria vigorously coming out of the infected cut portion, which turns the clear water to a whitish turbid or milky appearance.

2.3 BY STRING TEST

The string test is used to confirm the bacterial pathogen in disease sample of vascular wilt caused by bacterial pathogens. The test is particularly used to detect *Pseudomonas*/*Ralstonia* wilt pathogens in the roots of eggplant, tomato, tobacco, potato, and so on.

Material Required

Wilted plant along with root, distilled sterile water, beaker, razor blade, tissue paper, and so on.

Procedure

Collect the wilt diseased plant. Separate the root from the crown portion of the plant. Wash the root in tap water to remove the soil particles and dry it in blotter paper. Take distilled sterile water in a small beaker or test tube. Cut a portion of root at 45° angle and hang the cut end in water with the help of a thread and leave for 10 to 15 minutes. A string of bacterial ooze will be observed with the naked eye in the water. The string ooze confirms the involvement of bacteria as the disease causal agent. (See Figure 2.2.)

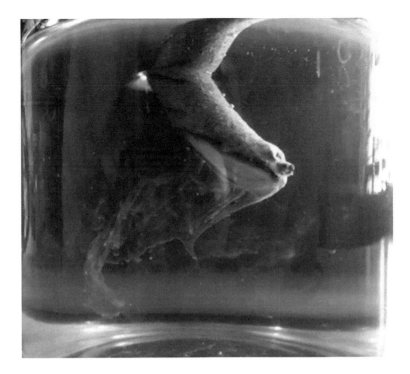

FIGURE 2.2 Oozing of bacteria in the form of string in water. (Courtesy of R. A. Yumlembam and S. G. Borkar, Department of Plant Pathology, Mahatma Phule Krishi Vidyapeeth, Rahuri.)

2.4 BY STAINING OF CRUSHED DISEASE PORTION

At times when the bacterial population is very low in the affected tissue, as in the case of young lesions of red stripe of sugarcane and *Xanthomonas* leaf spot of mung beans, distinct bacterial ooze may not be detected.

In such cases, staining of smears from diseased tissue may reveal the presence of bacterial cells.

Material Required

Infected leaf sample, glass slide, distilled sterile water, razor blade, crystal violet stain or carbol fuchsin stain, Bunsen burner for flame, tissue papers, microscope, cedar oil, and so on.

Procedure

Take an infected disease sample and wash in running tap water to remove dirt and external microbes. Dry the sample with blotter paper/tissue paper. Take a piece of the diseased tissue and crush it in a drop of water with the razor blade on one end of the glass slide. Allow it to stand for 1 minute and then tilt the slide so that the water suspension flows toward the other end of the slide, leaving most of the crushed host tissue behind. The smear so prepared is dried and fixed by passing over a flame. Flood the smear with crystal violet solution or carbol fuchsin and allow to stand for 1 minute. Drain the stain and wash the slide thoroughly in running tap water. Blot dry and examine under oil immersion objective for presence of bacterial cells.

Observation

If presence of the bacterium is not detected by the ooze test or by staining, there is no use trying isolation from such material.

3 Histopathology of Bacterial Infection

The presence of bacterial pathogens in plant host tissues and the change they cause in these plant tissues are studied by histopathology of infected plant material. The location of the bacterial pathogen as intercellular or intracellular or in water conducting tissues can be revealed with microtome sectioning and staining of the plant cellular material.

3.1 DETERMINATION OF THE LOCATION OF BACTERIAL PATHOGENS IN HOST

To determine the exact location of bacteria in diseased tissue, it is desirable that the material is fixed in a suitable killing and fixing solution. In sections of fresh material, considerable bacterial population may be lost when sections are transferred to water from the razor.

The most desirable killing and fixing solutions are formalin–acetic acid–alcohol (FAA) and the FAA saturated with mercuric chloride. The advantage of the latter is that it helps to bind the bacterial cells together and thus keeps the bacterial mass intact without leaching or dispersal during sectioning and staining. Even the bacterial exudates on the surface of the tissue are preserved in this fixing solution.

Composition of Fixing Solution (FAA)		
Alcohol (50 percent)	....	100 ml
Formalin	....	6.5 ml
Acetic acid	....	2.5 ml

Material Required

Disease sample, microtome, razor blade, fixing solution, distilled sterile water, glass slide, cover slip, tissue paper, differential stains, alcohol, xylol, balsam, microscope, and so on.

Procedure

1. Fixing the material: Wash the diseased sample in running tap water to remove dirt and external microbes. Dry with blotter paper/tissue paper. Cut the material into 5 mm pieces with microtome and transfer in FAA. The minimum time required for killing and fixing is 48 hours, but the material can be kept for any length of time. In FAA saturated with mercuric chloride, the tissue pieces are kept for 48 hours, washed in FAA to remove excess mercuric chloride, and then preserved in FAA.
2. Staining of free hand sections: The sections can be cut with a sharp razor or new shaving blade, and stained by the following methods.
   a. Staining with acid fuchsin: Mount the sections in 0.1 percent acid fuchsin in lacto phenol; allow to stand for 5 to 10 minutes and examine.

   Observation

   The bacteria are stained deep red and the host cell walls pink. In the affected tissues of coleoptiles, leaf sheet, stem, and so on, which contain little or no chlorophyll, the bacterial cells can be revealed prominently by this method.

b. Staining with differential stain: Stain the section with carbol thionin for 5 minutes (Thionin blue 0.1 gm, phenol crystals 5.0 gm, water 100 ml). Wash in water and transfer to 95 percent alcohol for a minute. Put in orange G for a few minutes (orange G saturated in absolute alcohol). Wash in absolute alcohol. Clear in xylol and mount in balsam.

Observation

The bacteria stain deep blue, the cellulose cell walls yellow or green, and the liquefied tissue light blue.

3.2 PREPARATION OF MICROTOME SECTIONS AND STAINING FOR HISTOLOGICAL STUDIES

Material Required

Disease sample, fixing reagent, dehydrating reagent, blotter paper, microscope, and so on.

Procedure

Wash the disease sample in running tap water to remove the dirt and dry with blotter paper/tissue paper.

1. Killing and fixing of plant material: Procedure is done in FAA or FAA-mercuric chloride as described in the preceding section.
2. Dehydration: Plant material affected by bacteria can best be dehydrated in tertiary butyl alcohol series (Jensen, 1962) as given below.

	Composition			
Sr. No.	Distilled Water (ml)	Ethyl Alcohol (ml)	Tertiary Butyl Alcohol (ml)	Plant Material to Be Kept for
1	50	40	10	2 hr
2	30	50	20	Overnight
3	15	50	35	1 hr
4	0	45	55	1 hr
5	0	75	25	1 hr
6	0	0	100	1 hr
7	0	0	100	1 hr
8	0	0	100	Overnight

3. Infiltration and embedding: Fill a vial three-quarters full of melted paraffin and allow to solidify but not cool completely and transfer the material from tertiary butyl alcohol liquid paraffin mixture on to the solidified paraffin in the vial. Cover the material with fresh tertiary butyl alcohol liquid paraffin mixture and place the vial in paraffin over for at least 1 hour or until the material has sunk to the bottom of the vial. Pour off the entire mixture leaving only the plant material in the vial and replace it with pure melted paraffin (m.p. 52°C). Give two more changes of paraffin at 2-hour intervals and allow the material to stay overnight embedded in wax.
4. Softening of paraffin-embedded tissue: Some of the plant material, such as stem, seed, and so on, becomes brittle during dehydration and it becomes difficult to obtain intact sections. Such material can be softened in the 1 percent aqueous solution of sodium lauryl sulfate (90 ml) in glycerol (10 ml; Alcorn and Ark, 1953). Trim the paraffin block so that the surface of the embedded material to be sectioned is exposed. Invert the block in a vial containing just enough

softening solution to cover the exposed surface. The time required for the softening is 24 to 28 hours, depending on material.

5. Sectioning: Section the embedded material at 10–12 μ thickness with a rotary hand microtome. The paraffin ribbons containing sections are fixed on slides with Haupt's Adhesive (Johansen, 1940).

    Composition of Haupt's Adhesive: Dissolve 1 g of fine quality gelatin in 100 ml of distilled water at 30°C and add 2 g of phenol crystals and 15 ml of glycerin. Mix and filter. Apply a very thin but uniform film of the adhesive to the slide and flood it immediately with 4 percent aqueous solution of formalin. Cut the ribbon into stripes containing desired number of sections and float over the formalin solution. Heat the slide gently so that the ribbon becomes straight and flat. Cool the slide and drain the formalin solution holding the ribbon in position with a camel hair brush and keep it at 30.0°C for thorough drying.

6. Staining: The Harris hematoxylin and orange G staining schedule are excellent to localize the bacterial cells in affected tissue. The dried slides are passed through following series (Johansen, 1940; Dickey, 1967).

Xylene	....	5 min
Xylene-absolute alcohol	....	5 min
Absolute alcohol	....	5 min
95 percent alcohol	....	5 min
70 percent alcohol	...	5 min
50 percent alcohol	....	5 min
Harris hematoxylin	....	5 min
Distilled water	....	Rinse to remove excess stain
Tap water to which a few drops of ammonia is added	...	5 min
Tap water	...	5 min
50 percent alcohol	....	5 min
70 percent alcohol	....	5 min
95 percent alcohol	....	5 min
Orange G solution	....	3 min
Wash off the excess stain with equal amounts of clove oil, absolute alcohol, and xylene. Agitate with equal amounts of clove oil, absolute alcohol, and xylene		15 sec
Xylene plus a few drops of absolute alcohol		5 min
Xylene		5 min
Xylene		5 min

Mount in Canada balsam

Observation

The bacterial cells stain deep blue and the host cell walls grayish to orange.

Preparation of Harris Hematoxylin		
Hematoxylin	....	5 g
Aluminum ammonium sulfate	....	3 g
50 percent ethyl alcohol	....	1000 ml
Mercuric oxide-red	....	6 gm

Dissolve the dye and the alum in the 50 percent alcohol with the aid of heat. Add the mercuric oxide and boil for 30 minutes. Filter and make up the volume to 1000 ml with 50 percent alcohol. Acidify in the proportion of 1 drop hydrochloric acid to each 100 ml, before use.

Preparation of Orange G Solution		
Orange G	....	1 g
Methyl cello solve	....	200 ml
95 percent alcohol	....	100 ml

Dissolve the dye in methyl cello solve and add the alcohol.

3.3 STUDY OF CHANGES IN CHEMICAL CONSTITUENTS OF THE AFFECTED PLANT TISSUES

To study the changes in chemical constituents of affected plant tissues, sections from fresh unfixed materials are preferred. Each section should contain about half infected and half healthy tissues.

3.3.1 DEGRADATION IN CELLULOSE

The following two methods are employed to determine degradation of cellulose by the pathogen (Rawlins and Takahashi, 1952).

Material Required

Tissue section, reagent, microscope, and so on.

1. Double refraction method: Mount a section in water and examine in polarizing microscope. Rotate the stage until the cell walls in healthy tissue appear brightest. Now examine the infected tissue. If the cellulose is degraded, the host cell walls around the bacterial mass would not appear bright.
2. Reaction of cellulose with zinc-chlor-iodine: Mount the section in several drops of zinc-chloriodine and observe under microscope. After a few minutes the cellulose-containing walls in parenchyma and meristematic tissue become blue. In case the bacterium degrades cellulose, the cell walls around the bacterial mass do not show blue color. (To prepare zinc-chlor-iodine, dissolve 50 g of zinc chloride and 16 g of KI in 17 ml of water. Add an excess of iodine and allow to stand for several days. The supernatant liquid is stored in a brown bottle.)

3.3.2 DEGRADATION OF PECTIC SUBSTANCES

Material Required

Microtome section, FFA, alcohol, Ferric chloride, potassium ferrocyanide, distilled water, wax, xylene, microscope, and so on.

Procedure

To stain pectic substances, microtome sections of FAA-fixed material can be used. Dissolve the wax in xylene and bring the slide down to water through 95, 70, and 50 percent alcohol. Place the slide in a 10 percent solution of $FeCl_2$ for 10 to 20 minutes and then wash in 4–5 changes of distilled water for 15 minutes. Add 1–2 drops of 2 percent potassium ferrocyanide solution on to the sections. After about 2 minutes, add a drop of 2 percent hot ferrocyanide solution and observe under microscope.

Observation

The cell walls with intact pectic material will give a Persian blue color, while those where pectin has been degraded by the bacterium remain unstained or stain only lightly.

3.3.3 DEGRADATION OF PROTEIN

Material Required

Tissue sections, distilled water, ninhydrin, microscope, and so on.

Procedure

Use sections of alcohol fixed or fresh material. Wash the sections in water and mount in a 0.25 percent solution of ninhydrin. Head the slide for 1 to 5 minutes until a blue color results in the healthy tissue. Examine the infected tissue sample in microscope.

Observation

If protein is degraded, the intensity of the blue color will be more in infected tissue than in healthy ones.

3.3.4 DEGRADATION OF LIGNIN

Material Required

Microtome section, phloroglucinol, microscope, and so on.

Procedure

Mount the sections in a few drops of saturated phloroglucinol in 18 percent HCR and observe under microscope.

Observation

The lignified cell walls become red within a few minutes. If lignin is degraded by the pathogen, cell wall will appear colorless.

The phloroglucinol test for lignin also holds good for wound gums and gives a red color when present mainly in cell walls of vascular tissue and sclerenchyma. The wound gums accumulate in the intercellular spaces and lumina of parenchyma and within the vessels adjacent to the bacterial mass.

4 Sterilization

Sterilization is a process by which the biological or nonbiological material is made free from living micro-organisms. It is a complete destruction of all living organisms by physical or chemical agents.

The material needed in a research laboratory, particularly glassware and other accessories such as forceps, inoculating needles, filter papers, and so on, need sterilization to make them microbe-free.

The common methods of sterilization includes dry heat sterilization, steam under pressure sterilization, filter sterilization, and UV radiation.

4.1 STERILIZATION OF GLASSWARE

Sterilization of glassware is achieved by dry heat sterilization. Hot air sterilization is done in a hot air sterilizer or hot air oven. The temperature of the sterilizer is maintained at 160–180°C for 60 to 90 minutes. If the temperature goes above 180°C, there will be charring of cotton plugs.

Procedure

Clean all the glassware perfectly. They should not have any scent or odor. Drain the water from these glassware by keeping them in a slanted position to dry them. The Petri dishes with cover, flasks, pipettes, beakers, and other glassware to be sterilized are kept in the hot air oven or sterilizer. Run the oven and increase the temperature slowly to 170°C. When the temperature of 170°C is reached, maintain the temperature for 1 hour and then put off the instrument. Open the oven or sterilizer when the temperature comes to ambient level (at room temperature) to remove the sterilized material.

Note: Under no condition should rubber goods, laboratory aprons, and culture media be sterilized in a hot air sterilizer. These will burn and catch fire in the instrument. Increase the temperature of the hot air sterilizer slowly till it reaches the desired temperature. Flasks and tubes containing water and solutions should not be kept for sterilization in the hot air oven. Do not open the hot air oven intermittently or when it is hot. In case of a material burn, turn off the instrument immediately.

4.2 STERILIZATION OF MEDIUM

For most types of media, cloth, rubber, and other material that would be destroyed by dry heat, steam under pressure sterilization is used. Such material is autoclaved at 121°C for 30 minutes using steam under 15-pound pressure in an autoclave.

4.2.1 BY AUTOCLAVE

Media, water blank, empty tubes with cotton plugs, and pipettes with cotton plugs (wrapped in butter papers) are sterilized in steam under pressure.

Certain sugars are destroyed in steam under pressure and therefore should not be sterilized in an autoclave under pressure.

Procedure

Prepare media and any other materials to be sterilized. Keep the material in the autoclave, close the lid, and tighten the screws. Start the autoclave. Keep the steam release cap open to release the steam. When the steam is formed in the autoclave and released through the steam release cap, close the steam release knob/cap so as to get the steam pressure. When the pressure reaches 15 pounds, adjust the pressure cap to maintain the pressure for 30 minutes. After this, turn off the autoclave button so as to bring the autoclave to normal pressure. When the pressure reaches zero, open the lid and remove the sterilized material.

Precautions

Check the water level before running the autoclave. Maintain the proper water level in the autoclave. Do not close the steam release cap in the beginning. Let some steam get released and then close the steam cap so as to avoid the escape of cotton plugs from the media flasks. Fill the media up to three-fourths capacity of the flask. Filling more than this capacity will touch your cotton plugs during sterilization and spoil the media. Do not open the autoclave unless the pressure reaches zero after sterilization. After opening the lid, do not put your hand in immediately to remove the sterilized material. Let the remaining steam come out.

4.2.2   BY FILTRATION

Materials such as certain sugars are destroyed by heating at temperatures normally used for sterilization. To sterilize such heat-labile materials that are liquid or substances dissolved in solutions, filtration is used.

The filters during filtration remove bacteria via the mechanical sieve-like action of the minute pores of the filter and via the adsorption of the microbes in the filter because of the differences in their electric charges.

A filter widely used is a membrane filter. The membrane filter is a cellulose or plastic membrane with a pore size sufficiently small (usually 0.45 μm) to trap and thereby remove bacteria from a liquid. Other filters used in sterilization include sintered glass filters.

Procedure

Sterilize the filter attached to the flask in the autoclave under steam pressure. Put the solutions to be sterilized in the filter cup. Attach the filtration unit to vacuum under pressure to start filtration. Complete the filtration and remove the filtered material in sterilized bottles or test tubes as required.

Precaution

Do not forget to put the cotton plug in the beak of the flask attached to the filter cup. This will reduce the risk of leakage of oil from the vacuum pump tube into the filter flask. This is also necessary to keep the filter flask sterilized.

When a few drops are left to be filtered, remove the vacuum tube attached to the beak of the filter flask in such a way that the cotton plug of the flask beak is retained in the beak itself. This will avoid the entry of outside air into the flask and keep the sterile condition in the flask.

4.3   STERILIZATION OF SOIL

4.3.1   STERILIZATION IN STEAM WITHOUT PRESSURE

For the microbes which are not destroyed by dry heat or moist heat at high temperatures, intermittent sterilization known as Tyndallization is used. The material is heated for 30 minutes in flowing steam (100°C) on each of three consecutive days, allowing it to remain at laboratory temperature between heating.

This method is generally followed to sterilize soil material. This technique of sterilization facilitates the heat-resistant spores of bacteria to be killed due to successive steaming.

Procedure

Keep the soil to be sterilized in an appropriate size tin box without a cover. Keep the soil contained in the tin box in the autoclave and steam it for 30 minutes. Place the soil in the autoclave and steam it again the next day. Follow the process for three consecutive days.

4.3.2 STERILIZATION WITH CHEMICAL VAPORS

Ethylene oxide and other gaseous vapors are important sterilizing agents. Soils of vegetable in raised bed nurseries are generally sterilized by this method.

Procedure

Loosen the soil of the raised bed and add ethylene into it. Mix the soil with ethylene and cover with a plastic sheet for 6 hours so that the ethylene vapors are released and kill the germs in the soil. Remove the plastic sheet and allow the fumes to release from the soil so that traces of the ethylene are not present in the soil. Use such soils for sowing in the nursery beds.

Precautions

Wear a mask, gloves, and goggles while mixing the ethylene in to the soil. Do not inhale the ethylene fumes or vapors.

4.3.3 STERILIZATION WITH SOLAR HEAT

This is also used to sterilize soil or nursery raised bed soils under solar heat.

Procedure

Loose the soil of the raised bed. Cover the soil or raised beds with black-colored polythene sheets and expose to sunlight or hot temperatures of 40°C or more. The process of heat treatment should be continued for at least five days or more to attain proper sterilization.

Note: Light irrigation of the soil during sterilization process enhances the degree of sterilization due to formation of hot air steam beneath the polythene sheet.

4.4 STERILIZATION OF THE WORKING PLACE

4.4.1 BY UV RADIATION

The working platform and the environment or space in the working platform is sterilized by UV radiation. In laminar flow benches the UV tubes are fitted for this purpose.

Procedure

Clean the working platform with methylated alcohol. Put on the UV light and run it for 30 minutes to kill the germs on the working platform and inside environment. Put on the laminar air flow to circulate the sterile air in the cabinet and close the UV radiation tube. Again clean the working platform with methylated alcohol before starting the actual work under laminar air flow.

Precaution

Do not work directly under UV radiation. Turn off the UV before starting your work. Do not look into the UV tube with the naked eye; this may irritate and damage your eyes.

5 Medium for Isolation of Plant Pathogenic Bacteria

Plant pathogenic bacteria, like all other living organisms, require basic nutrients, for the sustenance of life. The food material on which bacteria are grown in the laboratory is known as a *culture medium* (plural: media) and the growth itself is called a *culture*. In other words, the nutrient preparation on or in which a culture (i.e., a population of bacteria) is grown in the laboratory is called a culture medium. Although the bacterial pathogens have the same basic requirements, there is a diversity as to the use for organic and inorganic compounds. Thus culture media vary in form and composition, depending on the species to be cultivated. Some media contain solutions of inorganic salts and may be supplemented with the organic compounds while other media contain complex ingredients such as extracts or digests of plant and animal tissues. These gradients, except for the agar, are used to prepare broth or liquid media. Agar, which liquefies on heating to 96°C and hardens into a jelly at 40–45°C, is used to solidify liquid media. Media have been classified variously using different criteria, viz., chemical composition, physical state, and utility purpose.

On the basis of their composition, there are three main types of culture media:

1. Natural or empirical culture media
2. Semi-synthetic media
3. Synthetic or chemically defined culture media

The exact chemical composition of a natural medium is not known. Those media whose chemical composition is partially known are called semi-synthetic media. A medium that contains agar becomes a semi-synthetic medium, such as dextrose agar, Czapek–Dox agar, oatmeal agar, and beef peptone agar.

Synthetic or a chemical-defined culture medium is composed of special substances of known composition. The synthetic medium may be a general-purpose medium used for a wide variety of microorganisms; a selective medium used for a selected microbe; a differential medium used for differential isolation of a microbe; or an assay medium used for the assay of vitamins, amino acids, and antibiotics. Thus, several media are available and each formulation presumably offers some advantage for the isolation, maintenance, characterization, or growth of certain groups of organisms.

5.1 BASIC LIQUID MEDIA (BROTH) FOR THE ROUTINE CULTIVATION OF BACTERIA

Bacteria, in contrast to fungi, are often cultured in a liquid broth (i.e., media lacking agar). The most common constituents of basic media used in a routine bacteriological laboratory are beef extract (a beef derivative which is a source of organic carbon, nitrogen, vitamins, and inorganic salts) and peptone (a semi-digested protein). These may be modified in a variety of ways by supplementing with some specific chemicals or materials to provide a medium suitable for the cultivation or demonstration of a reaction for specific types or groups of bacteria.

Nutrient broth and glucose broth have been considered as basic liquid media for cultivation of bacteria.

Constituents for Nutrient Broth (pH 7.0)	
Peptone	5.0 g
Beef extract	3.0 g
Distilled water	1000.0 ml

Constituents for Glucose Broth (pH 7.3)	
Peptone	10.0 g
Glucose	5.0 g
Sodium chloride	5.0 g
Distilled water	1000.0 ml

Material Required

Ingredient of the respective medium, 1 N HCl, 1 N NaOH, pH meter, distilled water, hot plate/heater, autoclave, beaker (1 L), measuring cylinder, cotton, culture tubes, glass rod.

Procedure

Preparation of Nutrient Broth

Dissolve the given quantity of peptone and beef extract in 1 L of distilled water for preparation of nutrient broth and boil on the gas burner till all the ingredients are dissolved and mixed well. Adjust the pH to 7.0 and sterilize in an autoclave at 15 pounds of pressure for 30 minutes.

Preparation of Glucose Broth

Dissolve the given quantity of peptone, glucose, and sodium chloride in 1 L of distilled water for preparation of glucose broth and boil on the gas burner till all the ingredients are dissolved and mixed well. Adjust the pH to 7.3 and sterilize in autoclave at 15 pounds pressure for 30 minutes.

5.2   GENERAL MEDIUM FOR ROUTINE CULTIVATION OF BACTERIA

5.2.1   PREPARATION OF NUTRIENT SUCROSE AGAR (NSA) MEDIA

Material Required

Peptone (5 g), beef extract (3 g), sucrose (20 g), agar (20 g), one-liter beakers (2), 250-ml Erlenmeyer flasks (5), muslin cloth/cheese cloth, deionized water, measuring cylinder (1 L), non-absorbent cotton, heating arrangement.

Procedure

Take 500 ml of water in a one-liter sauce pan. Add 5 g of peptone and 3 g of beef extract to the water in the sauce pan and boil till they dissolve. Filter the nutrient extract through cheese cloth. Add 20 g of sucrose to the above nutrient extract. Take 500 ml of water in another beaker and heat it. Add 20 g agar bit by bit to the hot water (96°C) to dissolve it. Mix the agar with the nutrient extract. Bring the volume up to 1000 ml with the addition of distilled water. Dispense 200 ml each to five conical flasks or dispense 8 ml each in screw-capped tubes/test tubes when slants or deep tubes are to be prepared. Plug the flasks and test tubes containing the medium. Sterilize at 121°C and 15 pounds of pressure for 15 minutes in an autoclave. Allow the tubes to cool in slanting position (for agar slants) and upright position (for agar deep tubes). Allow the flasks to cool until the flasks can be held by hand.

Pour the medium into Petri dishes quickly under aseptic conditions. Allow the medium to gel to produce agar plates. The sterilized medium (in flasks or tubes) can be stored at room temperature (25°C) in a dust-free environment (if to be used within a week) or in a refrigerator (if to be stored longer).

Precautions

Do not fill the medium in excess of two-thirds of the capacity of the flasks/container used for autoclaving.

Cotton plugs should be kept loose when autoclaving. Do not pour the media into Petri plates that are too hot since this will produce much condensed water on the Petri plate lid and thus can fall onto the agar surface and may lead to culture contamination. Pour the medium quickly to avoid contamination by air spores and close the lid as soon as possible. Perform the pouring of the medium in a cabinet fitted with UV tubes or in laminar-filtered air flow. Pouring is to be performed near the flame or under aseptic conditions. Medium containing slants or deep tubes are to be stored always at low temperature in dust-free environments.

5.3 SELECTIVE AND DIFFERENTIAL MEDIA

The selective media are those that permit the growth of some specific group or type of organisms while preventing, retarding, or inhibiting the growth of others, thus facilitating bacterial isolation. This selective action is brought about by the addition of certain chemicals in the medium, for example, the dye crystal violet is selectively bacteriostatic for Gram-positive bacteria and is added into a selective medium for the isolation of Gram-negative bacteria.

A differential medium is that which will cause certain colonies to develop differentially (i.e., differentiated) from other organisms present by producing a characteristic change in the bacterial growth or the medium surrounding the colonies. These media are used for distinguishing among morphologically and biochemically related groups of organisms.

King B medium is a selective media for *Pseudomonas fluorescens* while tetrazolium agar medium is a differential media for differentiation of virulent and nonvirulent colonies of plant pathogenic bacteria.

5.3.1 SPECIAL MEDIUM

Most plant pathogenic bacteria grow on nutrient sucrose agar medium. However, some species and pathovars require certain ingredients in the medium and thus require a "special medium" for their growth. The growth of these bacterial pathogen is profuse on the special medium while it is scanty or negligible on nutrient agar sucrose medium.

5.3.1.1 Special Media for *Xanthomonas* (Dye, 1962)

Composition of Medium

Potato, 300 g; peptone, 2.0 g; sucrose, 20.0 g; KH_2PO_4, 0.2 g; Na_2HPO_4, 0.5 g; Ca (No3)$_2$, 0.5 g; KCl, 0.05 g; $FeSO_4$, 0.05 g; agar–agar, 20.0 g; water, 1 L.

5.3.1.2 Special Medium for *Pseudomonas fluorescens* (King's B Medium)

Composition of Medium

Protease peptone, 20 g; K_2HPO_4, 1.5 g; $MgSO_4.7H_2O$, 1.5 g; glycerol, 10 ml; agar, 20 g; distilled water, 1 L.

5.3.1.2.1 Special Medium for Detection of Fluorescent Pigment of
           Pseudomonas fluorescens

Composition of Medium

Pancreatic digest of casein, 10 g; peptic digest of animal tissue, 12 g; anhydrous K_2HPO_4, 1.5 g; $MgSO_4.7H_2O$, 1.5 g; agar, 15 g; pH after sterilization, 7.2 ± 0.2.
  Suspend the above constituents in 1000 ml of distilled water containing 10 ml of glycerol.

5.3.1.2.2 Special Medium for Detection of Pycocyanin Pigment of
           Pseudomonas fluorescens

Composition of Medium

Pectic digest of animal tissue, 20 g; potassium sulfate, 10 g; magnesium chloride, 1.4 g; agar, 15 g; pH after sterilization, 7.2 ± 0.2.
  Suspend the above constituents in 1000 ml of distilled water containing 10 ml of glycerol.

5.3.1.3 Special Medium for Isolation of Pectolytic *Erwinia*

Modified Miller–Schroth medium (Pierce and McCain, 1992).

Composition of Medium

Mannitol, nicotinic acid; L-asparagin (anhydrous); K_2HPO_4 dibasic powder, $MgSO_4.7H_2O$; sodium taurocholate; tergitol 7 (sodium heptadecyl sulfate); 2% nitrilotri acetic acid in 1.46% KOH; 0.5% bromothymol blue (Na salt); 0.5% neutral red; 1 N NaOH; 1% thallium nitrate; 0.33% $CaOCl_2$; MOPS[3]; 10% $CaCl_2.2H_2O$; agar; sodium polypectate.

Procedure

Heat 1 L of distilled water to 80–90°C. Add 10 g of mannitol, 0.5 g of nicotinic acid, 3 g of L-asparagines, 2 g of K_2HPO_4, 0.2 g of $MgSO_4.7H_2O$, and 2.5 g of sodium taurocholate into the water while magnetic stirring. Mix 0.1 ml of Tergitol 7 (sodium heptodecyl sulfate), 10 ml of 2 percent nitrilotriacetic acid (20 g + 14.5 g of KOH per liter), 9 ml of 0.5 percent bromothymol blue (Na salt), 2.5 ml of 0.5 percent neutral red, 5 ml of 1 N NaOH, 1.75 ml of 1 percent thallium nitrate, and 50 ml of 0.33 percent $CoCl_2$ in a container. Add this mixture to the hot medium. Add 15.30 ml of 1N NaOH, 4 g of MOPS (3-(4-morpholino) Propanesulfonic acid), 12 ml of 10 percent $CaCl_2.7H_2O$, and 4 g of agar separately to the heated stirring media. Finally add 18 g of sodium polypectate and stir until it dissolves. Maintain the temperature of the medium at 80–90°C while preparing and before autoclaving.
  Note: Pectate media will solidify at 70°C and will not form an acceptable gel if remelted. Poured plates must not have any surface water when used. Allow them to dry for several days. The NaOH must be added before the pectate is incorporated; otherwise precipitation will occur.
  The final pH should be 7.5 to 7.7. The color of the medium should be blue-green. 5 ml of 1 percent cyclohexamide may be added after autoclaving if fungal contamination is a problem.

Observation

Pectolytic *Erwinia* appear as pink to orange colonies located in deep pits. Non-pectolytic *Erwinia* from pink-orange colonies with no pits. *Pseudomonas* spp. either do not grow on this medium or form green colonies with no pits or pits that are very shallow.

5.3.1.3.1 Miller and Schroth Medium for Isolation of Erwinia

The medium is diagnostic for *Erwinia* spp. *Erwinia* colonies turn to medium yellow or orange in 1–3 days. The colonies then change black to blue-green in color.

Composition of Medium (for 1 L of medium):

Part I: Distilled water, 800 ml; agar, 20 g.

Part II: Tergitol 7 (Sodium heptadecyl sulfate), 0.1 ml; 2 percent nitrilotriacetic acid in 1.46 percent KOH, 10 ml; (20 g + 14.6 g KOH1L); 0.5 percent bromothymol blue (Na salt, 9 ml); 0.5 percent neutral red, 2.5 ml; 5N NaOH (do not use if precipitate from), 1 ml; 1 percent thallium nitrate (T1NO3), 1.7 ml; 14mM $COCl_2$ (0.83 g/250 ml H_2O), 50 µl.

Part III: Mannitol or sorbitol, 10 g; nicotinic acid, 0.5 g; L-asparagin (anhydrous), 3 g; K_2HPO_4 (dibasic powder), 2 g; $MgSO_4.7H_2O$, 0.2 g; sodium taurocholate, 2.5 g.

Procedure

Dissolve the Part II ingredients in 50 ml of water while stirring on a heated stir plate. Similarly, dissolve the Part III ingredient in 50 ml of water while stirring on a heated stir plate. Dissolve the agar in 800 ml of water while boiling, add Part II and Part III ingredients dissolved in water, and bring volume to 1 L. Autoclave the media.

Note: The pH should be about 7.3. It should be blue-green in color.

It is the best medium for isolation of *Erwinia* spp. Many *Pseudomonas* species can grow on MS media; however, these never turn yellow-orange but always remain blue-green. Plates should be observed every day for 3–4 days. After 3–5 days *Erwinia* and *Pseudomonas* spp. both appear blue-green. When selecting for soft rotters, use mannitol for rough colonies with convoluted margins. *Erwinia herbicola* and *Erwinia amylovora* develop smooth colonies with translucent margins. Use sorbitol instead of mannitol for *Erwinia amylovora*. Sorbitol is not readily used by *Erwinia herbicola* and may restrict some soft rot bacteria as well.

5.3.1.3.2 Selective Medium for Erwinia amylovora

To differentiate *Erwinia amylovora* from other *Erwinia*, a selective medium is used.

Composition of Medium (for 500 ml of medium)

Distilled water, 380 ml; sucrose, 160 g; agar, 12 g; crystal violet (0.1 percent in ETOH), 0.8 ml; 0.1 percent cycloheximide.

Autoclave, cool pour plates, and allow to dry for 2 hours prior to use or storage. After streaking, incubate at 28°C for 60 hours.

Observation

Examine the surface of each bacterial colony with a 30× magnification lens with aid of obliquely oriented light source. *Erwina amylovora* colonies develop in characteristic craters after 48–72 hours, but are most distinct after 60 hours.

Erwinia herbicola and other saprophytic isolates grow on this medium, but do not form craters.

5.3.1.3.3 Semi-Selective Media for Erwinia carotovora ssp.

Carotovora (Bdliya and Langerfeld, 2005)

It is a modified crystal violet sodium polypectate double layer medium with 2,3,6-triphenyltetrazolium chloride and L-asparagine in the basal medium and bromothymol blue in the upper, both containing tryptone.

Observation

Red to pink colonies of *Erwinia carotovora* ssp. *carotovra* appears on a blue background with cavity formation.

Note: Addition of 1.2 µg/ml polymyxin B sulfate to the upper layer inhibit the growth of *Erwinia carotovora* ssp. *atroseptica* while addition of 32 µg/ml crythromycin in the upper layer inhibit the growth of *Erwinia chrysanthemi*. The addition of 35 µg/µl novobiocin inhibit the growth of *Bacillus* sp.

Nontarget bacteria like *Pseudomonas fluorescens*, *Pseudomonas marginalis*, *Ralstonia solanacearum*, and *Erwinia herbicola* form pink, red, yellow, or white colonies without cavity formation.

5.3.1.4 Special Medium for *Agrobacterium*

5.3.1.4.1 For Isolation of *Agrobacterium tumefaciens (A Radiobacter Group) from Soil*

Composition of Medium

- Part I: Water, 1 L; mannitol, 10 g; sodium nitrate, 4 g; magnesium chloride, 2 g; calcium propionate, 1.2 g; magnesium phosphate, 0.2 g; magnesium sulfate, 0.1 g; sodium bicarbonate, 0.075 g; magnesium carbonate, 0.075 g; agar, 20 g.
- Part II: Berberine, 275 mg; sodium selinite, 100 mg; penicillin G, 60 mg; streptomycin sulfate, 30 mg; cycloheximide, 250 mg; tyrothricin, 1 mg; bacitracin, 100 mg.

Procedure

Dissolve the ingredients of Part I in hot water by boiling and autoclave. Add the ingredients of Part II. Adjust pH to 7.1 with 1.0 N HCl.

5.3.1.4.2 For Isolation of *Agrobacterium tumefaciens Species Comply (i.e., Biovar 1) and Rubi*

Composition of Medium

L-arabitol, 3.04 g; NH_4NO_3, 0.16 g; KH_2PO_4, 0.54 g; K_2HPO_4, 1.04 g; $MgSO_4.7H_2O$, 0.25 g; sodium taurocholate, 0.29 g; 0.1 percent crystal violet, 2.0 ml; agar, 15 g.

Adjust volume to 1 L with water and sterilized by autoclaving. Add K_2TeO_3, 0.08 g; and 2 percent cycloheximide, 1 ml.

5.3.1.4.3 For Isolation of *Agrobacterium rhizogenes (Biovar 2) 2E Medium*

Composition of Medium

Erythritol, 3.05 g; NH_4NO_3, 0.16 g; KH_2PO_4, 0.54 g; K_2HPO_4, 1.04 g; $MgSO_4.7H_2O$, 0.25 g; sodium taurocholat, 0.29 g; 1 percent yeast extract, 1 ml; 0.1 percent malachite green, 5.0 µl; agar, 15 g. Adjust volume to 1 L and sterilize by autoclaving. Add K_2TeO_3, 0.32 g; and 2 percent cycloheximide, 1 ml.

5.3.1.5 Yeast Glucose Chalk-Agar Medium for Maintenance of Bacterial Cultures

Composition of Medium

Yeast extract, 10 g; glucose, 10 g; CaCO3, 20 g; agar–agar, 20 g; water, 1 L.

5.4 PREPARATION OF SLANTS

The media slants are used to subculture and maintain the bacterial cultures. It is always easy and safe to maintain the bacterial cultures on media slants.

Procedure

Pour the media in test tube at one-third of test tube length or volume and plug with a cotton plug. Put the media test tubes in a sterilization basket in an upright position and sterilize in the autoclave under pressure at 15 pounds for 30 minutes. Remove the test tubes after sterilization and keep them in a slanted position by using a wooden bar or foot scale so as to get slants. When the media is solidified in the slanted positions, collect the slants and store in freezer or use as per requirements.

6 Isolation of Bacterial Plant Pathogens

Isolation of bacteria from infected disease plant material is an important step in the determination of the cause of the plant disease, and further, to prove Koch's postulate for confirmation of the cause of the disease as bacteria. This is also important to obtain a pure culture of the bacterial plant pathogen for its further identification up to genus and species level.

6.1 FROM DISEASED PLANT SAMPLE

For successful isolation of bacterial plant pathogens, the selection of material is important. From young developing lesions the bacteria can be easily isolated. In advanced stages of lesion development, isolation of true pathogenic bacteria becomes difficult due to the overwhelming population of saprophytes, which overgrow the pathogen in isolation plates. If the diseased specimen does not contain young lesions, it is desirable to inoculate macerate of the diseased tissue on to the healthy host plant and resulting young lesions be used for isolation.

Before proceeding for isolation, a part of the tissue from the lesion should be subjected to ooze test and staining to ensure the presence of the bacterium.

Material Required

Diseased plant sample, nutrient agar plates, distilled sterile water, mercuric chloride solution, sterile Petri plates, glass slide, razor blade, dissection needle, inoculating needle, spirit lamp, and so on.

Procedure

Rinse the selected lesion with the spirit and immediately dip in mercuric chloride solution (1:1000) for 15 seconds and then pass through three changes of sterile water. Rinse a slide in spirit, flame, and allow it to cool. Place the surface-sterilized lesion tissue on the slide in a few drops of water. In the initial ooze test, if the affected tissue shows abundant bacterial streaming, the lesion is cut into halves with a heat-sterilized razor blade and the slide is set aside for 2–3 minutes to permit the diffusion of bacteria into the water drops. The tissue showing feeble bacterial ooze should be teased apart with a sterilized blade to get the bacteria into suspension. The suspension thus prepared is used for isolation by any of the following methods.

1. **Streak plate method**

   Prepare nutrient agar medium with the following constituents:

   Nutrient agar (NA) medium: Peptone, 10 g; beef extract, 5 g; agar, 20 g; water, 1000 ml; pH, 7.0.

   Pour about 25 ml of sterilized lukewarm nutrient agar in Petri plates and allow to solidify for 1 hour; then invert the plates. After 2–3 hours these plates are used for isolation purposes.

   Streak a loopful of the suspension prepared as in the preceding section over the agar surface by the to-and-fro method with the help of the inoculation needle. Streak two more plates without recharging the wire loop with bacterial suspension. Label the plates and incubate in an inverted position at 25°C and examine daily. Most bacterial plant pathogens develop colonies within 3–5 days, but some of them may take as long as 10 days. Single colonies are usually obtained on second or third plate.

2. **Spreading diseased tissue macerate on NA medium**

In those plant samples where oozing is not visible, the isolation of bacteria is done by the tissue macerate method. Sterilize the disease sample in 0.1 percent mercuric chloride solution for 1–2 minutes, wash with distilled sterilized water thrice, and macerate with a pestle and mortar in sufficient sterilized water. Allow the macerate to settle down. Pipette out 0.2 ml of supernatant of macerate and place on NA medium on the previously divided plates as shown in Figure 6.1. Streak the supernatant in each compartment of the plate. Incubate the plate and observe for colony development.

3. **Spreading dilutions on solid agar**

Prepare nutrient agar medium plates as in the streak plate method. Dilute the bacterial suspension prepared from the infected tissue serially in 5 ml of sterile water blanks. Usually three dilutions will suffice. Put one drop of each of the original suspension and three serial dilutions in four agar plates separately. In each plate spread the drop uniformly with a sterile bent glass rod. Incubate the plates and observe for single colony development.

4. **Pour plate method**

The nutrient agar medium for this method should contain 1.5 percent agar. Sterilize the medium in test tubes in 20-ml quantities and allow to cool to 45°C. Inoculate one tube with

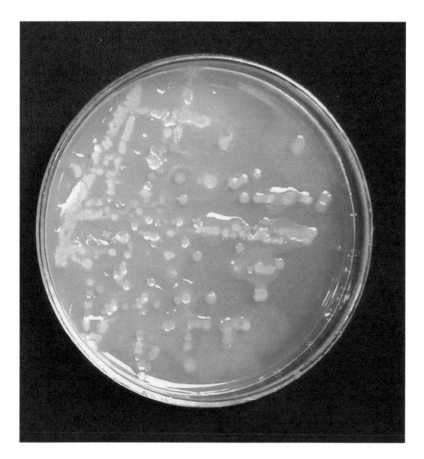

FIGURE 6.1   Isolation of plant pathogenic *Xanthomonas* from a grapevine plant sample. (Courtesy of Dr. S. G. Borkar and Swapanali Kadam, Department of Plant Pathology, Mahatma Phule Krishi Vidyapeeth, Rahuri.)

0.1 ml of the suspension from diseased tissue macerate. Mix thoroughly by rotating the tube between your palms. Transfer one loopful of this mixture to the second tube and mix thoroughly. Remove one loopful of the mixture from the second tube, transfer to the third tube, and mix thoroughly. Pour the above three dilutions in three separate Petri plates. Label the plates and incubate in inverted position after the agar has solidified. (See Figure 6.1.)

6.2 FROM DISEASE FIELD SOIL SAMPLE

Many plant pathogenic bacteria can be isolated from soil, too, as infected leaves and plant debris fall on the ground, are buried in the soil, and the decomposing infected material releases the bacteria in the soil of the plant canopy. Such soils can be used for the isolation of disease, causing bacterium when the symptoms are not available on the young growing leaves, for example, oily spot disease of pomegranate or bacterial wilt of solanaceous crops.

Material Required

Soil sample from diseased plant soil, nutrient agar plates, distilled sterile water, sterile water blank, sterile pipettes, inoculating needle, sterile glass rod, and so on.

Procedure

Collect the soil sample from 1 to 3 inches in depth of soil under the plant canopy in case of leaf spot disease and from rhizosphere in case of wilt disease. Add 10 g of the soil sample in 100 ml of distilled water, shake well, and allow to settle. Take 1 ml suspension from this to make serial dilution up to 10^{-8}.

Prepare the nutrient sucrose agar medium, sterilize, and add 500 ppm of aureofungin in the sterilized media before pouring into Petri plates. When media solidifies, keep the plates under UV radiation for 30 minutes. Use these plates for plating 0.1 ml suspension of individual serial dilutions in a separate plate. Replicate the plating of serial dilution for three replications. Incubate the plates at $28 \pm 2°C$ in a BOD incubator. Observe the plates for the development of bacterial colonies.

Note: Most of the plant pathogenic bacterial colonies appear in the media after 3 days. Do not pick up the colonies that appear within 24 or 48 hours. The colonies obtained and selected should be assessed for their pathogenic nature.

6.3 FROM FIELD WATER SAMPLE

Bacterial plant pathogens can also be isolated from field water and irrigation channel waters. This is specifically true in cases where the bacterial spread is through irrigation water or standing water in the field, for example, the bacterial blight of rice.

Material Required

Field water sample, Whatman filter paper, nutrient agar plates, sterile water blanks, sterile pipettes, inoculating needle, spreading glass rod, and so on.

Procedure

Collect the water sample from the field or irrigation channel in a plastic jar or bottle. Filter through a Whatman filter paper no. 42 to remove dirt, fungal, and algal structures, if any. Collect the filtrate and use for plating.

Prepare nutrient sucrose agar medium, sterilize, and add 500 ppm of aureofungin in the sterilized media before pouring in the Petri plates. When media solidifies, keep the plates under UV radiation for 30 minutes. Use these plates for plating 0.1 ml suspension of the water filtrate. Make five plates of this water filtrate. Incubate at $28 \pm 2°C$ in the BOD incubator. Observe the plates for development of bacterial colonies.

Note: Most of the plant pathogenic bacterial colonies appear in the media after 3 days. Do not pick up the colonies which appear within 24 or 48 hours. The colonies obtained and selected should be assessed for their pathogenic nature.

6.4 FROM INFECTED SEED MATERIAL

The bacterial plant pathogens which are seedborne or transmitted through seed (cotton bacterial blight, sesamum leaf spot, chili leaf spot, tomato leaf blight, halo bean blight, etc.) can be isolated from seed material.

6.4.1 ISOLATION OF EXTERNALLY SEEDBORNE BACTERIA

Material Required

Test seed material, nutrient agar plates, sterile forceps, blotter paper, and so on.

Procedure

Collect the seed samples. Put them in sterilized water-soaked blotter paper for 4–8 hours. Transfer these seeds to nutrient sucrose agar plates and incubate at $28 \pm 2°C$ in the BOD incubator. Bacterial growth formed on and around the seed is to be selected, subcultured, and assessed for pathogenicity.

6.4.2 ISOLATION OF INTERNALLY SEEDBORNE BACTERIA

Material Required

Test seed sample, nutrient agar plates, mercuric chloride solution, sterile pestle and mortar, sterile water blank, sterile pipettes, and so on.

Procedure

Collect the seed sample. Sterilize the seed in 0.1 percent $HgCl_2$ solution for 60 seconds followed by three washings of distilled sterilized water. Keep half of the seed on sterile NAS media in Petri plates. The other half of the seed is to be macerated with a sterilized pestle and mortar in 2 ml of distilled sterile water. Pipette out the macerate on NAS media in plate B. Incubate the plates at $28 \pm 2°C$ in the BOD incubator. Observe the plate and pick up the bacterial colonies which appear after 70 hours.

Note: The bacterial colonies that appear in plate B indicate the internally seedborne bacterial infection. If colonies appear in both the plates, this indicates both external and internal seedborne infection.

7 Purification of Plant Pathogenic Bacterial Cultures

During the isolation of bacterial plant pathogen from diseased plant samples, several nonpathogenic and saprophytic bacterial colonies also grow in the isolation plates. Generally, such colonies appear much earlier than plant pathogenic bacterial colonies. Sometimes the saprophytic bacterial colonies are intermingled with plant pathogenic bacterial colonies. Therefore, picking the pathogenic bacteria and their purification is an important step to obtain pure culture of bacterial plant pathogen.

Further, during the preservation of bacterial cultures, sometimes the cultures are contaminated with other microbes and therefore purification is a necessary aspect in the maintenance of these cultures.

Material Required

Plates of isolated bacteria, nutrient agar plates, inoculating needle, marker, incubator, and so on.

Procedure

Please see the following.

7.1 SELECTION OF SINGLE BACTERIAL COLONIES

With well-chosen diseased tissue, it is often possible that only the colonies of the pathogen develop. However, many times the colonies of saprophytic bacteria develop along with plant pathogenic bacteria. If more than one type of colonies are seen, it is preferable to select the representative colonies of the bacteria. Further, the colonies that came up more slowly are likely to be that of pathogenic bacteria. For instance, in isolation from neem leaves infected with *Pseudomonas azadiractae*, numerous white, bold, saprophytic colonies develop frequently within 48 hours and it is not until 4–5 days later that a few white, glistening, convex colonies of the pathogen become visible. In general, the colonies that appear before 48 hours are not likely to be that of a pathogen. In isolation of previously undescribed pathogens, sometimes it may be desirable to select two or three types of colonies. After multiplication on nutrient agar medium, these are tested for hypersensitive reaction and pathogenicity. The one which proves pathogenic is retained and the others discarded. (See Figure 7.1.)

7.2 PICKING OF SINGLE COLONIES AND SUBCULTURE

After selecting the right type of colonies, transfer them to the nutrient agar slants. Touch a small wire loop of a bacterial inoculation needle to a well-isolated colony and streak it on the agar slant in a tube. Incubate the tube at $28 \pm 2°C$ for 48 hours to obtain sufficient bacterial growth. (See Figure 7.2.)

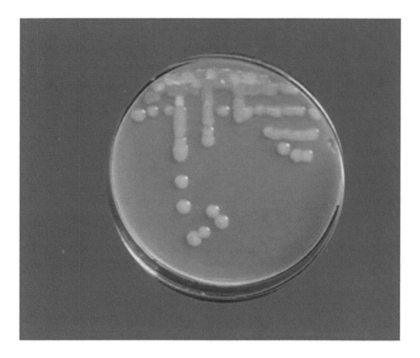

FIGURE 7.1   Isolation of single colonies of bacterial plant pathogens. (Courtesy of Dr. S. G. Borkar and Maria D'Souza, Department of Plant Pathology, Mahatma Phule Krishi Vidyapeeth, Rahuri.)

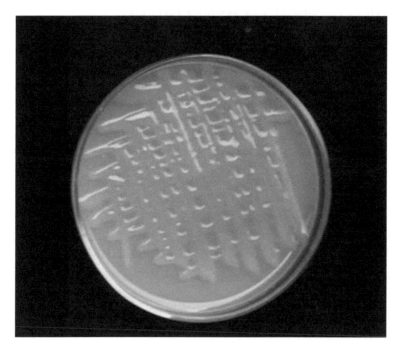

FIGURE 7.2   Pure culture obtained from single bacterial colony. (Courtesy of Dr. S. G. Borkar and Maria D'Souza, Department of Plant Pathology, Mahatma Phule Krishi Vidyapeeth, Rahuri.)

7.3 CHECKING THE PURITY OF THE ISOLATED CULTURE

The cultures obtained by single colony transfer are checked for purity. Make a dilute suspension of the culture in distilled sterilized water and streak on the nutrient agar plates. If the culture is pure only one type of colony with original characteristic should develop. Only the culture obtained from pure colonies should be maintained and used for further investigations. (See Figure 7.3.)

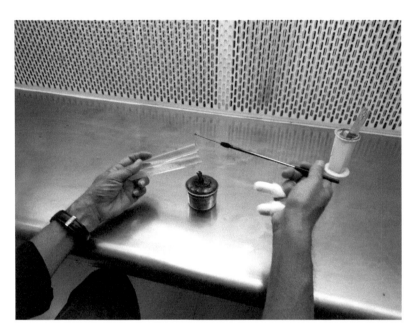

FIGURE 7.3 Pure culture colonies of single isolate. (Courtesy of Dr. S. G. Borkar, Department of Plant Pathology, Mahatma Phule Krishi Vidyapeeth, Rahuri.)

8 Rapid Assessment of Plant Pathogenic Nature of Bacterial Isolates

Once the bacterium is isolated from the diseased plant material, it is necessary to know whether the isolated bacteria is the actual cause of the disease and is pathogenic.

The determination of pathogenic nature of the isolated bacteria on its natural host generally takes 4–15 days or even longer, depending on the types of symptoms they induce on their natural host plant; per se, induction of water-soaking symptoms in natural host plants requires 4–8 days, induction of wilt symptoms requires 8–15 days, and induction of a tumor requires approximately a month or more.

Due to this, a rapid assessment method to determine the pathogenic nature of bacterial isolates is an important aspect.

Material Required

Test bacterial culture, tobacco plant, brinjal plant, stramonium plant, cowpea plant, hypodermal syringe, distilled sterile water, label tags, incubation humid chamber, and so on.

Procedure

Please see the following.

8.1 BY HYPERSENSITIVE REACTION ON A TOBACCO PLANT

The method is based on the fact that only plant pathogenic bacteria are able to induce a hypersensitive reaction (HR) in tobacco. Saprophytic bacteria do not induce HRs in tobacco.

Diluted bacterial culture suspension is inoculated by the injection infiltration method in tobacco leaves with a hypodermic syringe. Different cultures may be inoculated in different interveinal sections of the same tobacco leaf. The cultures, which are plant pathogenic, produce quick necrosis (HR) within 24 hours. Sometimes the HR is produced as early as 8 hours (Borkar, 1982). Those isolates that gives an HR in tobacco plants are surely plant pathogenic bacterial isolates.

If several different types of colonies are encountered in isolation plates, this method may be used to select pathogenic ones. The test is useful particularly in differentiating phytopathogenic xanthomonads and pseudomonads from saprophytic ones. (See Figure 8.1.)

8.2 BY HYPERSENSITIVE REACTION ON A NON-HOST PLANT

Most of the plant pathogenic bacteria produce hypersensitive browning reactions on tobacco plants. However, in most of the places and laboratories, the tobacco plant may not be available. In such cases, other non-host plants can be used to check the HR. Brinjal or eggplant and stramonium plant are the most suitable for this purpose (Borkar, 1982).

Inoculate a bacterial suspension (0.1 OD or 10^7 cfu/ml) in the dorsal side of the leaf of test plant by the syringe infiltration technique and keep the plant at $28 \pm 2°C$ with more than 85 percent atmospheric humidity.

Observe the HR reaction which develops within 18–24 hours on brinjal/stramonium leaves. (See Figure 8.2.)

FIGURE 8.1   Hypersensitive reaction induced by plant pathogenic bacteria on tobacco leaf. (Courtesy of Dr. S. G. Borkar and Pratiksha Mote, Department of Plant Pathology, Mahatma Phule Krishi Vidyapeeth, Rahuri.)

FIGURE 8.2   Reaction of *Xanthomonas vesicatoria* on potato and stramonium leaves. (Courtesy of Dr. S. G. Borkar and Pratiksha Mote, Department of Plant Pathology, Mahatma Phule Krishi Vidyapeeth, Rahuri.)

8.3   BY REACTION OF BACTERIA ON SOFT FRUITS

The reaction of plant pathogenic bacteria, particularly *Erwinia*, on soft pear fruit is very quick (Nagarale, 2010). The inoculation of bacteria into pear fruit produced soft rot of tissue and light to dark-brown coloration of rotted tissues within 24 hours of bacterial inoculation.

Infiltrate the bacterial suspension (0.2 ml) beneath the skin of the pear fruit or banana fruit and incubate at room temperature (at 27°C). Observe the development of soft rot in the fruit due to bacterium. (See Figure 8.3.)

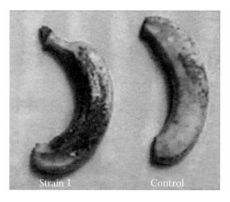
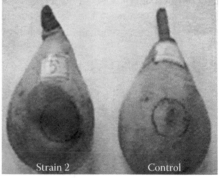

FIGURE 8.3   Reaction of *Erwinia chrysanthemi* on banana and pear fruit. (Courtesy of Dr. S. G. Borkar and D. T. Nagarale, Department of Plant Pathology, Mahatma Phule Krishi Vidyapeeth, Rahuri.)

9 Determination of Pathogenicity of the Isolated Bacterial Culture on Natural Host Plants

When testing a bacterial culture for pathogenicity, it is important that the same plant species and cultivar from which the bacterial culture is isolated is used. The young vigorously growing plants or shoots should be selected for inoculation. It is advisable to keep the plants before and after inoculation under high humidity in a moist chamber.

In carrying out inoculations by any method, check/control plants (treated in a similar way as of the test plant, but only with sterile water) should be maintained. The check plants should be treated first and then the test plant inoculated. This avoids chances of contamination and the labor involved in sterilization of instruments used for inoculation, which is necessary if the test plant is inoculated first.

Material Required

Hypodermic syringe, cotton, atomizer or plastic sprayer, vacuum pump, dissection needle, razor blade, carborandum powder, test bacterial suspension, tags, incubation humidity chamber, and so on.

Procedure

Preparation of bacterial inoculums for inoculation on/in plant: Suspend a 24–48 hour young bacterial growth in about 25–30 ml of sterile water. This bacterial suspension is used for inoculation purposes in the desired plant parts. Only freshly prepared bacterial suspension should be used.

The inoculation may be carried out with different methods depending upon the types of symptoms to be produced.

9.1 BY INFILTRATION METHOD

This method consists of injecting bacterial suspension into the intercellular spaces of leaves with a 25-gauge hypodermic syringe with a five-eighths size needle. Insert the hypodermic needle gently under the epidermis on the dorsal side of the leaf. The opening face of the needle should be toward the leaf. Inject the desired amount of inoculum so that the tissue becomes water-soaked. This method can also be used for inoculation of stem and other plant parts.

The method is generally used for determining the pathogenicity of leaf spot and leaf blight bacterial pathogens. The method produces pathogenicity symptoms (water soaking reaction) within a short period of time (4–5 days) as compared with other inoculation methods for producing leaf spot diseases. The method is appropriate for the inoculation of leaves with a rough texture. (See Figure 9.1.)

9.2 BY SWAB INOCULATION METHOD

It is the best alternative for spray inoculation in case of plant parts possessing high surface tension or waxy coating that do not allow the sprayed inoculums to stick to the plant's surface. Smear the inoculum soaked in muslin cloth on the surface of the plant part to be inoculated. Incubate the plant

Laboratory Techniques in Plant Bacteriology

FIGURE 9.1 Infiltration of bacterial suspension into intercellular spaces of leaves by hypodermic syringe. (Courtesy of Dr. S. G. Borkar, Department of Plant Pathology, Mahatma Phule Krishi Vidyapeeth, Rahuri.)

in a humid chamber or alternatively spray it with water thrice a day for a week or until symptoms develop.

The swab inoculation method is generally used for determining the pathogenicity of leaf spot and leaf blight bacterial pathogens. The method is more appropriate for the inoculation of leaves that have leaf hairs or trichomes on the plant surface.

9.3   BY SPRAY INOCULATION METHOD

In case of spot, blight, and canker diseases where the pathogens usually enter the tissue through stomata, hydathodes, or lenticels, this method of inoculation is the most natural.

Keep the plants in a humid chamber 24 hours prior to inoculation to allow the stomata to open and to create high intercellular humidity in the tissues around the natural openings. Spray the bacterial suspension with a hand atomizer or plastic sprayer to cover the plant surface thoroughly.

Put the plant in a humid chamber for 3 days and, later on, in natural environmental conditions. Spray with water thrice a day for a week or until symptoms develop.

The spray inoculation method is generally used for determining the pathogenicity of bacterial blossom blight, wild fire of apple and pear, and inflorescence diseases of bacterial pathogen or bacterial fruit spot diseases.

9.4   BY ROOT DIP INOCULATION METHOD

The method is probably the most natural way of inoculating bacterial pathogens that produce vascular wilts.

Uproot the young seedlings, wash the roots in water, clip the root tips, and immerse in the bacterial suspension for 10 minutes. Transplant the seedlings in soil and water the transplanted seedlings appropriately. Observe for the wilt symptoms' development.

This method is generally used for determining the pathogenicity of bacterial wilt pathogen (*Ralstonia*) or in case of soilborne bacterial diseases.

9.5   BY VACUUM INFILTRATION METHOD

In this method, the bacterial suspension is forced through the stomata into intercellular spaces. It reduces the incubation period for symptom development and is best applicable for stomata invaders. A very diluted bacterial suspension should be used to obtain the right type of symptoms. If a concentrated suspension is used atypical symptoms may result.

The method consists of applying the inoculums with an atomizer at 1.5 kg/cm^2 pressure on the lower surface of the leaf. The leaf is held with the hand in position and the nozzle of atomizer kept at about a 2-inch distance. The pressure is obtained by connecting the atomizer to the exhaust outlet of an electric motor pump.

The vacuum infiltration method is most appropriate where the plant leaf surface is smooth and waxy (e.g., banana leaf) where the inoculums are not retained on the leaf surface when applied with other inoculation methods.

9.6   BY PIN PRICK INOCULATION METHOD

This method is good for bacteria causing wilts, blights, soft rots, and galls. Prepare a pin bundle by fixing four to six fine insect pins on a piece of cork. Only the tips of the pins should project out of the cork piece. If a piece of cork is not available, one may use the disc of a carrot to make the pin prick bundle. Injure the plant part to be inoculated with the pin bundle and apply the bacterial suspension on the injured site.

For blights, the leaf is inoculated in the center, avoiding the mid rib. For wilts, inoculate the stem at a point between the cotyledon and first leaf. This method does not work well in many of the leaf spot diseases.

The pin prick method is most appropriate to test the pathogenicity of bacterial stalk rot, collar rot, or rhizome rot pathogen. (See Figure 9.2.)

Observations

The inoculated plants are examined for symptom development for at least for 1 month. In leaf spot or streak disease, which is parenchymatous, symptoms usually develop within 4–5 days. However, certain leaf spots, for example, mango canker, may take as long as 10–15 days. Wilts and blights usually take 7–10 days, while galls caused by *Agrobacterium tumefaciens* may take more than a month to develop.

FIGURE 9.2 Inoculation of bacteria with pin prick method in maize stalk. (Courtesy of Dr. S. G. Borkar, Department of Plant Pathology, Mahatma Phule Krishi Vidyapeeth, Rahuri.)

While taking observations on symptoms, it is important to record and describe different developmental stages of the symptoms rather than describing the final stage.

The inoculated bacterium is considered the true pathogen if it produces similar types of symptoms as seen on the original specimen from which it was isolated. In typical symptoms exactly the same alternations occur in an artificially inoculated plant as in nature and the development is slow. The susceptible cultivar of the plant produces typical symptoms while the resistant cultivar of the same plant produces atypical symptoms. In atypical symptoms, rapid (within 24 hours) necrosis of the tissue occurs at the site of inoculation without water soaking at any stage. These atypical symptoms are produced due to plant resistance, high inoculum concentration, or inappropriate methods of bacterial inoculation, even though the bacterium may be pathogenic.

10 Isolation and Enumeration of Microorganisms Associated with Bacterial Plant Pathogen from Soil, Rhizosphere, and Phylloplane

Microorganisms are abundant and ubiquitous in our environment. They occur on phylloplane, on/in soil and the rhizosphere, and some of them are generally associated with bacterial plant pathogens causing diseases of root systems or on above-ground plant parts. These microorganisms include fungi, bacteria, actinomycetes, yeast, and so on. They exert either synergistic or antagonistic effects on bacterial plant pathogens or the diseases they cause. Therefore, their isolation and enumeration is an important aspect in the studies of bacterial plant diseases. The associated microorganisms has to be isolated in pure form to study their effect/interaction.

A culture containing a single unadulterated species of cells is called a pure culture. Several different techniques are applied to isolate and study microorganisms in pure culture. For isolation, several media (solutions containing all the nutrients required for the growth of microorganisms) are employed.

10.1 ISOLATION OF MICROORGANISMS FROM CULTIVATED SOIL/ SOILS UNDER CROPPING SYSTEMS

Though various methods are available to isolate and enumerate microorganisms (bacteria, fungi, actinomycetes, protozoa, yeast, and algae) from soil, the serial dilution agar plating method or viable plate count method is one of the commonly used procedures for the isolation and enumeration of fungi, bacteria, yeast, and actinomycetes, which are the most prevalent microorganisms. This method is based upon the principle that when material containing microorganisms is cultured, each viable microorganism will develop into a colony.

Material Required

Soil sample, water blank, NAS and PDA media plates, pipettes, glass rod.

Procedure

Suspend 1 g of soil in 10 ml of distilled water and shake well. Allow to settle and make a serial dilution up to 10^{-7}. Pipette out 0.1 ml from each serial dilution and plate on NA and PDA plates individually and spread with the help of a glass rod. Incubate the plates in BOD incubator at $28 \pm 2°C$ temperature for 48 hours to observe the development and types of bacterial/fungal/actinomycetes/ yeast colonies on the concern media. Observe the same plates after 72 hours also.

Observation

Observe the plates for the appearance of fungal and bacterial colonies in all inoculated plates. Count the number of microorganisms and their types (bacteria and fungi) developed onto the nutrient agar and potato dextrose agar plates and calculate their population per gram of soil.

10.2 ISOLATION OF RHIZOSPHERE MICROFLORA

The term rhizosphere was introduced in 1904 by the German scientist Lorenz Hiltner to denote the region of soil that is under the influence of plant roots. It is defined as the region of the soil immediately surrounding the roots of a plant together with the root surfaces. Operationally, the rhizosphere can be defined as the region, extending a few millimeters from the surface of each root, where the microbial population of the soil is influenced by the chemical activities of the plant. The region provides certain characteristic conditions for the increased occurrence of microflora in it, which is attributed to the rich food materials provided by the added sloughed-off portions of root tissues and root exudates, which contain sugars, amino acids, vitamins, and other growth factors that serve as nutrients for microorganisms in the rhizosphere.

The number of bacteria in the rhizosphere usually exceeds the numbers in the neighboring soil by a factor of ten and often by a factor of several hundred. The rhizosphere to soil ratio (R:S), called the *rhizosphere effect*, can be calculated by dividing the number of microorganisms in the rhizosphere soil by the number in the soil free from plant growth. Greater rhizosphere effect is seen with bacteria (R:S values from 10 to 100 or sometimes more) than with actinomycetes and fungi; while with regard to protozoa and algae there are only negligible changes. The rhizosphere effect increases with the age of the plant and normally reaches its maximum at the stage of greater vegetative growth. Following the death of the plant, the microbial population reverts gradually to the level of that in the surrounding soil.

To study the rhizosphere microflora of the various techniques, soil dilution and plate count method is most widely used.

Material Required

Freshly collected roots, nutrient agar plates, Czapek–Dox agar plates, sterile water blanks, screw cap bottle, sterile 10 ml pipettes, sterile water, sterile Petri dishes, sterile polythene bags, Bunsen burner, glass marking pencil, and so on.

Procedure

Separate rhizosphere soil from five to six roots with the help of a brush in a Petri dish. Add 10 g of rhizosphere soil in 100 ml sterile water blank and shake it for 15 minutes on a magnetic shaker. Prepare serial dilutions 10^{-2} to 10^{-6} as outlined in an earlier experiment. Transfer 1 ml each of dilutions 10^{-4} to 10^{-6} to sterile Petri plates (three replicates for each dilution) and pour the melted and cooled (45°C) Czapek–Dox agar supplemented with streptopenicillin for isolation of fungi while adding nutrient agar media supplemented with aeurofungin for isolation of bacteria and actinomycetes, respectively, to various plates. Incubate the CDA plates at 25°C in an inverted position for 7 days and NA plates at 30°C for 24–48 hours.

Observation

Observe the plates after incubation period and count number of fungal/bacterial/actinomycetes/yeast colonies and calculate the population per gram of rhizosphere soil.

10.3   ISOLATION OF MICROORGANISM FROM ROOT SURFACE

Material Required

Roots of plant, nutrient agar plates, potato dextrose agar plates, water blank, agitator, pipettes, spreading glass rod, and so on.

Procedure

Take root pieces of primary, secondary, and tertiary roots of the plant and remove the attached soil particles with the help of a brush. Suspend 1 g of respective root pieces in 10 ml of distilled water and agitate on a magnetic stirrer until the root surface becomes clear. From this root wash, prepare a serial dilution up to 10^{-6} and plate on Czapek–Dox agar and nutrient agar plates. Incubate the plates for 7 days at 25°C in an inverted position for fungal colonies and up to 3 days for bacterial/actinomycetes colonies at 28 ± 2°C.

Observation

Count the number of fungal/bacterial/actinomycetes colonies and calculate the population for per centimeter of root length of respective root (primary, secondary, tertiary, etc.).

10.4   ISOLATION OF MICROORGANISMS FROM VIRGIN/ UNCULTIVATED/NON-RHIZOSPHERE SOIL

Material Required

Virgin soil, nutrient agar plates, potato dextrose agar plates, water blank, agitator, pipettes, spreading glass rod, and so on.

Procedure

Isolate microorganisms from the soil by the serial dilution technique as described in Section 10.1.

Observations

Observe the plates for the appearance of bacterial and actinomycete colonies after 2–3 days and for fungi after 4–5 days. Continue observing for 2 weeks.

Data Recording and Calculations

Record the data for the number of colonies per plate in various dilutions for bacteria, actinomycetes, yeast, and fungi in rhizosphere, rhizoplane, and non-rhizosphere soils. Identify the various isolates up to generic/species level and maintain the record for each.

Microbial counts in rhizosphere and non-rhizosphere soils per gram of soil (on a dry-weight basis) are calculated by applying the following formula:

$$\text{Number of microorganisms/g of soil} = \frac{\text{Number of colonies/plate} \times \text{dilution factor}}{\text{Dry wt. of the soil taken}}$$

Rhizosphere effect (R:S) can be calculated by applying the following formula:

$$\text{R:S} = \frac{\text{Number of microorganisms (i.e., fungi or bacteria) on the rhizosphere soil}}{\text{Number of microorganisms (i.e., fungi or bacteria) in the control (non-rhizosphere) soil}}$$

10.5   ISOLATION OF PHYLLOPLANE MICROFLORA BY THE SERIAL DILUTION METHOD

The term *phyllosphere* was introduced by F.T. Last in 1955 to denote leaf surface environment. In 1965, F.T. Last and F.C. Deighton suggested that phylloplane should be used while referring to leaf surface habitat. The terms *phyllosphere* and *phylloplane* are interchangeably used in literature. Phylloplane is a natural habitat on a leaf surface that supports a heterogeneous population, comprising both pathogen and nonpathogens. The phylloplane microbes cover a wide variety of microorganisms, including yeasts, filamentous fungi, bacteria, actinomycetes, blue green algae, and even ferns. The phylloplane microflora is of special interest from various viewpoints because some of them have antagonistic action against fungal parasites, degrade plant surface wax and cuticles, produce phytoalexins, act as a source of allergic airborne spores, and influence growth behavior and root exudation of plants.

Several methods are employed to study the phylloplane microflora. These are classified as direct (direct observations, impression films, clearing, scanning microscopy, phase contrast, fluorescent antibody, and infrared microscopy) and cultural (spore fall, plating, damp chamber, surface sterilization, leaf washing, leaf maceration and leaf impression) methods. Of these, serial dilution plate method and leaf impression method are the two commonly employed techniques.

Material Required

Fresh leaves, sterile polythene bags, potato dextrose agar medium, nutrient agar medium, 100-ml sterile water blank, 9-ml sterile water blanks, sterile 1-ml pipettes, sterile Petri plates, magnetic stirrer, cork borer, Bunsen burner, glass marking pencil.

Procedure

Collect fresh healthy leaves (of all ages) as well as diseased leaves (if available) in fresh sterile polythene bags and bring to the laboratory. Cut five discs each of 6-mm diameter from every leaf using a sterile cork borer. Transfer 50 discs to a 100-ml water blank and stir the discs for 20 minutes using a magnetic stirrer to wash out the phylloplane microflora from the leaf surface into the water. Prepare a serial dilution up to 10^{-4}. Transfer 1 ml aliquots from 10^{-4} and 10^{-2} dilutions to sterile Petri plates (six replicates for each dilution: three for fungi and three for bacteria and actinomycetes). Pour melted and cooled (45–50°C) potato dextrose agar medium and nutrient agar medium (three plates per dilution) to the above plates. Incubate the plates at 28 ± 2°C in an inverted position up to 7 days.

Observations

Observe the nutrient agar plates after 2–3 days and potato dextrose agar plates after 5–7 days of incubation for the appearance of colonies of bacteria, actinomycetes, and fungi, respectively.

Count the colonies of bacteria, actinomycetes, and fungi using the colony counter. Since one colony is represented by a single spore, hence all colonies of a particular species appearing on an agar plate are considered for the total number of spores, cells, or propagules of that particular species in 1 ml of the suspension. Microbial population/cm^2 of leaf area can be calculated by applying the following formulas:

$$\text{Propagules}/\text{cm}^2 = \frac{\text{Total no. of spores in one ml}}{\text{Total area of 50 discs} \times 2}$$

Area of one leaf disc = πr^2 (where r is the radius of disc in cms)

The microbial isolates are purified by applying hyphal tip (in case of fungi) or streaking methods (for bacteria). The purified cultures are then identified up to generic/specific level.

10.6   STUDY OF PHYLLOPLANE MICROFLORA BY LEAF IMPRINT METHOD

Material Required

Leaf samples, potato dextrose agar plates, nutrient agar plates, Bunsen burner, wax marking pencil.

Procedure

Take a leaf and press it from the dorsal surface momentarily against the agar surface of potato dextrose agar plates at three places. Press the same leaf from dorsal surface against the surface of nutrient agar at three places. Now press the same leaf from ventral surface against the surfaces of both the potato dextrose agar and nutrient agar plate, as above. Incubate the plates at 28 ± 2°C in an inverted position up to 7 days.

Observations

Observe the nutrient agar plates after 48–72 hours of incubation for the appearance of bacterial and actinomycetes colonies and potato dextrose agar plates for fungal colonies.

Record the data for number of colonies of bacteria, actinomycetes, and fungi on the leaf imprint per plate and make comparisons for the microbial populations on the dorsal and ventral surface of the leaves. Also note the position of particular microbial colonies on the leaf surface in print to determine their habitat affinity. (See Figure 10.1.)

FIGURE 10.1   Population density of bacterial ephiphytes on a hibiscus leaf. (Courtesy of Dr. S. G. Borkar and Rupert Anand Yumlembam, Department of Plant Pathology, Mahatma Phule Krishi Vidyapeeth, Rahuri.)

11 Preparation of Pure Cultures of Microorganisms

Microorganisms isolated from various substrates or sources (soil, air, contaminated water, disease plants, decomposing infected plant material), need to be obtained in pure form as cultures contain millions or billions of individual cells. A culture that contains only one kind of microorganism is called a *pure culture*. A culture that contains more than one kind of microorganism is called a *mixed culture*; if it contains only two kinds of microorganisms that are deliberately maintained in association with one another, it is called a *two-member culture*.

Pure cultures are essential in order to study colony characteristics, biochemical properties, morphology, staining reactions, and immunological reactions or the susceptibility to antimicrobial agents. Two kinds of operations, that is, isolation (the separation of a particular microorganism from the infected host, individually or as the mixed populations that exist in nature) and purification (the growth of a single cell into microbial populations as a culture in artificial environments [culture media] under laboratory conditions), are important irrespective of the kind of microorganism with which a microbiologist deals.

Pure cultures of microorganisms that form discrete colonies on solid media may be most simply obtained by one of the modifications of the plating method. This method involves the separation and immobilization of individual organisms on or in a nutrient agar medium. Each viable organism gives rise through growth to a colony from which transfer can be readily made.

The three most commonly used methods for obtaining pure cultures of microorganisms are (1) streak plate; (2) pour plate; and (3) spread plate. Other methods for obtaining pure cultures of microorganisms include: (1) use of enrichment media; (2) use of differential and selective media; (3) use of media containing antibiotics; and (4) selected techniques for the cultivation of anaerobes.

11.1 BY STREAK PLATE METHOD

The streak plate method offers a most practical method of obtaining discrete colonies and pure cultures. It was originally developed by two bacteriologists, Friedrich Loeffler and George Gaffkey, in the laboratory of Robert Koch. In this method, a sterilized loop or transfer needle is dipped into a suitable diluted suspension of organisms which is then streaked on the surface of an already solidified agar plate to make a series of parallel, non-overlapping streaks. The aim of this exercise is to obtain colonies of microorganisms that are pure, that is, growth derived from a single cell/spore.

Material Required

A 24–48-hour nutrient broth of mix or the contaminated colony culture, nutrient agar plates, inoculating loop, Bunsen burner, wax marking pencil.

Procedure

With a wax marking pencil, label all the plates on the bottom with the name of the organism(s) to be inoculated. Hold the tube containing the broth of mixed mixture in the left hand. Sterilize the loop holding in the right hand, remove the cotton wool plug using the little finger of the right hand, and immediately flame the mouth of the tube. Introduce the loop into the broth and withdraw one loopful of culture. Flame the mouth of the tube, replace the cotton wool plug, and place the tube in the test tube rack. Lift the Petri plate cover with the left hand and hold it at an angle of 60°. Place the inoculums (the loop containing the droplet of broth) on the agar surface at the edge farthest from you and streak the inoculums from side to side in parallel lines across the surface of area. Reflame and cool the loop and turn the Petri plate to 90°. Touch the loop to a corner of the culture in Area 1 and streak the inoculums across the agar in Area 2 as above (the loop should never enter Area 1 again). The rest of the agar surface is now used to complete the streaking. Replace the lid of the Petri plate after completing the streaking and sterilize the loop by flaming. Incubate all the plates at 28–30°C in an inverted position for 48–72 hours.

Observations

After incubation, examine each of the three plates for the growth of the colonies. A confluent growth will be seen where the initial streak was made, the growth is less dense away from the streak, and discrete colonies will be farthest away from the streak (i.e., end of the streak). Any colony not growing on the streak marks is regarded as a contaminant. Select a well-isolated colony from each plate and record their features. If their features agree with the original description, then the two colonies being examined are pure cultures.

Precautions

Avoid pressing the loop or needle too firmly against the agar surface as this will damage it. The inoculating loop should be cooled by touching the agar surface before lifting the inoculums for streaking. The Petri plate lid should never be lifted completely. Plating of the medium should be done 24 hours in advance of performing the exercise, so as to detect the contaminated plates to discard them. (See Figure 11.1.)

11.2 BY POUR PLATE METHOD

The forerunner of the present pour plate method was developed in the laboratory of the famous bacteriologist, Robert Koch. In this technique successive dilutions of the inoculums (serially diluting

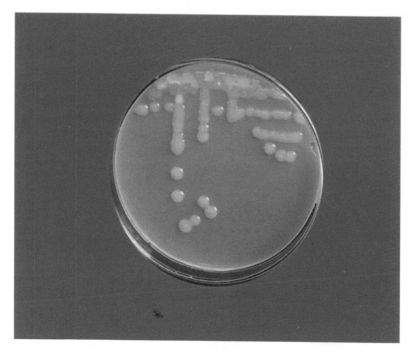

FIGURE 11.1   Pure culture colonies of bacteria in streak plate. (Courtesy of Dr. S. G. Borkar and Ashwini Bhosale, Department of Plant Pathology, Mahatma Phule Krishi Vidyapeeth, Rahuri.)

the original specimen) are added into sterile Petri plates to which melted and cooled (42–45°C) agar medium is poured and thoroughly mixed by rotating the plates which is then allowed to solidify. After incubation, the plates are examined for the presence of individual colonies growing throughout the medium. The pure colonies which are of different size, shape and color may be isolated/transferred into test tube culture media for making pure cultures.

Pour plates are also used as a means of determining the number of viable organisms in a liquid such as water, milk, urine, or broth culture as well as to determine the hemolytic activity of deep colonies of some bacteria, such as the streptococci, via an agar medium containing blood.

Material Required

Twenty-four hour nutrient broth cultures of mixed bacterium, nutrient agar medium, sterile 9-ml water blanks, sterile Petri plates, sterile 1-ml pipettes, test tube rack, water bath, Bunsen burner, wax marking pencil.

Procedure

Place the nutrient agar medium into the boiling water bath for melting. Allow this to cool to 45°C. Prepare a serial dilution of the mixed bacterial culture up to 10^{-4}. Also label the Petri plates and aseptically transfer 1 ml of the bacterial suspension from each dilution to the individual plates that are previously numbers. Remove a nutrient agar medium flask from the water bath (at 45°C), pour the medium (20 ml) into these plates, and rotate the plate gently to ensure uniform distribution of cells in the medium. Allow the medium to solidify. Incubate the inoculated plates for 24–48 hours at 28 ± 2°C in an inverted position.

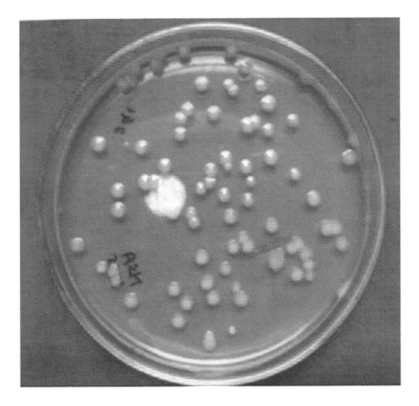

FIGURE 11.2 Pure distinct colonies obtained by pour plate method. (Courtesy of Dr. S. G. Borkar and Dr. Kunal Surywanshi, Department of Plant Pathology, Mahatma Phule Krishi Vidyapeeth, Rahuri.)

Observations

Examine the plates for the appearance of individual colonies growing throughout the agar medium. It will be observed that progressively poured plates will have fewer and fewer numbers of colonies that will be distributed more or less sparsely in the plates that may be transferred (subcultured) to other media (fresh plates) or agar slants for further study.

Precautions

The medium to be poured in the Petri plates should have a temperature of 45°C. The plates should be incubated in an inverted position to prevent collection of condensation on the agar surface. Unless the surface is dry it will be difficult to obtain discrete surface colonies.

11.3   BY SPREAD PLATE TECHNIQUE

The spread plate technique is used for the separation of a diluted, mixed population of microorganisms so that individual colonies can be isolated. In this technique microorganisms are spread over the solidified agar medium with a sterile L-shaped glass rod while the Petri dish is spun on a turntable. The theory behind this technique is that as the Petri dish spins, at some stage, single cells will be deposited with the bent glass rod on to the agar surface. Some of these cells will be separated from each other by a distance sufficient to allow the colonies that develop to be free from each other.

Material Requirements

Twenty-four hour nutrient broth cultures of known bacterial mixture, nutrient agar plates, lazy susan turntable, L-shaped bent glass rod, 95 percent alcohol, beaker (50 ml), Bunsen burner, wax marking pencil.

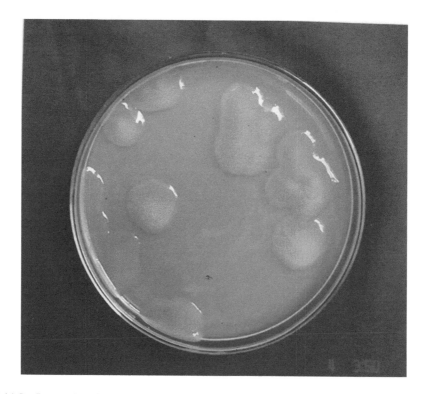

FIGURE 11.3   Pure and distinct bacterial colonies obtained by spread plate technique. (Courtesy of Dr. S. G. Borkar and Ashwini Bhosale, Department of Plant Pathology, Mahatma Phule Krishi Vidyapeeth, Rahuri.)

Procedure

Label nutrient agar plates with bacterial species with a wax pencil. Pour 95 percent alcohol into a beaker and dip the bent glass rod in it. Aseptically transfer a loopful culture of mix bacterium in the center of the appropriately labeled nutrient agar plate. Place the inoculated plate on the turntable. Remove the glass rod from the beaker and sterilize the bent portion in the Bunsen burner flame. Cool the rod for 10–15 seconds. Remove the cover of the Petri dish and spin the turntable. Lightly touch the sterile bent rod to the agar surface and move it back and forth while the turntable is spinning for spreading the culture over the agar surface. Replace the Petri dish cover when the turntable stops spinning. Immerse the bend rod in alcohol and reflame to sterilize it. Incubate all the three plates in an inverted position at $28 \pm 2°C$ for 24–48 hours.

Observations

Observe all the inoculated plates as to the distribution of colonies on each of the agar plates; some of the colonies will be free from each other. Select a discrete colony from first and second plate and record their form, elevation, pigmentation, and size of the colony.

12 Maintenance and Preservation of Bacterial Pure Cultures

Once a bacterium is isolated in a pure form, it is subcultured on plates or agar slants (a tube containing a solid medium prepared by keeping the tube tilted as agar solidifies; the resultant slope surface provides more area and is easier to streak than a horizontal surface) at regular intervals to maintain viability. The interval between subculturing, which varies between every two weeks or a month, depends on the storage conditions and on the growth rate of organisms. The bacterium can be subcultured on maintenance media especially designed to allow low growth rates and extend the culture's life. Storing cultures in a refrigerator at a temperature of 4°C, which slows growth, protects from damage due to evaporation of the medium and reserves the culture. Subculturing of refrigerated cultures is to be carried out at regular fortnightly intervals.

It is advisable to maintain two slants, one as the working culture to be used as a source for routine laboratory work and the other as stock culture from which new working cultures are prepared, whenever required. Such a system decreases the chances of contaminating the stock culture. Maintenance of stock cultures requires great care. To preserve bacteria, it is necessary to reduce their metabolism to a minimum. As a result, the processes that lead to aging and death are slowed down and the bacterium can be maintained in its inactive state for several years. Lyophilization (freeze-drying) or freezing in liquid nitrogen is used for preserving the bacterium where they can retain viability for several years.

There are several methods available for maintenance of pure cultures; the choice of the method depends upon the purpose, size of collection, and the laboratory. Some of the commonly used methods are as follows.

12.1 PRESERVATION ON YGCA MEDIA

Plant pathogenic bacterial cultures are maintained on yeast–glucose–chalk–agar medium rather than the routinely used nutrient–sucrose–agar medium. Due to the presence of calcium carbonate in the medium, the pH of the medium remains constant to sustain the bacterial viability. On YGCA media, the bacterial cultures can be maintained up to 3 weeks.

Composition of YGCA Media		
Yeast extract	:	10.0 g
Glucose	:	10.0 g
CaCO3	:	20.0 g
Agar–agar	:	20.0 g
Water	:	1 L

12.2 PRESERVATION IN REFRIGERATOR OR COLD ROOM STORAGE

Live cultures on a culture medium can be successfully stored in refrigerators or cold rooms maintained at 4°C. Generally the metabolic activities of the microorganisms will be greatly slowed down at this temperature, but not low enough to stop metabolism completely. Thus growth will occur slowly and nutrients will be utilized and waste products produced, which will eventually kill the microorganism. So regular subculturing is necessary, which ranges from an interval of 2–3 weeks.

12.3   PRESERVATION IN MINERAL OIL

This is a simple and economical method of preserving bacteria where they remain viable for several months at room temperature. In this method, sterile liquid paraffin is poured over a slope culture of the bacterium and stored upright at room temperature. The layer of paraffin prevents dehydration of the medium, and by ensuring anaerobic conditions, the microorganisms remain in a dormant state.

12.4   PRESERVATION AT –40°C IN GLYCEROL

Cultures can be preserved for several years in glycerol at –40°C in a deep freeze. In this method, approximately 2 ml of the glycerol solution is added to the agar slope culture; the culture is emulsified by shaking. The emulsified culture is transferred, 0.5 ml into each ampoule, which are placed in a mixture of industrial methylated spirit and carbon dioxide and are froze rapidly to –70°C. Ampoules are removed from the mixture and placed directly into a deep freezer at –40°C. To revive the cultures, the ampoules are placed in a water bath at 45°C for a few seconds or until the suspensions melt and are aseptically streaked onto agar plates.

12.5   PRESERVATION IN EDIBLE/NONEDIBLE OILS

In far-off places or remote area laboratories where mineral oils are not available, the edible/nonedible oils can be used for preservation of bacterial plant pathogen. Preservation of plant pathogenic bacterial cultures in edible/nonedible oils was reported from Professor Borkar's laboratory in 2007 (Birajdar, 2007).

Streak the pure bacterial culture on a nutrient agar slant in test tubes. After 48 hours of bacterial growth, edible oils like groundnut oil, sunflower oil, safflower oil, and sesame oil can be used for the preservation of bacterial culture. Any one of these oils can be added in the test tube till the culture slant is submerged in the oil. These tubes can be stored at room temperature where bacteria remain viable for up to a three-month period.

Among nonedible oils, castor oil and cottonseed oil were observed to be preservatives where bacteria remain viable for up to two months and one-and-a-half months, respectively.

12.6   PRESERVATION BY LIQUID NITROGEN METHOD (STORAGE AT LOW TEMPERATURE)

Freezing in liquid nitrogen at temperatures of –196°C suspends the metabolism of cells, and these survive unchanged for long periods. In this method, cell suspension in the presence of a stabilizing agent, such as glycerol or dimethyl sulfoxide (DMSO), that prevents the formation of ice crystals which may kill frozen cells, is sealed into small ampoules and stored in liquid nitrogen refrigerator (–196°C). Most species of bacteria can remain viable for 10 to 30 years or even more without undergoing change in their characteristics.

12.7   PRESERVATION BY PARAFFIN METHOD

This is a very simple, economical, and successful method for storing cultures of bacteria. The microbes remain viable for several years at room temperature. During revival or transfer from oiled cultures, a loopful of the culture is removed and inoculated on an agar slant or a broth tube.

Material Required

Slant culture of a bacterium, liquid paraffin (in flasks), oven (at 180°C).

Procedure

Sterilize the liquid paraffin at 180°C in an oven for 1 hour. Pour a sufficient amount of sterilized paraffin over a slope (slant) culture of bacterium so that the paraffin forms a layer about half an inch above the agar surface. Store the agar slant upright at room temperature. During transfer from a paraffin-preserved culture, draw a loopful of the culture and inoculate onto a fresh agar slant or into a tube of broth.

Precaution

While sterilizing the paraffin, the flasks should be kept in a metal box before placing them in the oven to avoid the pollution due to production of offensive odor when heated. A sufficient amount of the paraffin should be added to the slope culture so that it completely covers the agar surface; otherwise the tube will be dehydrated in due course of time.

12.8 PRESERVATION BY FREEZE-DRYING (LYOPHILIZATION) METHOD

Freeze-drying (lyophilization) is the rapid dehydration of organisms while they are in a frozen state. Most of the microbes are protected from the damage caused with water loss by this method. Because metabolism requires water, the organisms are in a dormant state and can retain viability for over 30 years unchanged in their characteristics. In this technique the culture is rapidly frozen at –70°C and then dehydrated by a vacuum and the tubes containing freeze-dried cultures are sealed and stored in the dark at 4°C in refrigerators.

The freeze-drying method is often used for storing of cultures in the national/international culture collection centers of the world as dehydrated (lyophilized) cultures retain their viability for several years.

Material Required

Corning or Borosil glass ampoules, small strip of filter papers, sterile 1-ml pipettes, cotton, oven, freeze-drying apparatus.

Procedure

Label the small strip of filter papers with the code number or name of the organisms with a pencil. Sterilize the ampoules in an oven at 180°C for 1 hour. Pour between 0.25 and 0.5 ml of a broth culture into a sterilized ampoule. Apply a cotton plug which is completely pushed into the ampoule. Connect the ampoule to a hole in the centrifuge head of freeze-drying apparatus. Centrifuging is continued for 7 minutes, which results in the rapid dehydration and freezing of the contents of the ampoules. Allow these ampoules for primary dehydration for 6 hours. The necks of the ampoules are drawn out and the ampoules are given secondary dehydration for 12 hours. Seal the ampoules under vacuum and store these at 4°C on the desk. Examine the ampoules for sealing using a vacuum tester. Production of a blue glow is indicative of sufficiently low pressure, that is, the proper sealing of the ampoules.

Revival of the Culture

During the opening of an ampoule, the ampoule neck is flamed, and a scratch is made with a file around the ampoule, corresponding to the equator of the cotton–wool plug. Open the ampoule by applying pressure at the scratch mark. Remove the cotton–wool plug and flame the ampoule neck. Using a sterile Pasteur pipette, add 0.2 ml of broth into the ampoule. Keep it for rehydration for a few minutes. Transfer the rehydrated culture to a tube of broth, using a fresh sterile pipette. Incubate the inoculated broth tube at 28 ± 2°C for 24 hours. Streak the agar plate from the incubated culture. Incubate the inoculated plate at 28 ± 2°C for 24 hours. Compare the colonies with the characteristics of the culture/ bacterium that was lyophilized.

13 Staining of Bacterial Cultures for Morphological Studies

Some microorganisms cannot be studied properly because they are transparent or colorless and therefore difficult to see when suspended in an aqueous medium. Specimens are, therefore, routinely stained to increase visibility and to reveal additional information to help identify microbes. Today several stains and staining procedures are available to study the properties of various microorganisms and their differentiation into specific groups/genera/species.

The chemical substances commonly used to stain bacteria are known as dyes. Dyes are classified as natural or synthetic; the former are used mainly for histological purposes while the latter are used mainly for the bacterial stain preparations. These are coal tar (aniline) dyes. Chemically a dye (stain) is defined as an organic compound containing a benzene ring plus a chromosphere and auxochrome group. Such dyes are acidic, basic, or neutral. The acidic dyes (e.g., picric acid, acid fuchsin, eosin) are anionic and stain the cytoplasmic components of the cells that are more alkaline in nature. On the other hand, the basic dyes (e.g., methylene blue, crystal violet, safranin) are cationic and combine with those cellular elements which are acidic in nature (e.g., nucleic acids). Neutral stains are formed by mixing together aqueous solutions of certain acidic and basic dyes. The coloring matter in neutral stains is present in both negatively and positively charged components.

Staining solutions are prepared by dissolving a particular stain in either distilled water or alcohol. The stain is applied to smears for 30–60 seconds, washed, dried, and examined under the microscope.

There are two kinds of staining procedures, simple and differential. Simple stains employ a single dye (e.g., methylene blue, crystal violet, or carbol fuchsin) and cells and structures within each cell will attain the color of the stain. Differential stains require more than one dye and distinguish between structures within a cell or types of cells by staining them different colors.

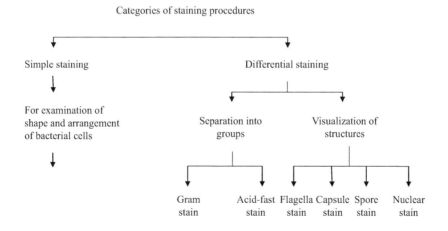

13.1 PREPARATION OF BACTERIAL SMEARS

Observations of microorganisms in the living state are the most difficult and limited. When light microscopy is used to study the morphology of bacteria, that is, size, shape, arrangement, and structure for detailed examination of cells, observation of internal cellular components, and for differentiation into specific groups, it is desirable that the cells are first fixed (killed) and then stained to make them more readily visible.

Bacterial cultures are grown either on solid (agar) media or in broth. Smears (films) of bacterial cultures are prepared by aseptically transferring a loopful of the material from the broth culture on a clean glass slide. For cultures on agar media, the organisms are first suspended and emulsified material is spread out into a thin film on clean glass by means of a platinum loop. This film is called the *smear*. The smear is air-dried and then passed with smear side up through a flame two or three times to heat fix the bacteria. Heat fixing denatures bacterial proteins (or enzymes), preventing autolysis, and also enhances the adherence of bacteria to the slide. The smear thus prepared is used for appropriate staining.

13.2 SIMPLE OR DIRECT STAINING OF BACTERIA

In simple staining, the cells (smear) are stained by the application of a single staining reagent. The purpose of the simple staining technique is to determine the cell shape, size, and arrangement of bacterial cells.

Most bacteria have a defined shape that falls into one of three morphological categories: (1) Spherical (cocci, singular coccus); (2) straight rods (bacilli, singular bacillus); and (3) spiral- or corkscrew-shaped organisms (spirilla, singular spirillum [rigid bacteria], or spirochetes [highly flexible]). A few bacteria change their shape and are called *pleomorphic* (e.g., arthorobacter).

Simple staining is performed by using basic stains that have different exposure times (e.g., crystal violet, 2–60 seconds; carbol fuchsin, 15–30 seconds; and methylene blue, 15–120 seconds).

Material Required

Twenty-four-hour-old cultures of test bacterium, staining solution of methylene blue/crystal violet/ carbol fuchsin, staining tray, glass slides, inoculating loop, Bunsen burner/spirit lamp, blotting paper.

Procedure

Take clean glass slides and wash and dry them. Prepare bacterial smears of all the bacterial cultures as described in Section 13.1. Keep a slide (heat fixed smear) on the staining tray and apply about five drops of a stain (any one of the above) for the designated period. Pour off the stain and wash the smear gently with slowly running tap water. Blot dry the slide using blotting paper (do not wipe the slide).

Observations

Examine all the preparations under oil-immersion objective. Make drawings for each bacterial organism. On the basis of the microscopic observations, write a description of the bacterium as shape and arrangement of cells in each. (See Figure 13.1.)

13.3 NEGATIVE STAINING OF BACTERIA

Negative staining provides the simplest and often the quickest means of gaining information about the cell shape, cell breakage, and retractile inclusions in cells such as sulfur and poly-ß-hydroxyl butyrate granules and about endospores.

A stain that stains the background and does not stain the bacteria is called a negative stain. In negative staining technique a simple stain, that is, nigrosin or India ink (an acidic stain), is used. The

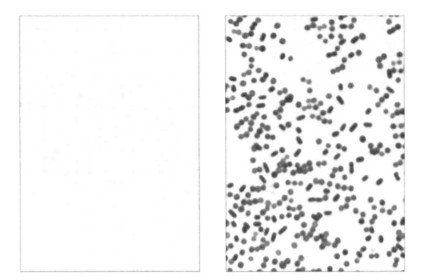

FIGURE 13.1 Bacterial cells in simple stain. (Courtesy of Dr. S. G. Borkar and Dr. Nisha Patil, Department of Plant Pathology, Mahatma Phule Krishi Vidyapeeth, Rahuri.)

acidic stain readily gives up a hydrogen ion and the chromophore of the dye becomes negatively charged. Since the surface of most of bacterial cells is negatively charged, the cell surface repels the stain, and therefore, bacterial cells appear transparent and unstained upon examination, whereas the background is stained.

Negative staining is advantageous for two reasons: (1) Cells appear less shriveled or distorted because no heat fixing is done; and (2) capsulated bacteria that are difficult to stain can be observed by this technique.

Material Required

Twenty-four-hour culture of test bacterium, nigrosin solution, clean glass slides, staining tray, inoculating loop, Bunsen burner, tissue paper, microscope.

Procedure

Place one drop of nigrosin at one end of a clean glass slide. With the help of a sterile inoculating loop, transfer a loopful of the inoculums from the broth culture in the drop of stain and mix gently with the loop. When the inoculums are taken from an agar medium, a bacterial suspension is made. Mix a drop of this suspension in the nigrosin. Take another clean slide, place it against the drop of suspended bacterium at an angle of 30° and allow the droplet to spread across the edge of the top slide. Spread the mixture of the stained inoculums out into a thin wide smear by pushing the top slide to the left along the entire surface of the bottom slide. Allow the smear to air dry.

Observations

Examine the preparation under an oil-immersion objective. Select the best microscopic fields and make drawings of a few cells. Note the shape, arrangement, and size of cells.

Precautions

Cultures used should be 24 hours old. Always place a small drop of stain close to one end of a clean slide. Do not spread the drop while mixing it with the culture. The thickness of the film should be uniform. Never heat fix the slide. Negatively stained preparations should not be used for cell length and width measurements. (See Figure 13.2.)

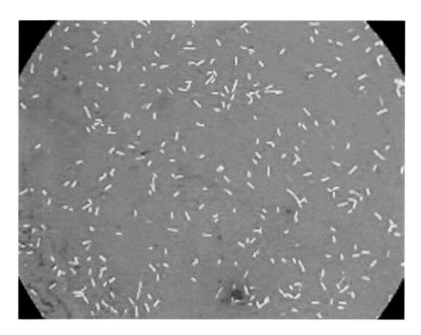

FIGURE 13.2   Bacterial cells in negative staining. (Courtesy of Dr. S. G. Borkar and Dr. Nisha Patil, Department of Plant Pathology, Mahatma Phule Krishi Vidyapeeth, Rahuri.)

13.4   GRAM STAINING OF BACTERIA

The Gram stain is a differential stain developed by Dr. Hans Christian Gram, a Danish physician, in 1884. The staining is called Gram staining, after Dr. Gram. It is a very useful stain for identifying and classifying bacteria into two major groups: Gram-positive and Gram-negative.

In this process, the fixed bacterial smear is subjected to four different reagents in the order listed: Crystal violet (primary stain), iodine solution (mordant), alcohol (decolorizing agent), and safranin (counterstain). The bacteria which retain the primary stain (appears dark blue or violet) are called Gram-positive, whereas those that lose the crystal violet and get counterstained safranin are referred to as Gram-negative.

The differences in staining responses to the Gram stain can be related to the chemical and physical differences in the bacterial cell walls. The Gram-negative bacterial cell wall is a thin, complex, multilayered structure and contains relatively high lipid content, in addition to protein and muco-peptides. The higher amount of lipids is readily dissolved by alcohol, resulting in the formation of large pores in the cell wall that do not close appreciably on dehydration of cell-wall proteins, thus facilitating the leakage of the crystal violet–iodine (CV-I) complex and resulting in the decolori-zation of the bacterium, which later takes the counterstain and appears red. In contrast the Gram-positive cell walls are thick and chemically simple, composed mainly of protein and cross-linked mucopeptides. When treated with alcohol, it causes dehydration and closure of cell wall pores, thereby not allowing the loss of the CV-I complex, and so the cells remain purple.

Material Required

Cultures of test bacterium that are 24 or fewer hours old, Gram staining reagents crystal violet, Gram's iodine solution, 95 percent ethyl alcohol, safranin, staining tray, wash bottle of distilled water, droppers, inoculating loop, glass slides, blotting paper/absorbent paper, tissue paper, Bunsen burner/spirit lamp, microscope.

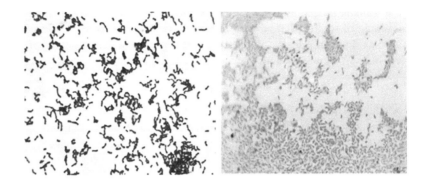

FIGURE 13.3 Gram staining of bacteria. (Courtesy of Dr. S. G. Borkar and Dr. Nisha Patil, Department of Plant Pathology, Mahatma Phule Krishi Vidyapeeth, Rahuri.)

Procedure

Make thin smears of test bacterium on glass slides. Let the smears air dry. Heat fix the smears. Hold the smears using a slide rack or clothes pin. Cover the smear with crystal violet for 30 seconds. Wash the slide with distilled water for a few seconds using the wash bottle. Cover the smear with Gram's iodine solution for 60 seconds. Wash off the iodine solution with 95 percent ethyl alcohol. Add ethyl alcohol drop by drop, until no more color flows from the smear. (The Gram-positive bacteria are not affected while all Gram-negative bacteria are completely decolorized.) Wash the slides with distilled water and drain. Apply safranin to smears for 30 seconds (counterstaining). Wash with distilled water and blot dry with absorbent paper. Let the stained slides air dry.

Observations

Examine the slides microscopically using oil-immersion objective.

Results

Those bacteria that appear purple are referred to as Gram-positive; those appearing pink are described as Gram-negative. (See Figure 13.3.)

Precautions

Always use fresh and young cultures (less than 24 hours old) to avoid misleading results. Excessive heat should be avoided during heat fixing. Overdecolorization of the smear should be avoided. Smears should be thin and uniform.

13.5 ACID-FAST STAINING OF BACTERIA

The acid-fast stain is a differential stain. It was developed by Paul Ehrilih in 1882 and was later modified by Franz Ziehl and Friedrich Neelsen. It is used by present-day microbiologists. Bacteria are classified as acid-fast if they retain the primary stain (carbol fuchsin) after washing with strong acid and appear red or as non–acid-fast if they lose their color on washing with acid and counterstained by the methylene blue.

The property of acid-fastness appears to be due to the presence of high contents of a lipid called mycolic acid in the cell wall, which makes penetration by stains extremely difficult. In the acid-fast procedure, bacterial smear is treated with carbol fuchsin, followed by heat fixing and treatment with acid alcohol and methylene blue.

The acid-fast staining procedure is useful for the identification of members of *Mycobacterium*, especially in staining of sputum that have *Mycobacterium tuberculosis*, the cause of tuberculosis, because this bacillus is the only acid-fast organism commonly found in the sputum. *M. smegmatis* is a nonpathogenic but normal inhabitant of the genitals and may be mistaken for *M. tuberculosis* in the urine.

Material Required

Nutrient agar slants or broth cultures of test bacterium, acid-fast staining reagents, carbol fuchsin, acid–alcohol (3% HCI + 95% alcohol), methylene blue, staining rack, hot plate/water bath, glass slides, Bunsen burner, inoculating loop, blotting paper/absorbent paper, tissue paper, microscope.

Procedure

Prepare smears of test bacterium on slides. Air dry and heat fix the smears. Flood smears with carbol fuchsin. Heat the slides to steaming for 3–5 minutes. From time to time, add more stain to prevent smears from becoming dry. Cool slides and wash with distilled water. Decolorize the smear with acid–alcohol for 10–30 seconds or until the smear is a faint pink color. Wash slides with distilled water. Counterstain smears with methylene blue for 1–2 minutes. Wash with distilled water. Blot dry with blotting paper.

Observations

Examine smears under oil-immersion objective and record the color of test organisms to classify the organisms as acid-fast or non–acid-fast.

Precautions

Do not allow the carbol fuchsin stain to evaporate and dry; add more stain, if necessary. Prevent carbol fuchsin stain from boiling. Overdecolorization of the smear is to be avoided.

13.6   BACTERIAL CELL WALL STAINING

All true bacteria possess a rigid cell wall. The cell wall is the structure that immediately surrounds the cell membrane. A bacterium from which the wall has been completely removed (usually by enzymatic digestion) is referred to as a *protoplast*. The enzyme, lysozyme, selectively dissolves the cell wall of Gram-positive bacteria. The chemical composition of the cell wall may vary in different species of bacteria since cellulose, hemicelluloses, and chitin have all been reported as the main constituents. In addition, peptidoglycan (also known as murein and mucopeptide), a substance found only in prokaryotes, is also present. Peptidoglycan is an enormous molecule composed of amino acids and sugars (peptide, amino acid + glycan, sugar). Possibly the cell wall is also differentially permeable. The cell wall can be visualized by special methods.

Material Required

Twenty-four-hour nutrient agar slant of test bacterium, aqueous Congo red solution, acetyl pyridinium chloride (0.34 percent), Loeffler's methylene blue solution, clear glass slide, dropper, inoculating loop, Bunsen burner.

Procedure

Prepare a smear of the test bacterium culture. Heat fix the smear. Apply three drops of acetyl pyridinium chloride and one drop of Congo red. Mix with a glass rod for 5 minutes. Flood the smear with Congo red. Wash the slide with distilled water. Let the smear air dry. Counterstain with methylene blue. Wash with distilled water. Blot dry the slide.

Observations

Examine microscopically using oil-immersion objective.

Results

The cell wall stains red and cytoplasm stains blue.

13.7 CYTOPLASMIC MEMBRANE STAINING

The outermost layer of the cytoplasm is called the cytoplasmic membrane. This layer is 75 Å (angstrom) (7.5 nm or 0.0075 μm) units thick. The cytoplasmic membrane in bacteria is a phospholipids–protein bilayer similar to that present in eukaryotic cells. The major difference is that there are no sterols in the cytoplasmic membranes of most prokaryotes. Major disruptions in the membrane result in the spilling of the cytoplasm from the cell and the death of the organism. A plasmolysis technique is used to separate cell wall from the cytoplasm and the membrane is then stained.

Material Required

Bacillus culture, potassium nitrate solution, Bouin fixative solution, Victoria blue solution, glass slide, inoculating loop, Bunsen burner.

Procedure

Prepare a smear of the bacterium on a clean slide. Heat fix the smear. Immerse the slide in potassium nitrate solution for 15 minutes for plasmolysis to take place. Keep the slide in Bouin fixative solution for 15 minutes. Wash the slide in tap water. Apply the Victoria blue solution for 3 seconds. Wash the slide in tap water. Blot dry the smear.

Observations

Examine the preparation microscopically under oil-immersion objective.

Results

The cytoplasmic membrane appears as a deep blue outer layer covering a contracted, irregularly shaped body, the cytoplasm.

13.8 BACTERIAL SPORE (ENDOSPORE) STAINING

Some bacteria are capable of forming or changing into dormant structures in unfavorable environments; these structures are metabolically inactive and do not grow or reproduce unless the favorable environment returns. Since these structures are formed inside the cells, hence these are called endospores. The German botanist Ferdinand Cohn (1828–1898) discovered the existence of endospores in bacteria. These are remarkably resistant to heat, radiation, chemicals, and other agents that are typically lethal to the organism. The heat resistance of spores has been linked to their high content of calcium and dipicolinic acid. A single bacterium forms a single spore by a process called *sporulation*. Sporulation takes place either by depletion of an essential nutrient or during unfavorable environmental conditions. During sporulation, a vegetative cell gives rise to a new, intracellular structure termed an endospore, which is surrounded by impermeable layers called *spore coats*. Complete transformation of a vegetative cell into a sporangium and then into a spore requires 6–8 hours in most spore-forming species. An endospore develops in a characteristic position within a cell, that is, central, subterminal, or terminal. Once an endospore is formed in a cell, the cell wall disintegrates, releasing the endospore that later becomes an independent spore. Endospores can remain dormant for long periods of time. One record describes the isolation of viable spores from a

3000-year-old archaeological specimen. However, a free spore may return to its vegetative or growing state with the return of favorable conditions.

Plant pathogenic bacteria do not form endospores. Endospores are formed by members of the seven genera, for example, *Bacillus*, *Clostridium*, *Coxiella*, *Desulfotomaculum*, *Sporolactobacillus*, *Sporomusa*, and *Thermoactinomyces*. These include nonpathogenic soil inhabitants (*Bacillus* and *Clostridium*) and pathogenic (*Clostridium tetani*, *C. Perfingens*, *C. Botulinum*, and *Bacillus anthracis*, the agents of tetanus, gas gangrene, botulism, and anthrax, respectively).

The spores are differentially stained by using special procedures that help dyes penetrate the spore wall. An aqueous primary stain (malachite green) is applied and steamed to enhance penetration of the impermeable spore coats. Once stained the endospores do not readily decolorize and appear green within red cells.

Material Required

Forty-eight-hour nutrient agar cultures of test bacterium, malachite green (5 percent aqueous), safranin (0.5 percent aqueous), staining tray, glass slides, inoculating loop, blotting paper, spirit lamp, microscope.

Procedure

Make smears of test bacterium on separate clean slides. Air dry and heat fix the smears. Flood the smears with malachite green. Heat the slides to steaming and steam for 5 minutes, adding more stain to the smear from time to time. Wash the slides under slowly running tap water. Counterstain with safranin for 30 seconds. Wash smear with distilled water. Blot dry slides with absorbent/blotting paper.

Observations

Examine the slide microscopically under oil-immersion objectives. In *Bacillus*, the endospores stain green and the vegetative cells stain red. Observe the position of spore, that is, central, subterminal, or terminal. The position of spore is useful for species identification in *Bacillum*. (See Figure 13.4.)

13.9   CAPSULE STAINING

Some bacterial cells are surrounded by a mucilaginous substance forming a viscous coat around the cell. This structure is referred to as a capsule when round and oval in shape and firmly bound to bacterium whereas it is referred to as slime layer when irregularly shaped and loosely bound to the bacterium.

Capsule that is external to the cell is also synthesized partially in the cytoplasm and is usually composed of polysaccharides but may contain other materials; for example, *Bacillus anthracis* has a capsule of poly-D-glutamic acid. The ability of the bacterium to form a capsule is genetically determined. The capsule is well-developed in some bacteria like *Streptococcus pneumonia*, *Clostridium perfringens*, and *Klebsiella pneumonia* while indistinct in other bacteria. In some bacteria (e.g., *Beijerinckia*) a capsule may enclose more than one cell. The diagnosis of pneumonia and other diseases is assisted by capsule staining.

Capsules help bacteria resist phagocytosis by phagocyte cells. Moreover, capsules contain a great deal of water and can protect bacteria against desiccation. They exclude bacterial viruses and most hydrophobic toxic materials such as detergents.

The capsules of specific pathogens can be displaced effectively (e.g., on pneumococcal, hemophilic influenza, and meningococcal) by the use of antisera specific for the capsule type, and this can provide a presumptive identification of bacterium. Capsules are clearly visible in the light microscope when negative stains and special capsule stains are employed.

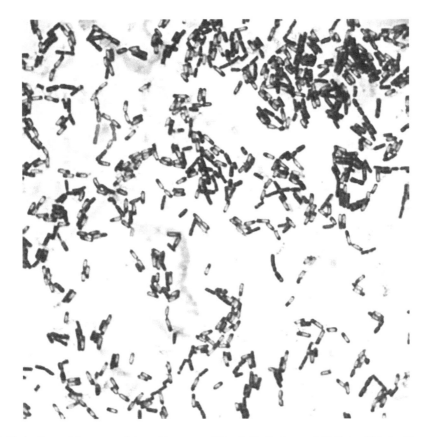

FIGURE 13.4   Bacterial spore staining. (Courtesy of Dr. S. G. Borkar and Kalindee Shinde, Department of Plant Pathology, Mahatma Phule Krishi Vidyapeeth, Rahuri.)

1. Capsule Staining by Negative Staining Technique

In negative staining, the background but not the bacterium is stained by the acidic stain india ink or nigrosin. The stain readily gives up a hydrogen ion and the chromophore of the dye becomes negatively changed. Since the surface of most of the bacteria cell is negatively charged, the cell surface repels the stain. In negative staining, the bacterial growth is mixed with a loopful of the stain (India ink or aqueous nigrosin) on a clean slide and overlaid with a cover slip. The capsule appears as a clear zone between the cell wall and the dark background under the high-dry and oil-immersion objective.

Material Required

A 36–48 hour milk culture of test bacterium, nigrosin/India ink, absorbent paper, clean slides (2), glass cover slips, inoculating loop, Bunsen burner.

Procedure

Take a clean glass slide. Put a drop of nigrosin or India ink close to one end of the clean glass slide. Add two loopfuls of a broth culture into the drop of the stain and mix with the loop. Prepare a smear of the suspended organism using edge of a second slide held at a 30° angle and pushed away to the other end of the slide. Air dry the smear.

Observations

Examine the slide under oil-immersion objective. The capsulated bacteria will appear as clear zones between the cell wall and dark background.

2. Capsule Staining by the Anthony Staining Method

In 1931, E. E. Anthony devised a simple method for capsule staining in bacteria. In this method, two stains, viz. crystal violet and copper sulfate, are used. Crystal violet (1 percent aqueous) is used as a primary stain. On application of crystal violet, both the capsular material and the cell wall will take the color of the stain and will appear dark blue in color. But the capsule being nonionic will not absorb the primary stain. Copper sulfate, which acts both as a decolorizing agent and a counterstain, on application, removes the excess primary stain and gives color to the decolorized capsular material. The capsule finally appears light blue in contrast to the deep purple color of the cell.

Material Required

A 36–48 hour milk culture of test bacterium, crystal violet aqueous (1 percent w/v), copper sulfate (20 percent), inoculating loop, glass slides (2), blotter paper.

Procedure

Take a clean glass slide. Put a loopful of the bacterial culture on one end of the slide. Prepare a heavy smear of the bacterium. Add crystal violet and allow the smear to react with crystal violet for 2 minutes. Wash off the dye with 20 percent copper sulfate. Drain copper sulfate. Gently blot-dry the slide.

Observations

Examine the slides under oil-immersion objective. In capsulated bacteria the vegetative cells are dark purple-colored, encircled by light blue colored capsules. No such distinctions are seen in a non-capsulated bacterium.

Precautions

Always prepare a heavy smear of the organism as some of the cells will slide off the slide during washing. Never heat fix the smears, as heating results in shrinkage which may create a clear zone around the cell that is an artifact and that can be mistaken for the capsule. Washing off the smear with water is to be avoided as the capsular materials are water-soluble and may be dislodged and removed with vigorous washing.

13.10   FLAGELLA STAINING

Many bacteria are motile due to the presence of one or more very fine threadlike, filamentous appendages called flagella. These are thin proteinaceous structures which originate in the cytoplasm and project out from the cell wall.

   Bacteria show four types of flagellation pattern: (1) Monotrichous, possessing a single flagellum at one end (or pole) of the cell; (2) lophotrichous, having many flagella in tufts or clusters at one end; (3) amphitrichous, possessing flagella at both ends, either singly or in tufts; and (4) peritrichous, possessing flagella all over the surface. If the flagella are present on the ends (either one or both) of the cell, they are called polar flagella. The presence, location, and the numbers possessed by the bacterium are some of the important criteria used in the identification and classification of bacteria. In order to see them with the light microscope, they are to be thickened first by the use of chemicals known as a mordant and later stained with a dye.

Material Required

An 18-hour nutrient agar culture of test bacterium, flagella mordant (i.e., Ziehl's carbol fuchsin), specially clean glass slides, 1 ml distilled water blank, dichromate solution, 95 percent alcohol, wash bottle, inoculating loop, Bunsen burner, microscope.

Procedure

Preparation of grease-free slides: Take new slides, dip them in dichromate solution, wash with water, rinse in 95 percent alcohol, wipe dry with cheese cloth, and pass the slide through flames of Bunsen burner. Allow the slides to cool.

Preparation of smear: Make suspension of the bacterium in distilled water. Incubate the suspension at room temperature for 10–15 minutes. Place one loopful of the suspension toward one end of the grease free slide.

Tilt the slide at 30° angle to allow the drop to spread to form a thin film on it. Allow the film to air dry at room temperature. Cover the slide with flagella mordant for 10 minutes. Wash gently with distilled water. Flood the slide with carbol fuchsin for 5 minutes. Wash the slide gently with distilled water. Air dry the slide.

Observations

Examine the slide microscopically under the oil-immersion objective. Record the number of flagella and various flagella arrangements. The cells appear as pink, straight rods surrounded by a deep-stained outer coat that bears pink-stained flagella. These flagella are fine, wavy threads of greater length than the cells. They may be peritrichous (i.e., in *Erwinia*), lophotrichous (in *Pseudomonas*), or monotrictium (in *Xanthomonas*).

Precautions

Always use especially cleaned, grease-free slides. Do not heat fix the smears. Do not blot dry the smears. Always use reduced illumination of the microscope to see flagella clearly.

13.11   STAINING OF NUCLEAR MATERIAL OF BACTERIA

Bacteria possess nuclear material consisting of a single circular molecule of DNA, in contrast to eukaryotes, where the genetic material is present in a membrane-bound structure, the nucleus. The nuclear material in prokaryotes is present in a region called the nucleotide, which is devoid of a nuclear membrane and does not divide by mitosis or meiosis.

In many bacteria, additional genetic information may be found on plasmids. These are small, circular pieces of DNA that can replicate independently of the chromosome and are usually less than one-hundredth of the size of the chromosome. Plasmid DNA may give a bacterium the power to synthesize new products and carry information providing the cell with resistance to antibiotics, with the ability to produce toxins, or with the ability to produce surface appendages essential for attachment and establishment of infection.

In nuclear material staining, the cytoplasm, which possesses a strong affinity for most stains and interfere with the observation of the nuclear material, is first hydrolyzed with hydrochloric acid and later stained with the Giemsa stain. Nuclear bodies appear purple-colored.

Material Required

Bacterial culture, 1N hydrochloric acid, Giemsa stain, clean glass slide, water bath, beaker, inoculating loop, blotting paper, Bunsen burner, microscope.

Procedure

Prepare a smear of the bacterial culture on a clean slide. Heat fix the smear. Keep the smear slide in a beaker containing HCl solution. Keep the beaker in a water bath at 60°C for 10 minutes. Wash the slide with tap water. Flood the smear with Giemsa stain for 2–3 minutes. Wash with tap water. Blot dry the smear.

Observations

Examine the preparation using oil-immersion objective. The nuclear bodies, one or two in each cell, stain purple surrounded by a colorless zone of cytoplasm. The cell membrane appears as a faintly purple layer.

13.12 VIABILITY STAINING OF BACTERIA

The distinction between living and dead cells in a culture is done by the viability staining technique. This method is based on the fact that certain changes take place in dead cells which result in these cells giving a different reaction to stains from that of living cells.

Material Required

Two-to-three-week-old *Bacillus* culture, Loeffler's methylene blue solution, dilute carbol fuchsin solution, glass slide, Bunsen burner, blotting paper, microscope.

Procedure

Prepare a smear of the culture. Heat fix the smear. Flood the smear with methylene blue for 10 minutes. Wash the smear thoroughly with tap water until the smear appears pale blue. Run carbol fuchsin down the slide for a very short time. Wash immediately with tap water. Blot dry the smear.

Observations

Examine the smear microscopically. Living bacteria appear purple while dead bacteria appear red or pink. Living spores stain faintly pink and dead spores stain blue.

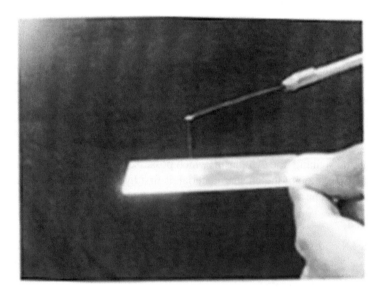

FIGURE 13.5 Formation of bacterial string by Gram-negative bacteria in KOH solution. (Courtesy of Dr. S. G. Borkar, Department of Plant Pathology, Mahatma Phule Krishi Vidyapeeth, Rahuri.)

13.13 SUSLOW REACTION TO DETERMINE GRAM-POSITIVE AND GRAM-NEGATIVE BACTERIA

Material Required

Test bacterial culture, 3 percent KoH solution, glass slide, streaking needle.

Procedure

Place two drops of 3 percent KoH solution on glass slide. Pick up a loopful of test bacterial culture by using inoculation needle. Put the culture in the KoH drop and stir clockwise till the culture is dispensed in the KoH solution or the mucoid suspension is formed. Lift the inoculating needle along with mucoid material to note the reaction.

Observation

If the culture forms the mucoid material and forms a string when inoculating needle is lifted, it is a culture of Gram-negative bacteria. If the culture is dispose in the KoH solution and does not form a string while lifting the needle, it denotes a culture of Gram-positive bacteria. (See Figure 13.5.)

14 Bacterial Mobility

Bacteria may or may not have mobility appendages, that is, flagella for their movement. The bacterial movement with the help of their flagella is called *true motility*. Some of the bacteria without flagella have gliding movement. Sometimes, the bacterial movement can be mistaken for the Brownian movement.

The plant pathogenic bacteria have the flagella appendages which help them to move in a thin film of water on the plant surface to reach the entry point to cause infection.

To demonstrate the mobility of bacteria due to presence of flagella, hanging drop technique is used.

14.1 HANGING DROP TECHNIQUE FOR DEMONSTRATING MOTILITY OF BACTERIA

Hanging drop preparation is useful for microscopic examination of living bacteria without staining them and to see their motility due to flagella.

Material Required

Twelve-hour-old broth culture of *Erwinia* spp., hanging drop (cavity) slide, cover slips, Vaseline/petroleum jelly, matchsticks, wax marking pencil.

Procedure

Clean and flame a hanging–drop slide and place it on the table with the depression uppermost. Spread a little Vaseline or petroleum jelly around the cavity of the slide. Clean a cover slip and apply petroleum jelly on each of the four corners of the cover slip, using a matchstick. Place the cover slip on a clean paper with the petroleum jelly side up.

Transfer one loopful of culture in the center of the cover slip. Place the depression slide on to the cover slip so that cavity is facing down the suspension. Press the slide gently to form a seal between the cover slip and the slide. Lift the preparation and quickly turn the hanging drop preparation with cover slip up so that the culture drop is suspended. Examine the preparation under low-power objective with reduced light.

Observations

Note the size, shape, and characteristics of motility of bacteria (i.e., whether true motility or Brownian movement; a vibratory or dancing movement of bacteria in suspension due to bombardment by the molecules of water). In case of true motility, the movement of bacteria is observed from one place to another.

14.2 BROWNIAN MOVEMENT

The random movement of microscopic particles suspended in a liquid or gas, caused by collisions with molecules of the surrounding medium, is known as Brownian movement.

It should not be confused with bacterial movement in the suspension.

Material Required

Zinc sulfate, distilled water, glass slide, microscope.

Procedure

Prepare a suspension of zinc sulfate in distilled water. Put a drop on glass slide, cover with cover slip, and observe in microscope.

Observations

Random movement of particles can be seen under the microscope at a fixed place while in case of bacterial movement, the movement can be observed in the microscopic field from one place to another.

14.3   RELATIONSHIP OF VIRULENCE WITH MOTILITY (IN *ERWINIA CAROTOVORA* VAR. *ZEAE*)

Medium (Motility Agar)

Composition	
Beef extract	3.0 g
Peptone	5.0 g
Agar	5.0 g
Distilled water (to make up)	1000 ml

Procedure

Inoculate *Erwinia carotovora* var. *zeae* by stabbing a loopful of inoculum in the medium vertically in a Petri plate and incubate at 25°C.

Observations

Virulent isolates of *Erwinia carotovora* var. *zeae* spread horizontally to a greater distance from the point of stabling on either side in Petri plate than nonvirulent isolates as evident by the formation of bacterial colonies.

15 Describing Bacterial Colony Morphology

Colony morphology is an important aspect in the detection of plant pathogenic bacteria in isolation plates. The plant pathogenic bacterial colonies usually develop after 48 hours of incubation while saprophytic non plant pathogenic bacterial colonies are observed after 24 hours or even in less incubation period.

Generally plant pathogenic bacteria produce round, raised, glistering nonpigmented or pigmented colonies on nutrient agar medium or specialized medium. The colony type with colony pigment is an important characteristic of the bacterial genus and species of plant pathogenic bacteria.

The different colony types, colony edge, internal colony structure, and colony pigment are illustrated below. (See also Figure 15.1.)

15.1 COLONY TYPE

Circular:	Round colonies
Irregular:	Without symmetry or even shape
Amoeboid:	Resembling or elated to amoebic
Rhizoid:	Root-like
Filamentous:	Very fine thread or thread-like structure
Curled:	To form into a spiral or curved shape coil
Myceloid:	Resembling mycelium
Toruloid:	Beaded, an aggregate of colonies like those seen in the budding of yeast

15.2 COLONY ELEVATION

Effuse:	To exude or flow out; spread out loosely
Raised:	Heighten or to set upright
Convex:	Having a surface that is curved or rounded outward
Concave:	Hollow and curved
Umbonate:	Having a rounded convex form
Pulvinate:	Cushion-shaped

15.3 COLONY EDGE

Entire:	Intact, not broken, without notches
Erose:	Having the margin irregularly incised as if gnawed, or uneven
Crenate:	Having the margin notched or scalloped so as to form rounded teeth
Undulate:	To have a wavy form or surface
Lobate:	Having the form of a lobe
Ciliate:	Having cilia on part or all of the body
Fimbriate:	Having a border of hairs
Lacerate:	To tear roughly; mangle
Ramose:	Branching

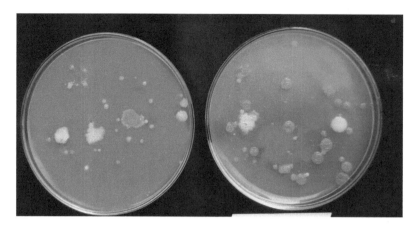

FIGURE 15.1   Colony types in bacteria. (Courtesy of Beatrice Leung, Genentech, Inc. Shijun Liu, Science Buddies, scibuddy@sciencebuddies.org.)

15.4   INTERNAL COLONY STRUCTURE (× 15)

Transparent:	Clear, permitting light to pass through so that objects on the opposite side are clearly seen
Translucent:	Permitting light to pass through but diffusing it so that objects on the opposite side are not clearly visible
Opaque:	Not transparent or translucent, not allowing light to pass through
Smooth:	Not rough, free of unevenness of surface
Finely granular:	Showing a granulated structure
Wavy interlaced:	To cross one another as if woven together; intertwine
Filamentous:	Thread-like structure
Arborescent:	Tree-like in form

15.5   COLONY PIGMENTS

The color of the colonies can be variable and the colonies appear as the following:

Yellow	White
Red	Gray
Dark brown	Pink
Blue	Milky

16 Physiological Characterization of Bacteria (Effect of Environmental Conditions on Growth of Bacteria)

The surrounding environment affects the growth and metabolism of bacteria and is therefore directly related to induction or suppression of disease initiation and symptoms development by the bacterial plant pathogen.

Temperature is one of the most important physical factors affecting microorganisms. Bacteria are different from higher plants and animals in the lack of homeostatic mechanism and cannot regulate heat generated by metabolism and are, therefore, directly and readily affected by temperature. Over a limited temperature range, there is a twofold increase in the rate of enzyme-catalyzed reactions for every 10°C rise in temperature.

Bacteria may be divided into three major groups with respect to their temperature requirements: (1) psychrophiles, those with optimum temperature between 0° and 20°C; (2) mesophiles, those with optimum temperature between 20° and 40°C; and (3) thermophiles, those with optimum temperature between 40° and 80°C. Thermophiles are of two types: (1) Facultative thermophiles, those with an optimum temperature of growth between 45° and 60°C; and (2) obligate thermophiles, those for which 50° and 100°C is in the lethal range for bacterial cells and spores. The range of temperature preferred by bacteria is genetically determined, which govern the metabolic process in bacterial cells.

Each organism grows within a particular temperature range (i.e., cardinal temperature points). The minimum growth temperature is the lowest temperature at which growth of a species will occur. The highest temperature at which a species can grow is its maximum growth temperature, and a species grows fastest at its optimum growth temperature.

Time of exposure is a vital factor in assessing the lethal effect of high temperature on bacterial cells. For this purpose, two methods are useful: (1) the thermal death point (TDP), the temperature at which an organism is killed in 10 minutes of exposure; and (2) thermal death time (TDT), the time required to kill cells/spores suspension at a given temperature.

16.1 EFFECT OF INCUBATION TEMPERATURES ON BACTERIAL GROWTH

Plant pathogenic bacteria, like all other bacteria, require certain temperature for growth. This is evident from the fact that certain plant pathogenic bacterial species are prevalent in temperate regions while some bacterial species are prevalent in tropical regions, indicating that temperature plays an important role in the growth, multiplication, and metabolic activities, including pathogenicity of the bacterium. Therefore, studies on requirements of temperature for a given bacterial species are important.

Material Required

Bacterial cultures; nutrient agar slants; Bunsen burner; inoculating loop; inoculating needle; refrigerator set at 4°C; four incubators set at 15°C, 35°C, 55°C, and 75°C.

Procedure

Label each of the nutrient agar slants with the test organisms (bacteria) and the temperature of incubation (0°C, 4°C, 15°C, 35°C, 75°C). Inoculate six nutrient agar slants/plates (labeled 0°C, 4°C, 15°C, 35°C, 55°C, 75°C) with test bacterium by streaking. Incubate one inoculated slant and plate from each set at 0°C (icebox of refrigerator) and 4°C (refrigerator) for 10–14 days and other slants and plates at 15°C, 35°C, 55°C, and 75°C (incubators set at these temperatures) for 24–48 hours.

Observations

Examine the cultures after two days of incubation (incubated at 15°C, 35°C, 55°C, and 75°C) and after 10–14 days of incubation (at 0°C and 4°C) for the presence (+) or absence (–) of growth and the degree of growth; that is, minimal growth (+), moderate growth (2+), heavy growth (3+), and very heavy (maximum) growth (4+). Express the results and determine the minimum, maximum, and optimum temperature ranges for test bacterium and classify these as psychrophile, mesophile, or thermophile.

16.2   DETERMINATION OF THERMAL DEATH POINT OF BACTERIA (TDP)

Thermal death point is an important aspect to be considered in the management of plant pathogenic bacteria. In the changing environment or climatic seasons the knowledge of thermal death point of bacteria will be helpful to determine the strategies of its management.

Enzymatic reactions proceed at maximum speed and efficiency at optimum temperature, which varies with the organism. The microbial enzymes become inactive beyond the maximum and minimum temperature extremes. High temperatures are harmful to the microorganisms because it denatures proteins, causing irreversible changes and total enzyme destruction, which results in the death of the cell. Low temperatures merely have an inactivating effect on the enzymes and are thus less damaging to the microorganisms. Temperatures in the range of 50–100°C are normally in the lethal range for bacterial cells but there may be considerable variation as to the length of time required to kill them since time of exposure is a vital factor in assessing the lethal effect of high temperatures on microorganisms. In other words, lethal action of heat is a time–temperature relationship. For determining the lethal effects of temperature on microorganisms, thermal death point (TDP; i.e., the temperature at which an organism is killed in 10 minutes of exposure) and thermal death time (TDT; i.e., the time required to kill a suspension of spores or cells at a given temperature) of an organism should be determined. The degree of heat tolerance of microorganism in the laboratory can be determined by exposing the cells to increasing temperature for a fixed period, that is, 10 minutes.

Material Required

Nutrient broth cultures of test bacterium (30–48 hours culture taken in flasks), nutrient agar plates, thermostatic hot water bath, sterile empty tubes with cap, sterile cotton-plugged 10 ml pipettes, culture tube with 10 ml water, centigrade thermometer, wax glass marking pencil, culture tube stand, inoculating loop/needle, Bunsen burner.

Procedure

Divide the bottoms of the two sterile nutrient agar plates into six sectors with the glass marking pencil. Label the six sectors of each plate as 40°C, 50°C, 60°C, 70°C, and 80°C. C (Control) stands for without any temperature treatment. Line up five sterile test tubes for the test organism and label with the temperatures 40°C, 50°C, 60°C, 70°C, and 80°C. Aseptically transfer 2 ml nutrient broth culture of test organism into its appropriately labeled tubes using sterile pipettes. Hold the tube of test bacterium labeled 40°C in the water bath (be sure that the water of the bath is above the broth level) where the water temperature is adjusted to 40°C, and keep the tube for 10 minutes at a constant 40°C temperature. Repeat the same steps for the remaining tubes for the other temperature (50°C, 60°C, 70°C, and 80°C) and make inoculations from the temperature-subjected broth and control broth immediately after temperature treatment into the respective temperature sectors. Incubate the bacterial inoculated plates at 30°C for 24–48 hours in an inverted position or until growth occurs in the control (C) sectors.

Observations

Examine the plates for the presence (+) or absence (–) of growth in all the sectors and compare with control.

From the above results find the thermal death point (at which death of an organism occurs) of the test bacterium.

Precautions

While making a streak inoculation of bacterium the inoculums should be at least 1–2 cm away from the plate center where the sectors converge. Inoculation of a bacterium may be made on the plate before cooling the broth. Constant attention should be given to the temperature of the water bath so that adjustment can be made before serious fluctuation in temperature occurs.

16.3 DETERMINATION OF THERMAL DEATH TIME OF BACTERIA

The time taken by the bacteria at a given temperature to cause its death is known as *thermal death time* (TDT). The determination of TDT, that is, the length of time that the bacteria is exposed to heat, contributes to the lethal effect. Their degree of heat tolerance can be assessed by exposing the bacterial cells to a fixed temperature (determined in the previous experiment) for increasing periods of time with subculturing being made at the end of each time interval.

Material Required

Nutrient broth cultures of test bacterium, nutrient agar medium plates, centigrade thermometer, inoculating loop/transfer needle, sterile empty tubes with cap, thermostatic hot water bath, marking pencil, and so on.

Procedure

Divide the bottoms of the nutrient agar Petri plate into six sectors with the glass marking pencil and mark these sectors as 0, 3, 6, 9, 12, and 18; on the other plate, mark the sectors as 21, 24, 27, and 30. Aseptically transfer 2 ml of nutrient broth culture of test bacterium into the appropriately labeled sterile tube using sterile pipettes. Place the first tube of test bacterium into the water bath set at a particular temperature (as determined in the previous exercise); say, 40°C for 4 minutes (1 minute for equalizing the temperature of the broth to the water and 3 minutes time of exposure). Remove the tube and cool quickly under tap water. Make streak inoculations on sector 3 of the NA plate. Now hold another tube, marked "6," in the water bath and allow 1 minute for the broth and water to become equalized before starting the next time interval, for instance, 6 minutes and repeat. Likewise for other time intervals of 9, 12, 15, 18, 21, 24, 27, and 30. Incubate the bacterial inoculated plates at 30°C in an inverted position for 24–48 hours or until growth occurs in the "0" time sector.

Observations

Observe the inoculated plates for the presence of growth in various sectors and compare with the "0" time sector. Record the growth as + (growth present) and – (no growth) and determine the TDT of test organism, that is, the time to kill the test organism at the given temperature.

Precautions

Water level of the water bath should be above the broth level. The temperature of the water bath should be kept constant. After a particular temperature treatment, the broth tubes should be immediately cooled to make the streaking.

16.4 EFFECT OF pH ON THE BACTERIAL GROWTH

The pH is the symbol for the logarithm of the reciprocal of hydrogen ion concentration ($logI/H^+$) in gram atoms per liter (i.e., (power) of $[H^+]_{10}$). For example, a pH 4 indicates a concentration of .0001 or 10^{-4} gram atoms of hydrogen ions in one liter of solution. In other words, the pH value is the negative exponent of the H^+ concentration. The measure of the relative acidity or alkalinity of a solution is called pH. pH is expressed as a number from 0 to 14. A more acidic solution at 0.01 (10^{-2}) M has a lower pH value (pH = 2). pH changes of the environment have a profound influence on the growth, activities, and survival of the bacteria. Each enzyme system of an organism has a particular pH range in which it can function. Like cardinal temperature points, an organism has a maximum pH (i.e., the concentration of H^+ [alkali] below which an organism fails to grow), optimum pH (i.e., the concentration of H^+ where growth of an organism is maximum), and minimum pH (i.e., the concentration of H^+ [acid] above which a microorganism fails to grow). In other words, the pH values above and below which an organism fails to grow are respectively termed as the minimum and maximum hydrogen ion concentrations. Enzymes differ greatly in their pH optima.

Each bacterial species has the ability to grow within a specific pH range that may be broad or limited, with the most rapid growth occurring within a narrow optimum range. Bacteria infecting the plant system multiplies in the plant tissues having plant sap that has specific pH. The plant sap of different plants may have some variation which either promote or inhibit the bacterial growth. Therefore, the requirement of pH is an important aspect in the studies of bacterial plant pathogen.

Although specific pH range for bacteria is between 4 and 9, the optimum growth usually occurs between 6.5 and 7.5. But there are some notable exceptions, including the acetic acid bacteria and *Thiobacillus thioxidans* which grows in acidic pH below 4. For example, *T. Thioxidans* tolerates normal acid, that is, pH 0. Therefore, organic acids are used to preserve foods such as fermented dairy products. Fungi are usually more tolerant of higher acid conditions than bacteria, with optimum activities at a pH of 4 to 6. These specific pH needs reflect the organism's adaptation to its natural environment. This property is used in selective media for the growth of fungi and bacteria.

Many bacteria produce organic acids from carbohydrate degradation and alkali from protein breakdown and these may inhibit growth due to shifts in pH. Buffers are frequently added to culture media to neutralize these acids and alkali and to maintain an equilibrium. Peptones, proteins, and amino acids, because of their amphoteric nature, in complex media act as natural buffers. Phosphate salts are often used in chemically defined media as buffers.

Material Required

Cultures of test bacterium; tubes containing sterile buffered nutrient broth adjusted to pH 2.5, 3.5, 4.5, 5.5, 6.5, 7.5, 8.5, and 9.5 (three tubes for each pH, i.e., total of 24 tubes); inoculating loop/transfer needle; Bunsen burner; sterile 10-ml pipettes.

Procedure

Label each of the nutrient broth tubes with the pH of the medium (i.e., 2.5, 3.5, ...) and name of the organism to be inoculated. Inoculate each of a series of the tubes (pH 2.5, 3.5, 4.5, 5.5, 6.5, 7.5, 8.5, and 9.5) with test bacterium by adding 0.1 ml of the culture to each tube. Incubate the bacterial inoculated tubes for 24–48 hours at 28 ± 2°C. After the incubation period, streak a loopful of inoculums from each pH tube on the nutrient agar media plate and incubate the plates for 48 hours at 28 ± 2°C to observe the growth *or* measure the bacterial growth in the nutrient broth of the individual pH in the Spectronic 20 at 620 nm after the incubation period of 48 hours and record the optical density values to denote the growth.

Observations

Record the growth of test organisms by using the symbols as follows: (–) no growth, (+) slight growth, (++) moderate growth, and (+++) best growth. From the results, state the optimum pH, minimum pH, and maximum pH for the test bacterium.

16.5 EFFECT OF SALT CONCENTRATION ON BACTERIAL GROWTH

Plants vary in their salt composition and content. The plant pathogenic bacteria may vary widely in their salt tolerance.

In a hypertonic (high-solute content) environment, all cells lose water by osmosis and become shrivelled, the phenomenon known as *cytolysis*. Its effect on cell reproduction is inhibitory. In a hypotonic (low-solute content) environment, water is taken by the cells and becomes swollen, a phenomenon called *plasmolysis*. There is no harmful effect on microorganisms of hypotonic environments; but in such environments, animal cells undergo lysis, which causes death. Most natural environments of high osmolarity contain high concentrations of salts, particularly sodium chloride. Microorganisms that grow in this type of environment are called halophiles. Bacteria can be divided into two groups depending on their ability to grow at various sodium chloride concentrations: Nonhalophilic (those capable of growing on media containing less than 2 percent sodium chloride) and halophilic (those capable of growth on a medium containing above 2 percent salt). Halophilic may further be of three types: Slightly halophilic (grow at salt concentrations of in between 2 and 5 percent), moderately halophilic (capable of growth at salt concentrations up to 10 percent), and extremely halophilic (capable of growth at salt concentration more than 20 percent).

To determine the effect of salt concentration, generally sodium chloride salt is used. However, one also may use other salts found in respective plants as its content.

Material Required

Broth cultures of test bacterium; nutrient agar plates containing respective sodium chloride concentrations (0.5 percent, 5 percent, 10 percent, 20 percent, 30 percent, and 40 percent); Bunsen burner; inoculating loop; wax marking pencil.

Procedure

Inoculate the test bacterium into the nutrient agar plates having different salt concentrations by making a single line loop inoculation. Incubate all the inoculated plates at 30°C in an inverted position for 2–5 days.

Observations

Observe all the nutrient agar plates of different salt concentrations for the growth of test bacterium. Indicate the degree of growth for test microorganism using these symbols: 0 for no growth, + for slight growth, 2+ for moderate growth, and 3+ for abundant growth. From the results, determine the salt concentration range at which growth occurred and optimal salt concentration for test species.

16.6 EFFECT OF METALS ON BACTERIAL GROWTH

Different metals are present in the soil and irrigation water on which crops are raised. These metals are absorbed through the root system and are present in the plant to exert its effect. The effects of metals are both stimulatory and inhibitory on bacteria. For example, iron is essential for growth of microorganisms, although desired in minute concentrations; some ions (e.g., mercury and silver) are toxic in small concentrations, while some metals (e.g., copper) are inhibitory at moderate concentrations. As the concentration of metal ions decreases, growth starts, and at a particular concentration, normal growth occurs. The ability of certain metals to exert a lethal action on bacteria is termed the *oligodynamic effect*.

A concentration gradient of the ions of metals is produced if a metal piece is placed in an agar medium, that is, a maximum concentration of ions will immediately surround the metal and there will be a gradual decline in an ion's concentration across the agar as the distance increases from the metal.

16.6.1 THE OLIGODYNAMIC ACTION OF METALS (ALUMINUM, COPPER, AND LEAD) ON BACTERIA

Material Required

Broth cultures of test bacterium, melted nutrient agar media at 50°C, sterile Petri plates, sterile 1-ml pipettes, sterile pieces of aluminium, copper and lead, forceps, Bunsen burner.

Procedure

Inoculate 0.25 ml of broth culture of test bacterium in 20 ml of sterilized lukewarm nutrient agar media and thoroughly mix the inoculums. Pour the contents into a sterile Petri plate. Add a piece of metal into each inoculated plate aseptically with the help of sterile forceps immediately after pouring the contents. Keep one plate as an uninoculated control where no metal has been added. Label each plate with the name of the metal added and test bacterium inoculated. After solidification of the medium incubate the plates at 30°C in an inverted position for 2 days.

Observations

Observe the inoculated plates for the growth of bacteria and compare with the control. Measure the zones of inhibition or zones of stimulation, if any, of the bacteria with different metals.

Precautions

A piece of metal should be added into the inoculated plate before the solidification of agar. Mixing of inoculums with the medium may be done by rotating the flask 3–4 times.

16.7   EFFECT OF DYES ON BACTERIAL GROWTH

Dyes are frequently used as an antiseptic. Crystal violet has long been employed therapeutically for fungus disease of the mouth, moniliasis, and thrush. Gentian violet and crystal violet in low concentrations inhibit the growth of Gram-positive bacteria, but has little effect on the Gram-negative bacteria. Inhibitory action of crystal violet on *Staphylococcus aureus* has been found to be on the cell wall synthesis. Dyes like acriflavine and proflavine inhibit action of Gram-negative bacteria but have little effect on Gram-positive bacteria. Practical application of these dyes is made in obtaining pure cultures and in the treatment of infectious processes.

Inhibitory action of crystal violet on bacteria can be demonstrated in the laboratory by growing a Gram-positive bacterium (*Corynebacterium*) and a Gram-negative bacterium (*Xanthomonas*) on nutrient agar medium supplemented with crystal violet.

Material Required

Broth cultures of *Xanthomonas* and *Corynebacterium*, nutrient agar medium, crystal violet nutrient agar medium, sterile Petri plates, inoculating loop, Bunsen burner/spirit lamp, wax marking pencil.

Procedure

Divide the bottom of a Petri plate into two sectors with a wax marking pencil. Label the two sectors of each plate for the two organisms to be inoculated. Melt the crystal violet–nutrient agar and the nutrient agar medium, allow these to cool to 50°C, and pour into individual sterile Petri plates. Inoculate *Xanthomonas* and *Corynebacterium* into the respective labeled sectors of the crystal violet–nutrient agar and nutrient agar plates by streak inoculation. Incubate the plates at 30°C for 24–48 hours in an inverted position.

Observations

Observe the inoculated plates for the growth of the test organisms on both the nutrient agar and crystal–violet nutrient agar plates.

Growth of *Xanthomonas* will only be observed in crystal violet nutrient agar plate while there will be no growth of *Corynebacterium* on this plate, indicating the bacteriostatic activity of crystal violet on the Gram-positive bacterium. Both the bacteria will grow on the nutrient-agar medium where no dye has been added.

16.8   EFFECT OF DIFFERENT WAVELENGTHS OF LIGHT ON BACTERIAL GROWTH

Photosynthesis is the process by which plants and certain bacteria (prototrophs) convert radiant energy in the form of light into metabolic energy and reducing power. Those organisms (plants, algae, and cyanobacteria) that utilize water are said to carry out oxygenic or plant photosynthesis, and others (purple and green bacteria) that utilize other electron sources are said to carry out an oxygenic or bacterial photosynthesis. Photoautotrophic microorganisms obtain their energy by converting visible light, radiant energy, into chemical energy. This reaction is catalyzed by a class of molecules termed *chlorophylls*, representatives of which nearly all prototrophs contain, using the light energy to break down inorganic molecules to produce hydrogen for the reduction of carbon dioxide. This results in the production of energy-rich organic compounds.

Visible light may be divided into a number of different colors (viz., violet-blue, green, yellow, orange, red) between 422 to 760 nm wavelengths of light. Photosynthesis does not utilize all parts of the visible spectrum with the same efficiency. The region of peak absorption within the longer wavelengths varies with the particular chlorophyll species and the proteins with which they are associated. In green microorganisms, the highest rate of efficiency is in the violet-blue portion and the minimum efficiency rate is in the green-yellow portion of the visible spectrum.

The effect of different wavelengths of light on the growth of bacteria in the laboratory can be studied by using colored celluloid filters to expose the bacteria to different regions of the visible spectrum.

Material Required

Culture of test bacterium, sterile slants of nutrient agar media, celluloid filters of different colors (blue, green, yellow, orange, red, and colorless), small open cardboard boxes (6), inoculating loop, Bunsen burner.

Procedure

Aseptically inoculate each plate with the test bacterial culture by the streak method. Place one inoculated culture plate in each box and cover the tops of all six with individual celluloid filters. Expose the boxes to light for three days under the normal light system.

Observations and Results

Observe all six plates exposed to different colors for the growth of bacterial cultures. Indicate the degree of growth using the following symbols: 0 for no growth, + for slight growth, 2+ for moderate growth, 3+ for abundant growth. Observe the changes in the bacterial growth pattern and coloration of bacterial culture.

From the results, find out the wavelength ranges of light most suitable for bacterial growth and multiplication.

16.9   LETHAL EFFECTS OF ULTRAVIOLET RADIATION (UV RAYS) ON BACTERIAL GROWTH

Different portions of the light spectrum have different effects on microbes. Radiation differs in wavelength and energy. The shorter wavelengths below 300 nm, such as ultraviolet (electromagnetic radiation), x-rays, gamma rays, and cosmic rays (termed ionizing radiations), possess more energy and are microbicidal in nature. Their principal effect is to ionize water into highly free radicals, which can break DNA molecules. X-rays and gamma rays, because of their high energy content and ability to penetrate matter, are used as a means of sterilization, especially for thermolabile materials. Cobalt radiations are commonly used to alter the DNA of malignant cells in an attempt to stop their uncontrolled multiplication as the radiation alters the gene that controls the speed of cell multiplication.

Ultraviolet means beyond the violet in the electromagnetic spectrum, corresponding to light having wavelengths shorter than 4000 Å (or 400 nm). Ultraviolet light produces a lethal effect in cells exposed to wavelengths in the range of 210–300 nm, 265 nm being the most lethal. It induces pyrimidine dimmers in the DNA, which result in the distortion of DNA molecules, interfere in the replication and transcription of DNA molecules during protein synthesis, and cause mutation. If mutation takes place in the critical genes, death of the cell results unless the damage is repaired. Ultraviolet lamps are used to sterilize some heat-labile solutions and to reduce the number of microorganisms in hospital operation theaters, inoculating chambers, and meat storage lockers. The effect of radiation is influenced by many variables, for example, the cell age, composition of media, temperature, distance, and to the time of exposure. Low-pressure mercury vapor lamps, which have a high output (90 percent) of 2537 Å (253.7 nm), make very effective bactericidal lamps.

This exercise deals with testing of effect of UV rays (254 nm) to determine the minimum amount of exposure required to effect a 100 percent kill of the organism.

Material Required

Nutrient broth cultures of test bacterium, sterile nutrient agar plates, ultraviolet lamp (254 nm), Bunsen burner, sterile serological 1-ml pipettes, wax marking pencil, glass spreaders.

Procedure

Take five nutrient agar plates and label the covers with the test organisms to be inoculated with exposure times to UV light as 0 (untreated control) and 1, 2, 4, 6, 8, and 10 minutes (with lid off or with cover). Inoculate all the labeled plates by the respective bacterium by pouring 0.1 ml of cell suspension from the stock cultures and spread the inoculums by using sterile glass spreaders, separate for each organism. Expose all the inoculated plates (except control) to ultraviolet light source according to the time schedule (i.e., 1, 2, 4, 6, 8, and 10 minutes) and conditions (with lid and without lid as) by placing them at a distance of 30 cm from the UV radiation source. Begin timing when you turn the UV lamp switch on. Replace the covers. Incubate all the plates (exposed and unexposed) at 25°C for 2–5 days in an inverted position.

Observation and Results

Observe all the nutrient agar cultures for the growth of bacterial species exposed for different periods of time. Record the degree of growth in tabular form using the following symbols: 0 for no growth, + for slight growth, 2+ for moderate growth, and 3+ for abundant growth.

From the results find out the killing efficiency of UV rays to the time of exposure of the bacterium tested and the effect of UV rays when they pass through glass.

Precautions

Do not look at the rays coming from the ultraviolet lamp because the rays can cause cataracts and other eye injuries. The UV lamp should be placed above 30 cm from the culture you wish to irradiate. The culture to be exposed must be in the direct path of the radiations. Avoid digging into the medium with your loop while streaking the surface of nutrient agar plate.

16.10 EFFECT OF RELATIVE HUMIDITY ON BACTERIAL GROWTH

Water is essential for the growth of all microorganisms. Some microorganisms are less sensitive to relative humidity (i.e., the amount of water vapor in the air, expressed as a percentage of the maximum amount that the air could hold at the given temperature) and can survive in its absence and recommence growth on the onset of favorable conditions. Others are more sensitive and soon die in a dry atmosphere.

The multiplication of bacterial plant pathogens, disease initiation, and symptoms developments are more vigorous under humid conditions as compared to dry conditions. Therefore, study on effect of relative humidity on bacterial plant pathogen is important.

Material Required

Plants sprayed with bacterial plant pathogens, incubation chambers with different humidities, hygrometer, nutrient agar plates, and so on.

Procedure

Spray the host plants with the bacterial plant pathogen and keep each plant separately in individual incubation chambers having different relative humidity ranging from 40 to 90 percent. Incubate the plants for three days in their respective chambers. Pluck the bacterium-sprayed leaves of inoculated plants and press them aseptically on nutrient agar plates to have imprints. Label the imprinted plates according to respective humidity and incubate the plates in a BOD incubator at 28°C.

Observations

Observe the development/nondevelopment of bacterial colonies of test bacterium in each plate and correlate with the humidity available during the growth of the bacterium in humidity chamber. Similarly count the bacterial population in each plate to assess the effect of relative humidity on bacterial growth and multiplication.

17 Biochemical Tests Used in Identification of Bacteria

Microbes differ in their metabolic reactions. Metabolic reactions that release energy (ATP) from the breakdown or degradation of a substrate (e.g., complex organic molecules) are called catabolic reactions, while the ones that use energy to assemble smaller molecules and produce biosynthetic building blocks are called anabolic reactions and when there is involvement of both the reactions, i.e., catabolism and anabolism in biochemical pathways, it is called as amphibolic reactions.

All these biochemical reactions that occur both outside and inside the cell are precisely controlled by some governing factors—the enzymes. An enzyme is a biological catalyst, a substance that accelerates the rate of a specific chemical reaction. The enzymes are either exoenzymes (i.e., extracellular) or endoenzymes (i.e., intracellular). Exoenzymes, which are a few in number, are released from the cell and act on the substrates. These are mainly hydrolytic enzymes that degrade, by the addition of water, high-molecular weight substrates (like polysaccharide, lipids, and proteins) into smaller components (e.g., glucose) that can enter into the cell and are later assimilated. Enzymes required for the hydrolysis of cellulose, starch, pectin, lipid, casein, and gelatin belong to the category of exoenzymes. Endoenzymes are utilized by the cell for further metabolic degradation of carbohydrates and are mainly responsible for synthesis of new protoplasmic requirements and production of cellular energy from assimilated materials. These enzymes function inside a cell. All of the end products of these metabolic reactions are used in hydrolysis. Endoenzymes are also involved in carbohydrate fermentation, litmus milk reactions, hydrogen sulfide production, nitrate reduction, catalyzed reaction, urease test, oxidase test, IMViC test, and the triple sugar iron test.

These reactions are examples of biochemical characteristics, the detection of which aid in the identification and classification of microorganisms that appear morphologically identical. A series of biochemical tests can provide a microbial "fingerprint."

17.1 AMYLASE PRODUCTION TEST (OR DEMONSTRATION OF STARCH HYDROLYSIS)

Amylase is an exoenzyme that hydrolyzes (cleaves) starch, a polysaccharide (a molecule that consists of eight or more monosaccharide molecules) into maltose, a disaccharide (double sugar, i.e., composed of two monosaccharide molecules), and some monosaccharides such as glucose. These disaccharides and monosaccharides enter into the cytoplasm of the bacterial cell through the semi-permeable membrane and are thereby used by the endoenzymes. Starch is a complex carbohydrate (polysaccharide) composed of two constituents—amylose, a straight-chain polymer of 200–300 glucose units, and amylopectin, a larger branched polymer with phosphate groups.

Amylase production is known in some bacteria. The ability to degrade starch is used as a criterion for the determination of amylase production by a microbe. In the laboratory it is tested by performing the starch test to determine the absence or presence of starch in the medium by using iodine solution as an indicator. Starch in the presence of iodine produces a dark blue coloration of the medium and a yellow zone around a colony in an otherwise blue medium indicates amylolytic activity.

Material Required

Test bacterial slant cultures, starch agar medium (60 ml), Gram's iodine solution, sterile Petri dishes (3), dropper, inoculating loop, Bunsen burner, wax marking pencil.

Procedure

Sterilize the starch agar medium, cool to 45°C, and pour into the sterile Petri dishes. Allow it to solidify. Label each of the starch agar plate with the name of the test isolate to be inoculated. Make a single streak inoculation of each bacterial isolate into the center of the appropriately labeled plate. Incubate the bacterial inoculated plates for 72–96 hours at 28°C in an inverted position. After the incubation period, flood the surface of the plates with iodine solution with a dropper for 30 seconds. Pour off the excess iodine solution. Examine the plates for the starch hydrolysis around the line of growth of each isolate, that is, the color change of the medium.

Observations

A typical positive starch hydrolysis due to amylase production is shown as a clear zone surrounding the microbial colonies which has diffused into the medium surrounding the growth by the addition of test reagent Gram's iodine solution. A negative reaction is indicated by no change in the dark blue coloration of the medium.

17.2　CELLULASE PRODUCTION TEST (DEGRADATION OF CELLULOSE)

A prominent carbonaceous constituent of higher plants and probably the most abundant organic compounds is cellulose. Cellulose is a polysaccharide composed of glucose units in a long linear chain linked together by β-1, 4 glycosidic bonds. Degradation of cellulose is brought about by microbes by the secretion of extracellular enzyme, cellulase. It is a complex enzyme composed of at least three components, viz., endoglucanase (endo-1, 4-β-D-glucanase), exoglucanase (1,4-β-D-glucancellobiohydrolase), and a β-glycosidase. The cooperative action of these three enzymes is required for the complete hydrolysis of cellulose to glucose. Utilization of cellulose can be detected by using hexadecyltrimethyl ammonium bromide. This reagent precipitates intact carboxymethyl cellulose (CMC) in the medium and thus clear zones around a colony in an otherwise opaque medium indicating degradation of CMC.

Material Required

Test cultures of bacteria, modified Czapek mineral salt medium, carboxymethyl cellulose (CMC), hexadecyltrimethyl ammonium bromide (1 percent solution), sterile Petri plates, inoculating loop/needle, glass rod, Bunsen burner, wax marking pencil.

Procedure

Prepare Czapek mineral salt agar medium of following constituents: sodium nitrate (NaNO$_3$), 2.0 g; potassium phosphate (K$_2$HPO$_4$ or KH$_2$PO$_4$), 1.0 g; magnesium sulfate (MgSO$_4$.7H$_2$O), 0.5 g; potassium chloride (KCl), 0.5 g; carboxymethyl cellulose (CMC), 5.0 g; peptone, 2.0 g; agar, 20.0 g; distilled water, 1000 ml; pH, 6.5. Pour the autoclaved medium into sterile Petri plates. Allow the medium to solidify. Label the plates each with the test isolate to be inoculated. Inoculate the appropriately labeled plates with the test isolate. Incubate inoculated plates at 28 ± 2°C in an inverted position for 2–5 days. Flood the plate with 1 percent aqueous solution of hexadecyltrimethyl-ammonium bromide. Observe the plates for the formation of a zone around the growth.

A clear zone formed around the colonies indicate degradation of CMC by the production of extracellular enzyme, that is, cellulase (or CMCase).

Precautions

While preparing the Czapek mineral salt medium, phosphate should always be dissolved separately and be added to the mixture last.

17.3 PRODUCTION OF PECTOLYTIC ENZYMES (DEGRADATION OF PECTIN)

Pectic substances are the primary constituents of the middle lamella, a tissue constituent located between individual cells, and are the structural elements of the primary wall. The pectic carbohydrates are complex polysaccharides composed mainly of galacturonic acid units bound to one another in a long chain linked by α-1, 4 glycosidic bonds in linear chains. The pectic substances include four types: (1) protopectin, (2) pectin, (3) pectinic acid, and (4) pectic acids.

Decomposition of pectic substances is brought about by the microbial enzymes which can be divided into three major categories: (1) hydrolytic that carry out a hydrolytic cleavage of the polymer; (2) transeliminative that brings about transeliminative cleavage of glycosidic links by action of a lyase; and (3) pectinesterases which hydrolyze off the methoxyl groups from the pectins, thereby converting pectin or pectinic acids to pectic acid. The pectinase which splits or hydrolyzes the α-1, 4 glycosidic bonds of the pectic substances can be classified into eight types, on the basis of whether they are exo or endo, whether they are hydrolytic or transeliminative, and whether their specificity is toward methylated or unmethylated acid polymers. The pectic substances degrading enzymes produced by microbes are important in host infection by the pathogens. The pectic enzymes are usually extracellular and can be detected by various techniques, that is, the use of solid media, viscosity loss method, and spectrophotometric method. One of the simplest techniques for detection of extracellular pectolytic enzymes is by growing a microbe on a solid medium containing pectin or sodium polypectate as a carbon source and then flooding the inoculated plates with hexadecyltrimethyl ammonium bromide solution after 2–5 days incubation. This reagent precipitates intact pectin of the medium and thus clear zone around the colony in an otherwise opaque medium indicates degradation of pectin.

The medium at pH 7 is used to detect pectate lyase production (pectate transeliminase) and at pH 5 to detect polygalacturonase activity (pectin depolymerase, pectinase).

Material Required

Cultures of *Erwinia carotovora* a bacterial soft rot plant pathogen, Hankin's medium, hexadecyltrimethyl ammonium bromide (1 percent solution), sterile Petri plate, inoculating loop/needle, Bunsen burner, pH meter, wax marking pencil.

Procedure

Prepare Hankin's medium with following constituents: $(NH_4)_2SO_4$, 2 g; KH_2PO_4, 4 g; Na_2HPO_4, 6 g; $FeSO_4 \cdot 7H_2O$, 0.2 g; $CaCl_2$, 1 mg; H_3BO_3 10 mg; $MnSO_4$, 10 mg; $ZnSO_4$, 70 g; MoO_3, 10 mg; in 100 ml of distilled water, pH 7 or 5 as needed, yeast extract, 1.0 g; agar, 15.0 g; pectin (citrus or apple), 5.0 g. Make volume to 1 L. Pour the autoclaved medium into sterile Petri plates and allow to solidify. Inoculate two labeled plates (pH 7 and pH 5) with *E. carotovora*. Incubate inoculated plates with *E. carotovora* at 30°C for 48–72 hours in an inverted position. Flood the plates with 1 percent aqueous solution of hexadecyltrimethyl-ammonium bromide. Observe the plate for the formation of a clear zone around the growth on both the media.

Observations

A clear zone around the colonies in the plate is observed due to the precipitation of intact pectin by the hexadecyl-trimethyl ammonium bromide, indicating the degradation of pectin due to the secretion of extracellular polygalacturonase and pectate lyase.

17.4   HYDROLYSIS OF GELATIN, A PROTEIN (PRODUCTION OF GELATINASE)

Proteins are organic molecules composed of amino acids; in other words, proteins contain carbon, hydrogen, oxygen, and nitrogen, though some proteins contain sulfur, too. Amino acids are linked together by peptide bonds to form a small chain (a peptide) or a large molecule (polypeptide) of protein. Gelatin is a protein produced by hydrolysis of collagen, a major component of connective tissue and tendons in humans and other animals. It dissolves in warm water (50°C), exists as a liquid above 25°C, and solidifies (gels) when cooled below 25°C.

Large protein molecules are hydrolyzed by exoenzymes, and the smaller products of hydrolysis are transported into the cell. Hydrolysis (liquefaction) of gelatin is brought about by microorganisms capable of producing a proteolytic exoenzyme known as gelatinase, which acts to hydrolyze this protein to amino acids.

Hydrolysis of gelatin in the laboratory can be demonstrated by growing microorganisms in nutrient gelatin. Once the degradation of gelatin occurs in the medium by an exoenzyme, it can be detected by observing liquefaction (i.e., even very low temperatures like 4°C will not restore the gel characteristic) or testing with a protein-precipitating material (i.e., flooding the gelatin agar medium with the mercuric chloride solution and observing the plates for clearing around the line of growth) because gelatin is also precipitated by chemicals that coagulate proteins while the end products of degradation (i.e., amino acids) are not precipitated by the same chemicals.

Material Required

Twenty-four- to forty-eight-hour-old slant bacterial cultures, gelatin agar medium, nutrient gelatin deep tubes (4), mercuric chloride solution in a dropping bottle (20 ml), sterile Petri dishes (2), refrigerator, inoculating needle, Bunsen burner, wax marking pencil.

Procedure

Sterilize the gelatin–agar medium, cool to 45–50°C, and pour into sterile Petri dishes (approximately 15 ml in each) and test tube as per requirement and allow it to solidify. Label each of the nutrient–gelatin deep tubes and gelatin agar–medium plates with the name of the bacterial isolate to be inoculated. Using inoculating loop, make a stab inoculation (i.e., puncture the agar column from top to bottom with withdrawal of the needle through the same path) from each culture into its appropriately labeled deep tube of nutrient gelatin. Uninoculated deep tube should be used as a control. Similarly, make a single streak inoculation from each culture into its appropriately labeled Petri plate across the surface of the medium. Incubate all the inoculated tubes, uninoculated deep tube, and plates at 28°C for 4–7 days.

After incubation, place the tubes into a refrigerator at 4°C for 15 minutes. Flood the incubated agar plates with mercuric chloride solution and allow the plates to stand for 5–10 minutes.

Examine the refrigerated gelatin tubes to see whether the medium is solid or liquid and the flooded plates for any clearing zone around the line of bacterial growth.

Observations

Deep gelatin-inoculated tubes that remain liquefied indicate the production of gelatinase and show positive testing for gelatin hydrolysis while those tubes that remain solid demonstrate a negative reaction for gelatin hydrolysis.

A clear zone observed around the growth of the bacterium in the presence of mercuric chloride solution in the inoculated Petri plates demonstrates proteolytic hydrolysis of gelatin. (See Figure 17.1.)

17.5   CASEIN HYDROLYSIS

Casein is a major protein found in milk. It is a macromolecule composed of amino acids linked together by peptide bonds CO-NH. Some microorganisms have the ability to degrade the protein

Gelatin liquefication

FIGURE 17.1   Hydrolysis of gelatin by bacterial isolate. (Courtesy of Dr. S. G. Borkar and Jyoti Mergerwar Department of Plant Pathology, Mahatma Phule Krishi Vidyapeeth, Rahuri.)

casein by producing proteolytic exoenzyme, called proteinase (caseinase), which breaks the peptide bond CO-NH by introducing water into the molecule. This liberates smaller chains of amino acids, called peptides, which are later broken down into free amino acids by extracellular or intercellular peptidases, which are transported through the cell membrane into the intracellular amino acid pool for use in the synthesis of structural and functional cellular proteins.

Casein hydrolysis can be demonstrated by supplementing nutrient agar medium with milk. The medium is opaque due to the casein in colloidal suspension. Formation of a clear zone adjacent to the bacterial growth, after inoculation and incubation of agar plate cultures, is evidence of casein hydrolysis. Pathogenic bacteria, for example, pseudomonas, are known to produce extracellular proteinases.

Material Required

Twenty-four-hour-old slant bacterial cultures, skim milk agar plates, inoculating loop, Bunsen burner, wax marking pencil.

Procedure

Prepare skim milk agar of the following constituents (per liter of medium): skim milk powder, 100.0 g; peptone, 5.0 g; agar, 15.0 g; pH, 7.2. Pour the autoclaved and cooled (45–50°C) medium into sterile Petri plates (20 ml each) and allow to solidify. Label the skim milk agar plates with the name of the bacterial organism to be inoculated and keep the uninoculated plate as control. Make a single line streak inoculation from each culture into its labeled Petri plate across the surface of the medium. Incubate the plates (inoculated and uninoculated) for 24–48 hours at 28°C in an inverted position. Observe all the inoculated plates for any clearing zone around the line of growth.

Observations

A clear area surrounding the bacterial growth is a positive reaction for extracellular caseinase secretions while absence of a clear zone around the growth of an organism is a negative reaction. (See Figure 17.2.)

17.6   UREASE TEST

Urea is a major organic waste product of protein digestion in most vertebrates and is excreted in the urine. Some microorganisms have the ability to produce the enzyme urease. The urease is a

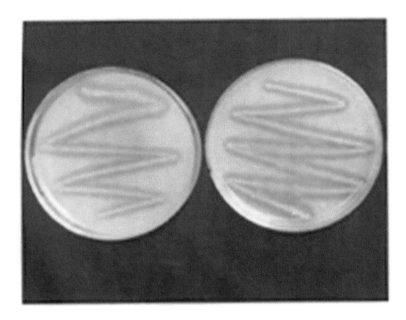

FIGURE 17.2 Casein hydrolysis by bacterial isolate. (Courtesy of Dr. S. G. Borkar, Department of Plant Pathology, Mahatma Phule Krishi Vidyapeeth, Rahuri.)

hydrolytic enzyme that attacks the carbon and nitrogen bond in amide compounds (e.g., urea) with the liberation of ammonia. It is a useful diagnostic test for identifying bacteria, especially to distinguish members of the genus from the Gram-negative pathogens.

Urease test is performed by growing the test organisms on urea broth or agar medium containing the pH indicator phenol red (pH 6.8). During incubation, microorganisms possessing urease will produce ammonia that raises the pH of the medium/broth. As the pH becomes higher, the phenol red changes from a yellow color (pH 6.8) to a red or deep pink (cerise) color. Failure of the development of a deep pink color due to no ammonia production is evidence of a lack of urease production by the microorganisms.

Material Required

Cultures of test bacterium, urea agar medium/urea broth tubes, inoculating loop, Bunsen burner, wax marking pencil.

Procedure

Prepare urea agar medium of the following constituents (per liter): peptone, 1.0 g; sodium chloride, 5.0 g; potassium monohydrogen (or dihydrogen) phosphate, 2.0 g; agar, 20.0 g; distilled water, 1000 ml (dissolve the ingredients by heating, adjust the pH to 6.8, and autoclave at 121°C for 15 minutes and cool to 50°C); glucose, 1.0 g; phenol red (0.2 percent solution), 6.0 ml (add to the molten base and steam for 1 hour, cool to 50°C); urea, 20 percent aqueous solution 100.0 ml (sterilize by filtration and add aseptically to the basal medium). Mix well and distribute into sterile containers, that is, flasks and culture tubes, and allow the medium to solidify in a slanting position to form slopes. Label tubes of the urea agar medium/urea broth tubes with the name of the bacterial organism to be inoculated. Inoculate urea agar slant/broth with the test bacterium. Incubate inoculated slants/ broths for 24–48 hours at 37°C.

Examine the slants/broths for their color (the presence of urease gives a red or cerise color while no urease is indicated by a yellow color).

Precautions

Urea should never be sterilized in an autoclave, along with other constituents, as it is unstable and breaks down at 15 psi steam pressure.

17.7 HYDROGEN SULFIDE PRODUCTION TEST

Hydrogen sulfide (H_2S), commonly known as "rotten egg" gas, releases copious amounts of gas (when eggs decompose by certain bacteria, such as *Proteus vulgaris*) through reduction hydrogenation of sulfur containing amino acids (e.g., cystine, cysteine, and methionine) or through the reduction of inorganic sulfur compounds such as thiosulfates ($S_2O_3^{2-}$), sulfates (SO_4^{2-}), or sulfites (SO_3^{2-}).

The hydrogen sulfide production can be detected by incorporating a heavy metal salt containing (Fe^{2+}) or lead (Pb^{2+}) ion as H_2S indicator to a nutrient culture medium containing cysteine and sodium thiosulfate as the sulfur substrates. When produced, hydrogen sulfide, a colorless gas, reacts with the metal salt (ferrous sulfate), forming visible insoluble black ferrous sulfide precipitates.

Material Required

Nutrient agar slant cultures of test bacterium and *Pseudomonas*, SI (sulfide indole motility) agar tubes, inoculating needle (loop), Bunsen burner, wax marking pencil.

Procedure

Prepare SIM agar medium of the following constituents (per liter): peptone, 30.0 g; beef extract, 3.0 g; ferrous ammonium sulfate, 0.2 g; sodium thiosulfate, 0.025 g; agar, 3.0 g; distilled water, 1000.0 ml. Label SIM agar deep tubes with the name of the organism to be inoculated. Inoculate the test organism into appropriately labeled tube by means of stab inoculation. Incubate the inoculated tubes at 28°C for 48 hours. Examine the tubes for the presence or absence of black coloration along the line of stab inoculation.

Observations

Blackening of the culture medium is positive test for H_2S production. This black color is due to the production of H_2S from an ingredient of the medium (i.e., sodium thiosulfate) that then combines with another ingredient of the medium (ferrous ammonium sulfate), resulting in the formation of the black insoluble compound, ferrous sulfite. (See Figure 17.3.)

17.8 CARBOHYDRATE CATABOLISM BY MICROORGANISMS (OXIDATION AND FERMENTATION OF GLUCOSE)

Carbohydrates are organic molecules that contain carbon, hydrogen, and oxygen in the ration (CH_2O) n. These are of three types, for instance, monosaccharides (simple sugars containing three to seven carbon atoms), disaccharides (composed of two monosaccharide molecules), and polysaccharides (containing eight or more monosaccharide molecules). Chemical reactions which release energy from the complex organic molecules decomposition are referred to as catabolism. The power to break down carbohydrates is possessed by a large number of microbes. This microbial activity is of great significance in the carbon cycle in nature and industrially. Organisms use carbohydrate differently depending upon their enzyme complement. The pattern of fermentation is characteristic of certain species, genera, or groups of organisms, and for this reason this property has been extensively used as a method for biochemical differentiation of microbes.

Intercellular enzymes are utilized by the cell for metabolic degradation of carbohydrates. Two examples of such enzymes are maltase and lactase. The enzyme maltase acts upon maltose to yield

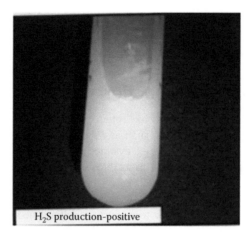

H₂S production-positive

FIGURE 17.3   H$_2$S production by bacterial isolate. (Courtesy of Dr. S. G. Borkar and Dr. Nisha Patil, Department of Plant Pathology, Mahatma Phule Krishi Vidyapeeth, Rahuri.)

two molecules of glucose, while lactase decomposes lactose and produces one molecule of glucose and one of galactose. All of the end products of these reactions are used in glycolysis.

Glucose, after entering a cell, can be catabolized either aerobically (biooxidation in which molecular oxygen (O$_2$) can serve as the final electron acceptor, i.e., oxidative metabolism) or anaerobically (biooxidation in which inorganic ions other than oxygen, e.g., NO$_3^-$ or SO$_4^{2-}$ can serve as the final electron acceptor, i.e., fermentative metabolism) or both aerobic and anaerobic pathways and some organisms lack the ability to oxidize glucose by either. The metabolic end products of carbohydrate fermentation can be either organic acids (e.g., lactic, formic, or acetic) or organic acid and gas (e.g., hydrogen or carbon dioxide).

Microorganisms that require organic carbon can bound as an energy source are called chemo-organotrophic. Whether an organism is oxidative or fermentative can be determined by using Hugh and Leifson's medium (OF medium), tryptone, and bromothymol blue (an indicator). Two tubes are used, one open to the air and the other sealed with paraffin (to keep air out). Growth of the micro-organisms in this medium is either by utilizing the tryptone, which results in an alkaline reaction (dark blue color), or by utilizing glucose, which results in the production of acid (turning bromothymol blue to yellow). Those microorganisms which produce acid in both closed and open tubes are described as fermentative while those which produce acid only in the open tube are called oxidative.

Material Required

Broth cultures of test pathogen, medium tubes each containing 5 ml of the medium, sterile liquid paraffin (or mineral oil), sterile 10 percent glucose solution, inoculating needle, Bunsen burner.

Procedure

Prepare OF glucose–agar medium of following composition: peptone, 2.0 g; sodium chloride, 5.0 g; dipotassium hydrogen phosphate, 0.3 g; bromothymol blue solution, 15.0 ml; agar, 3.0 g; distilled water, 1000.0 ml. (Dissolve the weighed constituents, make volume to 1000, adjust pH to 7.1, and add bromothymol. Pour the basal medium into tubes and flasks; sterilize by autoclaving at 121°C for 15 minutes; melted medium taken in the tubes and add 1 ml of sterile glucose solution [10 percent] to produce a final concentration of 1 percent). Allow the tubes to cool. Label each of the OF medium tubes with the name of the test organism (two tubes per organism) to be inoculated. Inoculate two tubes of OF glucose medium with test bacterium by stabbing with a straight wire. Two uninoculated tubes may be used as control. Pour liquid paraffin over the medium to form a layer about 1 cm in depth into one of the tubes of each pair. Incubate the tubes at 28°C for 24–48 hours.

Observation

Observe the tubes and record for the presence of growth of the bacteria and color of the medium and record the type of metabolism. The bacterium which acts on glucose and produces an acid in turn changes the blue color of the indicator to yellow and is thus fermentative in reaction. The bacterium lacking the capacity to attack glucose but attacks the tryptone which in turn produces a deep blue color, that is, an alkaline reaction, is thus oxidative in reaction.

Precautions

Glucose should never be autoclaved along with other constituents of the medium. Sufficient liquid paraffin should be added to the inoculated tubes to form a layer about 1 cm in depth to create anaerobic conditions.

17.8.1 GROUPING OF BACTERIAL ISOLATES OF *XANTHOMONAS MALVACEARUM* ON THE BASIS OF LACTOSE UTILIZATION

Variation in utilization of lactose is observed in the isolates of *Xanthomonas malvacearum*. Some isolates utilize lactose as sugar or carbohydrate source while other isolates of the same pathogen does not. On the basis of lactose utilization the isolates can be grouped as Lac + or Lac −.

Procedure

Prepare the basal medium for *Xanthomonas* (Dye, 1962): $NH_4H_2PO_4$, 0.5 g; K_2HPO_4, 0.5 g; $MgSO_4$ $7H_2O$, 0.2 g; NaCl, 5.0 g; yeast extract, 1.0 g; agar–agar, 20 g; H_2O, 1 L; and add carbon compound (lactose) to give 1 percent concentration. Add bromocresol purple 0.7 ml (of 5 percent alcoholic solution) into this medium. Adjust the pH to 6.8 and sterilize the medium by steaming for 30 minutes for three successive days. Streak the 48-hour young growing cultures of test isolates on the medium and incubate the tubes at 27 ± 2°C temp for 3 days.

Observation

Observe the growth of the bacterium on carbon medium. Utilization of lactose will change the color of the medium with bacterial growth while nonutilization of lactose will not change the color of the medium without any growth.

17.9   FERMENTATION OF CARBOHYDRATES

Fermentative degradation of various carbohydrates such as glucose (a monosaccharide), sucrose (disaccharide), and cellulose (polysaccharide) by microbes under anaerobic condition is carried out in a fermentation tube. A fermentation tube is a culture tube that contains a Durham tube (i.e., a small tube placed in an inverted position in the culture tube) for the detection of gas production, as an end product of metabolism. The fermentation broth contains ingredients of nutrient broth, a specific carbohydrate (glucose, lactose, maltose, sucrose, or mannitol) and a pH indicator (phenol red), which is red at a neutral pH (7) and turns yellow at or below a pH of 6.8 due to the production of an organic acid.

Material Required

Broth/slant of test bacterial cultures, sterile fermentation tubes (of glucose broth, lactose broth), inoculating loop, Bunsen burner, wax marking pencil.

Procedure

Prepare fermentation medium of the following constituents: Trypticase/peptone, 10.0 g; carbo-hydrate, 5.0 g; sodium chloride, 15.0 g; phenol red, 0.018 g; distilled water, 1000.0 ml; pH, 7.3

(a specific carbohydrate such as glucose, sucrose, and lactose is added). Broth is taken into fermentation tubes and is autoclaved at 12 pounds pressure for 15 minutes. Label each of specified fermentation tubes of media with the name of the organism to be inoculated, inoculate the three types of sugar fermentation broths with the bacterium (three per culture), and keep one uninoculated tube of each fermentation broth as a comparative control. Incubate all the inoculated and uninoculated tubes at 28°C for 24–48 hours.

Observation

Observe the reactions that develop in the three fermentation media by comparing with the uninoculated tubes (control), that is, change in color (due to production of acid) or change in color and appearance of bubbles (due to production of acid and gas) findings.

17.10   MICROBIAL REACTION IN LITMUS MILK

Milk is an excellent medium for the growth of microorganisms because it contains the milk protein casein, the milk sugar lactose, vitamins, minerals, and water. Litmus, a pH indicator, is incorporated in the medium for the detection of production of acid or alkali and oxidation–reduction activities. A variety of different chemical changes occur in milk, depending upon which milk ingredients are utilized by the bacteria. Once again this is dependent upon the type of enzymes that the organism is able to produce.

Litmus milk medium consists of 10 percent powdered skim milk and the dye molecule litmus. Litmus, upon addition to rehydrated skim milk, changes the colloidal milk suspension from white to lavender (pale bluish-purple).

Some bacteria ferment the lactose of milk with the production of lactic acid, which is detected when litmus medium changes from blue or purple (neutral pH) to red or pink (acidic pH). The accumulation of acid acts on the milk protein, casein, resulting in the formation of curd (clot) due to the precipitation of casein as calcium caseinate, which is very firm. As the acidity increases, the curd becomes so solid that there is a squeezing out of a clear liquid (whey) from the curd. Acid curd is identified by the clot, which remains immobile when the tube is inverted. Some bacteria produce an enzyme, rennin, which acts on casein and in the presence of calcium ions forms calcium paracaseinate that is insoluble and is called rennet curd. Unlike acid curd it is a soft semisolid clot that will flow slowly upon tilting the tube. Acid or rennet curds are quite palatable as dairy products such as cottage cheese.

Some bacteria possess proteolytic enzymes (e.g., caseinase) and hydrolyze casein (called proteolysis, peptonization) resulting in the release of large quantities of ammonia that makes the medium alkaline have a foul smell and the litmus turn a purplish-blue. With further incubation of the medium, an opaque clearing of the milk occurs as the casein is hydrolyzed to peptides and amino acids. The opaque (turbid) liquid supernatant (whey-like appearance) turns brown in color.

Some bacteria ferment lactose of milk with the production of acid and gases (like $CO_2 + H_2$). Gas production is detected by bubbles in the acid curd or by the development of tracks or fissures within the curd.

Reduction of the litmus by oxidation–reduction activities of bacteria can be shown by loss of litmus color. In the litmus milk test, litmus (purple in the oxidized state) acts as such an acceptor of hydrogen from the substrate and becomes reduced and turns white- or milk-colored.

An alkaline reaction due to partial degradation of casein into shorter polypeptide chains with the simultaneous release of alkaline end products, is evident when the color of the medium either does not change or changes to a deeper blue.

Material Required

Trypticase soy agar slant cultures of test bacterium, sterile tubes of litmus milk medium (10 ml tube) autoclaved at 7 psi for 20 minutes, Bunsen burner, inoculating loop, wax marking pencil.

Procedure

Label each of the litmus milk medium tubes with the bacterium to be inoculated. Inoculate tubes of litmus milk, using loop inoculation, with bacterial species and keep one tube as an uninoculated comparative control. Incubate the inoculated tubes at 28°C for 24–48 hours. Refrigerate the uninoculated control tube of litmus milk. Make frequent observations of the incubated tubes for the color and consistency of the medium and record the observations.

Conclusions

Acidic reaction (pink to red): Demonstrates the production of a considerable quantity of acid.

Alkaline reaction (blue to purple): Indicates utilization of milk proteins as a source of carbon and nitrogen by these species.

Litmus reduction: Indicates the oxidation reduction activity. This reaction begins at the bottom of the tube and spreads upward.

Peptonization: Reduction in curd size occurs and a brownish supernatant, called whey, is formed.

17.11 NITRATE REDUCTION

Composition of Medium

Peptone, 10 g; NaCl, 5 g; KNO_3 (nitrite-free), 2 g; agar, 3 g; distilled water, 1 L.

Procedure

Dispense the medium in the test tube to a depth of 5 cm and autoclave. Inoculate the sterilized medium tubes with test bacterium and incubate at 27°C.

Reduction of nitrate can be tested up to 15 days at regular intervals in different slant culture by adding few drops of sulfanilic acid (0.8 percent in 5 N acetic acid) and dimethyl alpha naphthylamine (0.5 percent in 5 N acetic acid).

Observations

Nitrate is reduced if the mixture becomes pink or red. No color means nitrate is present as such or has not been reduced to ammonia and free nitrogen.

17.12 OXIDASE TEST

Composition of Trypticase Soy Agar Medium

Trypticase, 17 g; phytone, 3 g; NaCl, 5 g; K_2HPO_4, 2.5 g; glucose, 2.5 g; agar–agar, 15 g; distilled water, 1 L; pH, 7.3.

Procedure

Streak the overnight grown culture of test bacterium on the trypticase soy agar plate and incubate the plates at 27°C for 24 hours. After incubation, add 2 to 3 drops of tetramethyl-phenylene diamine dihydrochloride on the growth surface of test isolate.

Observations

The color change to maroon indicates the oxidase positive reaction.

17.13 PIGMENT PRODUCTION

Production of Fluorescein and Pycocyanin Pigments

Production of these pigments can be demonstrated by using special agars particularly Tech (*Pseudomonas* P agar) and Flo (*Pseudomonas* F agar). Fluorescein production is enhanced by bacterial cultures on Flo agar because of this medium's limiting levels of ferric ions. Fluorescein producing cultures (*Pseudomonas fluorescens*) on Flo-agar fluoresce when the plates are open and viewed under UV radiation.

Pycocyanin is a blue-green water soluble pigment produced by *P. aeruginosa*. On Tech agar this pigment appears within 6–7 days.

Incubate the test bacterial cultures on Flo and Tech agar plates for 48 hours at 28 + 2°C temp. Observe the plates under UV light for fluorescent pigment production. (See Figure 17.4.)

17.14 THE IMViC TESTS

This consists of four different tests: (1) indole production, (2) methyl-red, (3) Voges–Proskauer, and (4) citrate utilization. The name IMViC stands for the first letter of the name of each test in the series, with the lower case "i" included for ease of pronunciation. The IMViC tests were designed to differentiate Gram-negative intestinal bacilli (family Enterobacteriaceae), particularly *Escherichia coli* and the *Enterobacter–Klebsiella* group, on the basis of their biochemical properties and enzymatic reactions in the presence of specific substrates.

17.14.1 INDOLE PRODUCTION TEST

Tryptophan, an essential amino acid, is oxidized by some bacteria by the enzyme tryptophanase, resulting in the formation of indole, pyruvic acid, and ammonia. The indole test is performed by inoculating a bacterium into tryptone broth; the indole produced during the reaction is detected by adding Kovac's reagent (dimethylaminobenzaldehyde) which produces a cherry-red reagent layer.

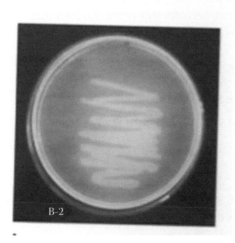

FIGURE 17.4 Production of fluorescein pigment by *Pseudomonas*. (Courtesy of Dr. S. G. Borkar and Jyoti Mergerwar, Department of Plant Pathology, Mahatma Phule Krishi Vidyapeeth, Rahuri.)

Material Required

Nutrient broth cultures of test bacterium, tubes containing 1 percent tryptone broth (5 ml/tube), Kovac's reagent, dropper bottle, 1 ml pipette, Bunsen burner, inoculating needle.

Procedure

Prepare (1 percent) tryptone broth: Dissolve 10 g of tryptone in 1 L of distilled water. Sterilize in the autoclave at 15 psi (121°C) for 15 minutes. Inoculate tryptone broth with test bacterium and keep one tube as an uninoculated comparative control. Incubate inoculated and uninoculated tubes at 28°C for 48 hours. After 48 hours of incubation, add 1 ml of Kovac's reagent to each tube including control. Shake the tubes gently after intervals for 10–15 minutes. Allow the tubes to stand to permit the reagent to come to the top. Examine the tubes as to the color in the reagent "layer."

Results

Development of a cherry (deep) red color in the top layer of the tube is a positive test for indole production. Absence of red coloration is indole negative.

Precautions

Use fresh tryptone broth for best results. Repeated gentle shaking of the tubes is necessary for 10–15 minutes after addition of Kovac's reagent for the development of a deep red color. Do not incubate the inoculated tyrptone broth tubes more than 48 hours because after 48 hours of incubation, indole itself may be attacked and further degraded. If this happens, the indole will eventually disappear, and you may get a misleading negative test result.

17.14.2 Methyl-Red and Voges–Proskauer Tests

The methyl red (MR) and the Voges–Proskauer (V-P) tests are used to differentiate two major types of facultative anaerobic enteric bacteria that produce large amounts of acid and those that produce the neutral product acetone as an end product. Both of these are performed simultaneously because they are physiologically related and are performed on the same medium MR-VP broth. Opposite results are usually obtained for the methyl red and Voges–Proskauer tests, that is, MR+, VP–, or MR–, VP+. In these tests, if an organism produces a large amount of organic acids, that is, formic, acetic, lactic, and succinic as end products, from glucose, the medium will remain red (a positive test) after the addition of methyl red a pH indicator (i.e., pH remaining below 4.4). In other organisms, methyl red will turn yellow (a negative test) due to the elevation of pH above 6.0 because of the enzymatic conversion of the organic acids (produced during the glucose fermentation) to nonacidic end products such as ethanol and acetoin (acetylmethylcarbinol).

MRVP tests are of value in the separation of *Escherichia coli* and *Enterobacter aerogenes* (both coliform bacteria) which appear virtually identical except for certain physiological differences that are used as indicators of the sanitary quality of water, foods, food production, and eating establishments.

Material Required

Nutrient broth cultures of test bacterium, MRVP broth tubes (5 ml/tube), methyl red pH indicator (dropper bottle), V-P reagent I (napthol solution), V-P reagent II (40 percent potassium hydroxide), clean empty test tubes, Bunsen burner, inoculating loop.

Procedure

Prepare MRVP broth (pH 6.9) tubes of the following composition: peptone, 7.0 g; dextrose/glucose, 5.0 g; potassium phosphate, 5.0 g; distilled water, 1000.0 ml. Pour the 5 ml broth in each tube and sterilize by autoclaving at 15 pounds of pressure for 15 minutes. Inoculate MRVP tubes with test bacterium as set I and set II and keep one tube as uninoculated comparative control. Incubate all tubes at 28°C for 48 hours. Add 5 drops of methyl red indicator to the tube of set I. Observe the change in color of methyl red for MR test. Add 12 drops of V-P reagent I and 2–3 drops of V-P reagent II to the tubes of set II as well as to uninoculated control tube. Shake the tubes gently for 30 seconds with the caps off to expose the media to oxygen. Allow the reaction to complete for 15–30 minutes. Observe the tubes for change in color for the VP test.

Observations

In the MR test, the methyl red indicator in the pH range of 4 will remain red (throughout tube) which is indicative of a positive test, while the turning of methyl red to yellow is a negative test.

In the VP test, the development of a crimson to ruby pink (red) color, which may be most intense on the surface, is indicative of positive VP test while no change in coloration is a negative test.

17.14.3   CITRATE UTILIZATION TEST

Citrate test is used to differentiate among enteric bacteria on the basis of their ability to utilize/ferment citrate as the sole carbon source. The utilization of citrate depends on the presence of an enzyme citrase produced by the organism that breaks down the citrate to oxaloacetic acid and acetic acid.

These products are later converted to pyruvic acid and carbon dioxide enzymatically.

The citrate test is performed by inoculating the microorganisms into an organic synthetic medium, Simmons citrate agar, where sodium citrate is the only source of carbon and energy. Bromothymol blue is used as an indicator. When the citric acid is metabolized, the CO_2 generated combines with sodium and water to form sodium carbonate an alkaline product, which changes the color of the indicator from green to blue and this constitutes a positive test. Bromothymol blue is green when acidic (pH 6.8 and below) and blue when alkaline (pH 7.6 and higher).

Material Required

Nutrient broth cultures of test bacterium, Simmon's citrate agar slants, Bunsen burner, inoculating needle.

Procedure

Prepare Simmons' citrate agar (pH 6.9) slants of the following composition: ammonium dihydrogen phosphate, 1.0 g; dipotassium phosphate, 1.0 g; sodium chloride, 5.0 g; sodium citrate, 2.0 g; magnesium sulfate, 0.2 g; agar, 15.0 g; bromothymol blue, 0.8 g; distilled water, 10000.0 ml. Dissolve all the constituents, except phosphates, which are to be dissolved separately in 100 ml of water. Mix these and make volume to 1 L. Adjust the pH to 6.9. Pour the medium in the culture tubes and sterilize by autoclaving at 15 pounds pressure for 15 minutes and prepare the slants.

Inoculate Simmons' citrate agar slants, with test bacterium by means of a stab-and-streak inoculation. The other tube is kept as an uninoculated comparative control. Incubate all the slants at 28°C for 48 hours. Observe the slant cultures for the growth and coloration of the medium.

Observations

In inoculated slant, when growth is visible on the surface and the medium color is blue the test is citrate positive while when there is no growth and there is no change in the color of the medium (i.e., green) the test is citrate negative test.

Precautions

Always use a light inoculums to minimize carry-over of nutrients from the medium to grow the inoculums on Simmons' citrate agar slants.

17.15 CATALASE TEST

During aerobic respiration in the presence of oxygen, microorganisms produce hydrogen peroxide (H_2O_2), which is lethal to the cell. The enzyme catalase present in some microorganisms breaks down hydrogen peroxide to water and oxygen and helps them in their survival.

Catalase test is performed by adding H_2O_2 to trypticase soy agar slant culture. Release of free oxygen gas ($O_2\uparrow$) bubbles is a positive catalase test.

Material Required

Cultures of test bacterium, trypticase soy agar slant, hydrogen peroxide (3 percent), Bunsen burner, inoculating loop.

Procedure

Prepare trypticase soy agar (pH 7.3) slants of the following composition: trypticase, 15.0 g; phytone, 5.0 g; sodium chloride, 5.0 g; agar, 15.0 g; distilled water, 1000.0 ml. Pour the medium in culture tubes and flasks and sterilize by autoclaving at 15-pound pressure for 15 minutes. Inoculate the trypticase soy agar slants with the test bacterial culture. Keep an uninoculated trypticase soy agar slant as control. Incubate the cultures at 28°C for 24–48 hours. While holding the inoculated tube at an angle, allow three to four drops of hydrogen peroxide to flow over the growth of each slant culture. Observe each culture for the appearance or absence of gas bubbles.

Observations

A catalase positive culture will produce bubbles of oxygen within 1 minute after addition of H_2O_2. (See Figure 17.5.)

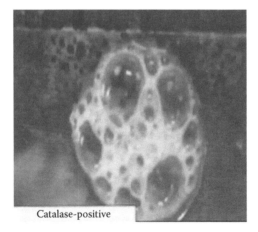
Catalase-positive

FIGURE 17.5 Bacterial culture showing catalase positive reaction. (Courtesy of Dr. S. G. Borkar and Dr. Nisha Patil, Department of Plant Pathology, Mahatma Phule Krishi Vidyapeeth, Rahuri.)

18 Characterization of Phytopathogenic Bacteria up to Genus

The cellular, morphological, and biochemical characters used for the characterization and determination of the genus of the plant pathogenic bacterial isolates as given by Garden and Luseitte are illustrated in Table 18.1.

These characters/tests include Gram reaction of bacterial cells, the number of flagella and their positions, bacterial colony morphology, and colony pigmentation. The major biochemical tests include reduction of nitrate; production of enzymes; catalase, oxidase, and ureas; carbohydrate metabolism, that is, oxidative, inactive, or fermentative; and hypersensitive reaction on tobacco plants. Some of the bacterial isolates do not produce HR on tobacco; therefore, one should not consider them nonplant pathogenic.

Once the pathogenicity of the bacterial isolates is tested on its natural host (from where it was isolated) and found pathogenic or induced the disease symptoms, the above characters/tests are to be carried out for the determination of bacterial genus of the concern isolates.

Other biochemical tests given in previous chapters are useful for the determination of group/species/pathovars of the various genus. Similarly these tests are also useful for identification of the nonplant pathogenic bacterial isolates associated with the plant pathogenic bacteria and do exist in rhizosphere or phylloplane with the bacterial plant pathogens. These may have a synergetic or inhibitory effect on the bacterial plant pathogens in/on plant system and therefore their identification is also necessary.

Thus, most of the morphological, physiological, and biochemical tests given in previous chapters are required in the identification of one or another of the bacterial isolates in the plant bacteriology laboratory. The detailed characteristics of the plant pathogenic bacteria of each genus are illustrated in Table 18.1.

18.1 AGROBACTERIUM

The bacterial cells are rod shaped, $0.6–1.0 \times 1.5–3.0$ μm in size and occur singly or in pairs. They are nonspore-forming and Gram-negative. Motility occurs by one to six pertrichous flagella. They are aerobic, possessing a respiratory type of metabolism with oxygen as the terminal electron acceptor.

Some strains are capable of anaerobic respiration in the presence of nitrate. Most strains are able to grow under reduced oxygen tensions in plant tissues. Optimum temperature is 25–28°C. Colonies are usually convex, circular, smooth, and nonpigmented to light beige. Growth on carbohydrate-containing media is usually accompanied by copious extracellular polysaccharide slime. Catalase positive and usually oxidase and urease positive. 3-Ketoglycosides are produced by the majority of strains belonging to *A. tumefaciens* biovar 1 and *A. radiobacter* biovar 1. chemoorganotrophs utilize a wide range of carbohydrates, salt of organic acids, and amino acids as carbon sources but not cellulose, starch, agar of glucose, d-galactose, and other carbohydrates. Ammonium salts and nitrates can serve as nitrogen sources for strains of some species and biovars; others require amino acids and additional growth factors. With the exception of *A. radiobacter*, members of this genus invade the crown, roots, and stems of a great variety of dicotyledonous and some gymnospermous plants via wounds, causing the transformation of the plant cells into autonomously proliferating tumor cells.

TABLE 18.1

Characterization of Phytopathologenic Bacteria up to Genus

Gram Reaction	No. of Flagella and Its Position	Colony Morphology and Pigmentation	Major Biochemical Test							Infection Site	Symptoms Induced	Genus
			HR Test on Tobacco	Nitrate	Catalase	Oxidase	Urease	Carbohydrate Metabolism				
Negative	1–6 peritrichrous	Convex, circular smooth, nonpigmentated to light-beige	HR ±	+	+	+	+	Respiratory	Crown, root and stem	Tumor or gall former	Agrobacterium
Negative	1 or more polar flagella	Fluorescent differential pigment	HR	+	+	±	D	Respiratory oxidative	All plant parts	Leaf spot and blight	Pseudomonas
Negative	One or more polar flagella	Pearly cream white, flat, irregular, fluidal colonies often with characteristic whorls in center, does not produce fluorescent pigment but produce diffusible brown pigment	HR	+	+	+	D	Oxidative	Root xylem vessel	Wilt	Ralstonia
Negative	Single polar flagella	Smooth, butyrous, and viscid; yellow	HR	+	+	+	D	Respiratory oxidative	All plant parts	Leaf spot, canker, blight	Xanthomonas
Negative	Nonmotile	Convex to pulvinate smooth, opalescent entire margin or umbonate rough, nonpigmented	d	d	+	−	D	Strictly aerobic oxidative	In the xylem of plant tissue	Leaf scorching	Xylella

(Continued)

TABLE 18.1 (CONTINUED)

Characterization of Phytopathogenic Bacteria up to Genus

Gram Reaction	No. of Flagella and Its Position	Colony Morphology and Pigmentation	Major Biochemical Test					Carbohydrate Metabolism	Infection Site	Symptoms Induced	Genus
			HR Test on Tobacco	Nitrate	Catalase	Oxidase	Urease				
Negative	Single polar flagella	Circular, semitranslucent slightly raised, glistering pale yellow, entire margin	d	–	+	–	+	Aerobic oxidative	Grapevine specific	Bacterial necrosis and canker	*Xylophilus* (formerly *Xanthomonas*)
Negative	Peritrichrous flagella	Mucoid, raised, circular, entire, pink, blue or yellow pigment depending on sp.	d	–	+	–	D	Fermentative or oxidative	Soil borne root, rhizome	Soft rot	*Erwinia*
Negative	Peritrichrous flagella	Most strains produce yellow pigment	d	+	+	–	D	Fermentative or oxidative	Soil borne	Rot	*Pantoea*
Positive	Polar flagella or nonmotile	Convex, semi opaque with mat surface, nonpigmented or yellow or orange pigment	HR ±	+	+	–	D	Inactive or fermentative	All plant part	Leaf spot, fruit spot, wilt, canker, hypertrophy	*Clavibacter* (formerly *Corynebacterium*)
Positive		Smooth, convex colonies, yellow or orange	d	d	+	d	D	Oxidative	All plant parts	Stem canker, leaf spot, bulb spot	*Curtobacterium*

Note: d = not known.

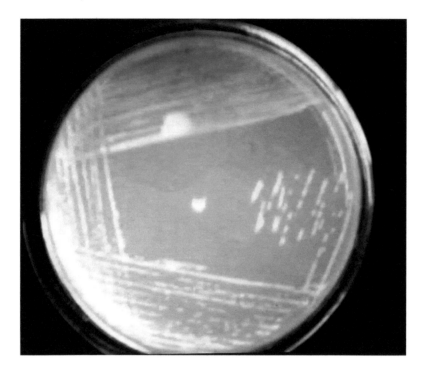

FIGURE 18.1   Bacterial colonies of *Agrobacterium*. (From Z.-Q. Luo and S. K. Farrand, *J. Bacteriol.*, 181, 618–626, 1999. Copyright ©1999, American Society for Microbiology, stephenf@life.illinois.edu.)

The induced plant diseases are commonly known as crown gall, hairy root, and cane gall. Some strains possess a wide host range, whereas others (e.g., grape vine isolates) possess a very limited host range. The tumors are self-proliferating and graftable. The tumor induction by *Agrobacterium* is correlated with the presence of a large tumor-inducing plasmid (Ti-plasmid) in the bacterial cells.

Agrobacteria are soil inhabitants. Oncogenic strains occur mainly in soils previously contaminated with diseased plant material. Some nonocogenic *Agrobacterium* strains have been isolated from human clinical specimens. Type species: *Agrobacterium tumefaciens*. (See Figure 18.1.)

18.2   *PSEUDOMONAS*

The bacterial cells are straight or slightly curved rods, 0.5–1.0 × 1.5–5.0 μm in size. Many species accumulate poly-β-hydroxybutyrate as carbon reserve material, which appears as sudanophilic inclusions. They do not produce prosthecae and are not surrounded by sheaths. Cells stain Gram-negative. Motility occurs by one or several polar flagella; they are rarely nonmotile. In some species lateral flagella of shorter wavelengths may also be formed. They are aerobic, having a strictly respiratory type of metabolism with oxygen as the terminal electron acceptor; in some cases nitrate can be used as an alternate electron acceptor, allowing growth to occur anaerobically. Xanthomonadins are not produced. Most, if not all, species fail to grow under acidic conditions (pH 4.5). Most species do not require organic growth factors. Oxidase positive or negative. Catalase positive and chemo-organotrophic; some species are facultative chemolithotrophs, able to use H_2 or CO as energy sources. Widely distributed in nature. Some species are pathogenic for humans, animals, or plants. Type species: *Pseudomonas aeruginosa*. (See Figure 18.2.)

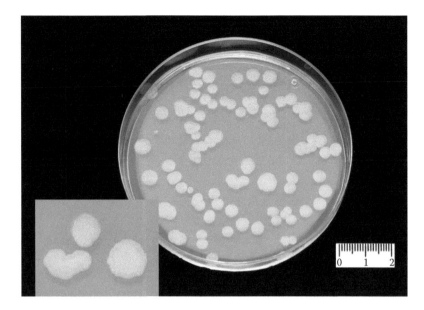

FIGURE 18.2 Bacterial colonies of *Pseudomonas*. (From http://www.microbiologyinpicture.com /pseudomonasaeruginosa.html.)

18.3 RALSTONIA

Ralstonia is a Gram-negative bacterium with rod-shaped cells, 0.5–1.5 μm in length, with a single or tuff of polar flagellum. The positive staining reaction for poly-β-hydroxybutyrate granules with Sudan Black or Nile Blue distinguishes. *R. solanacearum* from many other (Phytopathogenic) Gram-negative bacterial species. The colonies often produced nonfluorescent but diffusible brown pigment. On a general nutrient media, virulent isolates of *R. solanacearum* develop pearly cream white, flat, irregular, and fluidal colonies often with characteristic whorls in the center, while the avirulent forms of *R. solancearum* form small, round, nonfluidal, butyrous colonies that are entirely cream white. On Kelman's tetrazolium and SMSA media, the colonies of virulent isolates are blood red in color while avirulent *R. solanacearum* are entirely deep red.

Ralstonia does not produce fluorescent pigment like *Pseudomonas*. The most important plant pathogen is *Ralstonia solanacearum*, a soilborne bacterium colonizing xylem vessels of root causing wilt in a very wide range of potential host plant.

The important plant pathogenic species are *Ralstonia solancearum* which infect potato, tomato, eggplant, banana, geranium, ginger, tobacco, capsicum, rose, and soybean plants, while the species *Ralstonia syzgii* is pathogenic on clove plants in Indonesia. The other species of the genus are not plant pathogenic and are reported to be clinical pathogens. (See Figure 18.3.)

18.4 XANTHOMONAS

The bacterial cells are straight rods, usually 0.4–0.7 × 0.7–1.8 μm in size, predominantly single. Do not produce poly-β-hydroxybutyrate inclusions. Do not have sheaths or prosthecae. No resting stages are known. Cells stain Gram-negative. Motility occurs by a single polar flagellum (except *X. maltophilia*, which has multitrichous flagella). They are aerobic, having a strictly respiratory type of metabolism with oxygen as the terminal electron acceptor. No denitrification or nitrate reduction

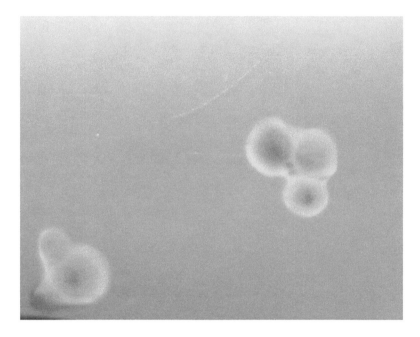

FIGURE 18.3 Bacterial colonies of *Ralstonia*. (Photo courtesy of Dr. S. G. Borkar, T. S. Ajaysree and R. A. Yumlembam, Department of Plant Pathology, Mahatma Phule Krishi Vidyapeeth, Rahuri.)

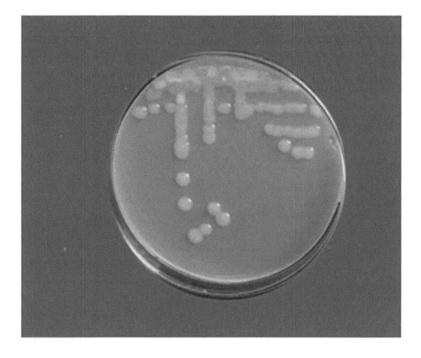

FIGURE 18.4 Bacterial colonies of *Xanthomonas*. (Courtesy of Dr. S. G. Borkar and Maria D'souza, Department of Plant Pathology, Mahatma Phule Krishi Vidyapeeth, Rahuri.)

occurs (except *X. maltophilia*, which reduces nitrate to nitrite). Optimum temperature is 25–30°C. Colonies are usually yellow, smooth, and butyrous or viscid. The pigments are highly characteristic brominated early polyenes, or "Xanthomonadins" (except for *X. maltophilia*, which does not produce xanthomonadins). Oxidase negative or weakly positive. Catalase positive. Chemoorganotrophic, able to use a variety of carbohydrates and salts of organic acids as sole carbon sources. Small amounts of acid are produced from many carbohydrates. Acid is not produced in purple milk or litmus milk. Grow on calcium lactate but not on glutamine. Asparagine is not used as a sole source of carbon and nitrogen. Growth is inhibited by 0.1 percent (and usually by 0.02 percent) triphenyltetrazolium chloride. Growth factors usually required include methionine, glutamic acid, nicotinic acid, or a combination of these. Plant pathogens occur in association with plants except for *X. maltophilia* (which is an opportunistic pathogen of humans). Type of species: *Xanthomonas campestris*. (See Figure 18.4.)

18.5   *XYLELLA*

The genus was created in 1987 by Wells et al. It includes only one species, *X. fastidiosa*.

The bacterial cells are single straight rods, 0.25–0.35 × 0.9–3.5 µm in size, with long filamentous strands under some cultural conditions. Colonies are of two types: Convex to pulvinate, smooth, opalescent with entire margins and umbonate, rough with finely undulated margins. Cells stain Gram-negative. Nonmotile. Oxidase negative and catalase positive. Strictly aerobic, nonfermentative, non-halophilic, nonpigmented. Nutritionally fastidious, requiring a specialized medium such as BCYE containing charcoal or glutamine—peptone medium (PW) containing serum albumin. Optimal temperature for growth is 26–28°C. Optimum pH is 6.5–6.9. Habitat is the xylem of plant tissue. Type (and only) species: *Xylella fastidiosa*.

Characteristics of the species: Hydrolyzes gelatin and utilizes hippurate. Most strains produce β-lactamase. Glucose is not fermented. Negative in tests for indole, H_2S, β-galactiosidase, lipase, amylase, coagulase, and phosphatase. The species has been isolated as a phytopathogen from tissues of a number of host plants. The type strain was isolated from grapevine with Pierce's disease.

18.6   *XYLOPHILUS*

The genus was created in 1987 by Willems et al. It includes only one species, *X. ampelinus* (formerly *Xanthomonas ampelina*).

The bacterial cells are straight to slightly curved rods, 0.4–0.8 × 0.6–3.3 µm in size. Filamentous cells (length: 30 µm or more) may occur in older cultures. Cells occur singly, in pairs, or in short chains. Motility occurs by a single polar flagellum. Cells stain Gram-negative. Oxidase negative, catalase positive. Aerobic, chemoorganotrophic. Oxidative carbohydrate metabolism. Even at the optimal temperature of 24°C, growth is generally very slow and poor. Growth occurs on L-glutamine but not on calcium lactate (in contrast to *Xanthomonas* strains). Plant pathogens, causing bacterial necrosis and canker. Type (and only) species: *Xylophilus ampelinus*.

Characteristics of the species: On nutrient agar, colonies are circular, semi-translucent, slightly raised, glistening, and pale yellow with entire margins. After 6 days the colonies are 0.2–0.3 mm in diameter; after 15 days the colonies are 0.6–0.8 mm in diameter. Better growth is obtained on GYAC medium (Williams et al., 1987), and the best growth is obtained on a medium containing 1 percent yeast extract, 2 percent D-galactose, 2 percent $CaCO_3$, and 2 percent agar at 24°C. On this medium, colonies are yellow, and a brown diffusible pigment is produced. Some strains may produce two colony types on GYCA medium. Average colony diameters of 1.4 × 0.6 and 0.6 × 0.3 mm are attained after 15 days at 24°C by the fast and slow growing types, respectively. Minimal and maximal growth temperatures are 6 and 30°C, respectively. The maximal NaCl concentration tolerated is 1 percent.

Growth is very slow in the presence of 0.01–0.02 percent 2,3,5-triphenyltertrazolium chloride; L-glutamate (0.1 percent) is required for growth while L-methionine is not required. Acid is produced from L-arabinose and D-galactose but not from D-xylose, D-ribose, L-rhamnose, D-glucose, D-mannose, D-fructose, sucrose, trehalose, cellobiose, lactose, maltose, D-melibiose, raffinose, glycogen, insulin, dextrin, adonitol, dulcitol, D-mannitol, D-sorbitol, l-sorbose, salicin, α-methylglycoside, arbutin, and meso-inositol. Grows on acetate (0.2 percent but not 0.5 percent), citrate, DL-malate, succinate, DL-tartrate, and fumarate but does not grow on formate, propionate, malonate, maleate, oxalate, benzoate, or calcium gluconate. Grows on Dye's asparagines medium. No hydrolysis of gelatin, esculin, starch, casein, arbutin, and sodium hippurate; lipolysis of Tween 80. Potato soft rot test is negative. H_2S is formed from cysteine and weakly from thiosulfate. Urease positive. The following features are lacking in all strains that have been studies: Nitrate reduction, production of indole and ammonia, Voges–Proskauer, arginine dihydrolase, lysine and ornithine decarboxylase, lecithinase, and acid from glucose.

The bacterium has been isolated from *Vitis vinifera*, on which it causes bacterial necrosis and canker, mainly of the woody parts, in the Mediterranean region and South Africa. Similar symptoms have been reported in Switzerland, Austria, Bulgaria, the Canary Islands, and Argentina.

18.7 *ERWINIA*

The bacterial cells are straight rods, 0.5–1.0 × 1–3 µm in size, occur singly, in pairs, and sometimes in short chains. Gram-negative. Motile by peritrichous flagella (except *E. stewartii*). Facultatively anaerobic. Chemoorganotrophic, having both a respiratory and a fermentative type of metabolism. Optimum temperature is 27–30°C. D-Glucose and other carbohydrates are catabolized with the production of acid; most species do not produce gas. Oxidase negative and catalase positive; they are lysine decarboxylase, arginine dihydrolase, and ornithine decarboxylase negative. Nitrates are not reduced by many species. Ferment fructose, galactose, β-methyl-glycoside, and sucrose (usually D-mannitol, D-mannose, ribose, and D-sorbitol), but rarely adonitol, dextrin, dulcitol, and meleitose. Utilize acetate, fumarate, gluconate, malate, and succinate, but not benzoate, oxalate, or propionate as carbon and energy-yielding sources. Associated with plants as pathogens, saprophytes, or constituents of the epiphytic flora. Very rarely isolated from humans. Type species: *Erwinia amylovora*.

18.8 *PANTOEA*

The genus was created in 1989 by Gavini et al. It includes two species, *Pantoea agglomerans* (synonyms: *Enterobacter agglomerans*, *Erwinia herbicola*, *Erwinia milletiae*) and *Pantoea dispersa*.

The bacterial cells are straight rods, 0.5–1.0 × 1–3 µm in size. Gram-negative. Motile by peritrichous flagella. Most strains produce a yellow pigment. Facultatively anaerobic. Chemoorganotrophic, having both a respiratory and a fermentative type of metabolism. Optimal temperature is 30°C. D-glucose and other carbohydrates are catabolized with production of acid, but not gas. Oxidase negative, catalase positive, indole negative, Voges–Proskauer and Simmons citrate positive, and methyl red variable. Negative for lysine and ornithine decarboxylase and for arginine dihydrolase (Gavini et al., 1989; found 30 percent of *P. agglomerans* to be ornithine positive; previous studies reported all strains negative). No production of H_2S; urea is not hydrolyzed. Most strains grown on KCN. Malonate utilization varies among species. Reduce nitrates. Carbohydrates fermented by all or most strains include L-arabinose, D-galactose, maltose, D-mannitol, D-manose, L-rhamnose, sucrose, trehalose and D-xylose. Isolated from plant surfaces, seeds, soil, and water as well as from animals and human wounds, blood, and urine. An opportunistic human pathogen. Type of species: *Pantoea agglomerans*.

P. agglomerans frequently referred to by its synonyms, especially *Enterobacter agglomerans*. (See Figure 18.5.)

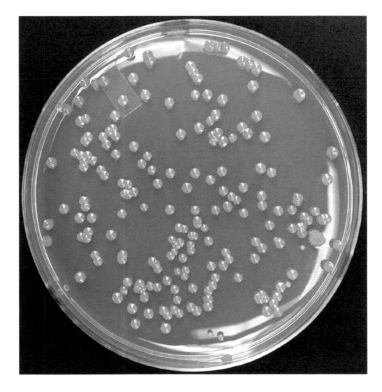

FIGURE 18.5 Bacterial colonies of *Pantoea*. (From Arias, R. S., Power, I. L., Culbreath, A. K., Sobolev, V. S., and Lamb, M. C., *Plant Health Progress*, Published 29 July 2013, doi:10.1094/PHP-2013-0729-04-BR, renee.arias@ars.usda.gov.)

18.9 *CLAVIBACTER (CORYNEBACTERIUM)*

It was noted in *Bergey's Manual of Systematic Bacteriology* that the traditional genus *Corynebacterium* was heterogeneous and that there were proposals to divide it into several genera. A number of such proposals have now received general assent, and in particular the aerobic plant pathogens containing 2,4-diaminobutyric acid in the cell wall are treated under the new genus *Clavibacter*.

The bacterial cells are straight or slightly curved, slender rods have tapered or sometimes clubbed ends and are 0.3–0.8 × 1.5–8.0 μm in size. One species (*C. Matruchotii*) has a whip handle–shape. Cells are usually arranged singly or in pairs, often in a V formation or in palisades of several parallel cells. Gram-positive, though some cells stain unevenly, giving a beaded appearance. Metachromatic granules of polymetaphosphate are commonly formed within the cells. Nonmotile, nonsporing, non–acid-fast. Facultative anaerobes, commonly requiring nutritionally rich media such as serum or blood media, on which colonies are usually convex and semi opaque, with a mat surface. Chemo-organotrophs with fermentative metabolism, most species produce acid without gas from glucose and some other carbohydrates. Catalase positive, often reduce nitrate and tellurite. Rarely acidify lactose or raffinose or liquefy gelatin. Primarily obligate parasites of mucous membranes or skin of mammals, but occasionally they are found in other sources; some species are pathogenic for mammals. Type species: *Corynebacterium diphtheria*.

18.10 *CURTOBACTERIUM*

The bacterial cells are small, short, irregular rods in young cultures, 0.4–0.6 × 0.6–3.0 μm in size, become coccoid in old cultures. Arranged singly or sometimes in pairs, often in a V formation; no

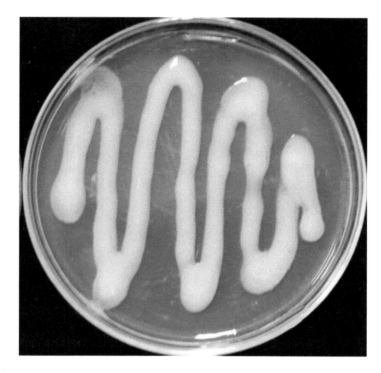

FIGURE 18.6   Bacterial colonies of *Curtobacterium*. (From Moreira Soares, R. et al., 2013 *Tropical Plant Pathology*. 38(5), 452–454. Copyright by the Brazilian Phytopathological Society, www.sbfito.com.br, rafael.soares@embrapa.br.)

branching is found. Gram-positive but cells from old cultures are easily decolorized. Generally motile by peritricchous flagella. Nonsporing, non–acid-fast. Metachromatic granules are absent. Obligately aerobic, they yield smooth, convex colonies on nutrient agar; the colonies are usually yellow or orange. Chemoorganotrophic, not especially exacting nutritionally. Metabolism is respiratory, yielding small amounts of acid from glucose and some other carbohydrates. Catalase positive. The optimum growth temperature is 25–30°C. Occur on plants, in soil and in oil brine; *C. flaccumfaciens* is a plant pathogen. Type species: *Curtobacterium citreum*. (See Figure 18.6.)

19 Differentiation of Bacterial Genus into Group

The differentiation of the bacterial genus into group is based on various biochemical tests and utilization of carbohydrates. The differentiation of the genus into groups is necessary because the bacterium of each group has a certain host range and does not infect the plants infected by other group of bacterium.

19.1 DIFFERENTIATION OF GENUS *XANTHOMONAS* INTO GROUPS

Test	Species				
	X. campestris	*X. fragariae*	*X. albineans*	*X. axonopodis*	*X. ampelina (3)*
Hydrolyse esculine	+	+	+	+	−
Gelatinolyse	+	+	d	−	−
H$_2$S	+	−	−	+	D
Tween esterase	+	+	d	D	−
Urease	−	−	−	−	+
Amidon	+	+	d	D	−
Utilisation of					
Arabinose	+	−	−	−	+
Glucose	+	+	+	+	−
Galactose	+	−	d	−	+
Cellobiose	+	−	−	−	−
Sorbitol	−	−	−	−	−
Inositol	−	−	−	−	−

Source: Bergey's Manual of Systematic Bacteriology, 2nd ed., vol. 2, December 14, 2007. The Proteobacter, Part B: The Gammaproteobacter, Xanthomonas, p. 323.

19.2 DIFFERENTIATION OF SOME WELL-STUDIED SPECIES OF THE GENUS *PSEUDOMONAS* INTO GROUP/SPECIES AND BIOVARS

Characteristic	*P. aeruginosa*	*P. caryophylli*	*P. cepacia*	*P. cichorii*	*P. fluorescens* Biovar I	Biovar II	Biovar III	Biovar IV	Biovar V	*P. gladioli*	*P. pseudoalcaligenes*	*P. putida* Biovar A	Biovar B	*P. solanacearum*	*P. syringae* pathovars	*P. viridiflava*
Number of flagella	1	>1	>1	>1	>1	>1	>1	>1	>1	>1	1	>1	>1	>1	>1	1-2
Fluorescent, diffusible pigments	+	–	–	+	+	+	+	+	+	–	–	+	+	–	+	+
Diffusible nonfluorescent pigments	+ (blue-green[e])	+ (yellow-green)	+ (various colors)	–	–	–	–	–	–	+ (yellow-green)	–	–	–	d (brown)	–	–
Nondiffusible nonfluorescent pigments	–	–	–	–	–	–	–	+ (blue)	–	–	–	–	–	–	–	d (blue-green)
Poly-β-hydroxybutyrate accumulation	–	+	+	–	–	–	–	–	–	+	–	–	–	+	–	–
Extracellular poly-β-hydroxybutyrate hydrolysis	–	–	–	–	–	–	–	–	–	–		–	–	–	–	–
Autotrophic growth with H$_2$	+	+	D	–	–	–	–	–	–	+	+	–	–	–	–	–
Growth at 41°C	+	+	D	–	–	–	–	–	–	+	+	–	–	+	–	–
Growth at 4°C	–	–	–	–	+	+	+	+	d	–	–	d	+	–	d	–
Organic growth factors required (pantothenate, biotin, cyanocobalimin, methionine, or cystine)	–	D	–	–	–	–	–	–	–	–	–	–	–	–	–	–
Levan formation form sucrose	–	+	–	–	+	+	+	+	+	–	–	–	–	–	D	–
Arginine dihydrolase	+	+	–	–	+	+	+	+	+	+	d	+	+	–	–	–
Oxidase reaction	+	+	+	+	+	+	+	+	+	+	+	+	+	+	–	–
Denitrification	+	+	–	–	–	d	+	+	+	+	+	–	–	+	–	+
Gelatin hydrolysis	+	–	D	–	+	+	+	+	+	+	d	–	–	–	D	+
Starch hydrolysis	–	–	–	–	–	–	–	–	–	–	–	–	–	–	–	–

(Continued)

(Continued)

Characteristic Utilization of	P. viridiflava	P. syringae pathovars	P. solanacearum	P. putida Biovar B	P. putida Biovar A	P. pseudoalcaligenes	P. gladioli	P. fluorescens Biovar V	Biovar IV	Biovar III	Biovar II	Biovar I	P. cichorii	P. Cepacia	P. caryophylli	P. Aeruginosa
Glucose	+	+	+	+	+	−	+	+	+	+	+	+	+	+	+	+
Trehalose	−	−	+	−	−	−	+	+	+	+	+	+	−	+	D	−
2-Ketogluconate	−	−	−	+	+	−	+	+	+	+	+	+	−	+	+	+
Meso-Inositol	+	D	d	−	−	−	+	+	+	+	+	+	D	+	+	−
Geraniol	−	−	−	−	−	−	−	−	−	−	−	−	−	−	−	+
L-Valine	−	−	d	+	+	d	+	+	+	+	+	+	−	D	D	D
B-Alanine	−	−	−	+	+	−	+	+	+	+	+	+	−	+	+	+
L-Arginine	+	D	−	+	+	+	+	+	+	+	+	+	+	+	+	+
D-Xylose							+							D	+	
D-Ribose			d				+							+	+	
L-Rhamnose			−				−							D	+	
Saccharate			+				+							+	+	
Levulinate			d				−							+	−	
Citraconate			−				+							+	−	
Mesaconate			−				+							−	−	
D(−)-Tartrate			d				+							−	+	
mesco-Tartrate			−				+							+	−	
Erythritol			−				−							−	−	
Adonitol			−				+							+	+	
2,3-Butylene glycol							−							+	−	
m-Hydroxybenzoate			−				−							+	−	
Tryptamine			−				−							+	−	
α-Amylamine			−				−							+	−	

Characteristic	P. aeruginosa	P. caryophylli	P. cepacia	P. cichorii	P. fluorescens Biovar I	P. fluorescens Biovar II	P. fluorescens Biovar III	P. fluorescens Biovar IV	P. fluorescens Biovar V	P. gladioli	P. pseudoalcaligenes	P. putida Biovar A	P. putida Biovar B	P. solanacearum	P. syringae pathovars	P. viridiflava
Sucrose																
Malonate																
Nitrate used as a nitrogen source	+	+	+	+	+	+	+	+	+		+	+	+		+	+
Lecithinase (egg yolk reaction)	–	D	D	+	+	Weak	+	+	d	+		–	–	–	D	d
Catechol, ortho cleavage	+	+		+	+	+	+	+	+	+		+	+			+
Protocatechuate																
Ortho cleavage	+	+	+	+	+	+	+	+	+	+	+	+	+	+	+	+
Meta cleavage	–	–	–	–	–	–	–	–	–	–	+	–	–	–	–	–
Saprophytic or opportunistic animal pathogens	+	–	+	–	+	+	+	+	+	–	+	+	+	–	–	–
Phytopathogens	–	+	–	+	–	–	–	–	–	+	d[h]	–	–	+	+	+
Parasite of animals and humans	–	–	–	–	–	–	–	–	–	–	+	–	–	–	–	–

Source: Bergey's Manual of Systematic Bacteriology, 2nd ed., vol. 2, December 14, 2007. The Proteobacter, Part B: The Gammaproteobacter, Xanthomonas, p. 323.

Note: 1. Symbols: See standard definitions.

2. Two phytopathogenic subspecies of *P. pseudoalcaligenes* have been proposed. *P. pseudoalcaligenes* subsp. *Citrulli*, pathogenic for watermelon (Citrullus lanatus; see Schaad, N. W. et al., *Int. J. Syst. Bacteriol.*, 28, 117–125, 1978) and *P. pseudoalcaligenes* subsp. *Konjaci*, pathogenic for the leaves of konjac (Amorphophalus konjac; see Goto, M., *Int. J. Syst. Bacteriol.*, 33, 539–545, 1983).

19.2.1 Characteristics Useful for Differentiation of Various Denitrifying Pseudomonas

Characteristic	P. aeruginosa	P. alcaligenes	P. caryo-phylli	P. fluorescens II, III P. aureofaciens	P. mallei	P. mendocina	P. pickettii	P. pseudoalcaligenes	P. pseudomallei	P. solanacearum	P. stutzeri
Number of flagella	1	1	>1	>1	0	1	1	1	>1	>1	1
Poly-β-hydroxybutyric acid accumulation	−	−	+	−	+	−	+	D	+	+	−
Growth at 40°C	+	+	+	−	+	+	+	+	+	−	+
Pyoverdin production	+	−	−	+	−	−	−	−	−	−	−
Pyocyanin production	+	−	−	−	−	−	−	−	−	−	−
Yellow cellular pigment	−	d	−	−	−	+	−	−	D	−	−
Arginine dihydrolase	+	+	+	+	+	+	−	D	+	−	−
Starch hydrolysis	−	−	−	−	D	−	−	−	+	−	+
Poly-β-hydroxybutyrate hydrolysis	−	−	−	−	D	−	−	−	+	−	−
Growth on											
D-xylose	−	−	+	d	+	−	+	−	−	−	−
Maltose	−	−	−	−	D	−	−	−	+	−	+
Saccharate	−	−	+	d	−	+	+	−	−	+	d
Mannitol	+	−	+	+	+	−	−	−	+	d	d
Ethylene glycol	−	−	−	−	−	+	−	D	−	−	+
2,3-Butylene glycol	+	−	+	d	−	−	d	−	−	−	d
Geraniol	+	−	−	−	−	+	−	−	−	−	−
Azelate	+	−	−	d	−	+	+	−	+	d	−
Levulinate	+	−	−	d	−	+	+	−	+	d	−
Glycolate	−	−	+	−	−	+	+	−	−	d	+
L-Serine	d	−	+	d	D	+	+	D	+	d	d
L-Arginine	+	+	+	+	+	+	−	+	+	−	−
L-Histidine	+	d	+	+	+	+	d	D	+	+	−
Betaine	+	−	+	+	+	+	−	+	+	−	−
Sarcosine	+	−	−	+	D	+	−	D	D	d	−

Source: Bergey's Manual of Systematic Bacteriology, 2nd ed., vol. 2, December 14, 2007. The Proteobacter, Part B: The Gammaproteobacter, Xanthomonas, p. 323.

Note: 1. Symbols: See standard definitions.

2. Not all strains of *P. pseudoalcaligenes* are denitrifiers.

19.2.2   SOURCES AND CHARACTERISTICS OF ADDITIONAL *PSEUDOMONAS* SPECIES (ISOLATED FROM DISEASED PLANTS AND MUSHROOMS)

Source	Species	Characteristics and References
Cultivated mushrooms	*P. agarici*	One, rarely two, flagella. Fluorescent pigment produced. Acid is produced from glucose and various other sugar. Oxidase positive. Causes drippy gill of mushrooms. One of the main differences with another mushroom pathogen, *P. tolaasii*, is in the utilization of benzoate. (See *Bergey's Manual of Systematic Bacteriology*, vol. 1, p. 188.)
	P. tolaasii	One to five flagella. Fluorescent pigment produced. Acid is produced from glucose. Gelatin is liquefied. Isolated from brown spot of cultivated mushrooms. (See *Bergey's Manual of Determinative Bacteriology*, 7th ed., p. 136.)
Bird's nest fern	*P. asplenii*	One to three flagella. Fluorescent pigment produced. Temperature range: 1–34°C. Acid is produced from glucose and various other sugars. Gelatin is liquefied. Isolated from lesions of the bird's nest fern (*Asplenium nidus*). (See *Bergey's Manual of Determinative Bacteriology*, 7th ed. p. 124.)
Pawpaw	*P. caricapapayae*	Three to six flagella. Fluorescent pigment produced. Temperature range: 7–45°C. Acid is produced from glucose and various other sugars. Gelatin is liquefied. Isolated from water soaked, angular spots on leaves of pawpaw. (See *Bergey's Manual of Systematic Bacteriology*, vol. 1, p. 188.)
Almond tree	*P. amygdali*	One to six flagella. No fluorescent pigment produced. Temperature range: 3–32°C. Acid produced from glucose and various other sugars. Gelatin is not hydrolyzed. Produces a hyper plastic bacterial canker in the almond tree (*Prunus dulcis*, fam. *Rosaceae*). Not pathogenic for other fruit trees. (See *Bergey's Manual of Systematic Bacteriology*, vol. 1. pp. 188–189.)
Ficus erecta tree	*P. ficuserectae*	Motile by one to five flagella. Poly-β-hydroxybutyrate is accumulated, pigments are not produced. Similar to *P. armygdali* in many properties but differs by forming larger colonies on nutrient agar, utilizing raffinose and glycerol, and failing to hydrolyze Tween 80, to produce H_2S and to utilize ribose, mannitol, and sorbitol. Causes dark brown, water-soaked spots on the leaves and stems of *Ficus erecta*, resulting either in defoliation or shoot blight on severely infected plants. (For other characteristics, see Goto, *Int. J. Syst. Bacteriology*, 33, 546–550, 1983.)
Sorghum, corn, clover, and velvet bean	*P. andropogonis*	One, rarely two, flagella. Sheathed flagella have been reported in some strains. No fluorescent pigment produced. Poly-β-hydroxybutyrate is accumulated. Most strains are oxidase negative. Glucose and various other sugars are utilized. Gelatin is not hydrolyzed. The species may be divided into two specialized pathovars, namely, pv. andropogonis, the agent of a stripe disease of sorghum, and pv. stizolobii, which has been described as the cause of leaf spot of velvet bean (*Stizolobium deeringianum*). (See *Bergey's Manual of Systematic Bacteriology*, vol. 1, p. 189.)

(Continued)

Source	Species	Characteristics and References
Oats (*Avena sativa*) and foxtail (*Chaetochloa lutescens*)	*P. avenae*	No fluorescent pigment produced. Poly-β-hydroxybutyrate is probably accumulated. Oxidase negative. Acid is produced from glucose and various other sugars. Gelatin liquefaction is variable. Pathogenic for oats (*Avena sativa*) and foxtail (*Chaetochloa lutescens*). (See *Bergey's Manual of Systematic Bacteriology*, vol. 1, p. 189.)
Orchids	*P. cattleyae*	One or two bipolar flagella. No fluorescent pigment produced. Acid produced from glucose and various other sugars. Gelatin is not liquefied. Pathogenic for *Cattleya* sp. and *Phalaenopsis* sp. (fam, *Orchidaceae*). (See *Bergy's Manual Determinative Bacteriology*, 7th ed., p. 148.)
Cissus plants	*P. cissicola*	Nonmotile immediately after isolation, but motile clones appear after subculturing, the cells of which have a polar flagellum. Poly-β-hydroxybutrate is accumulated. No fluorescent pigment is produced. No growth below 5°C and above 37°C. Acid produced from glucose and various other sugars by most isolates. Gelatin is liquefied. Starch is hydrolyzed. Arginine dihydrolase is negative. Pathogenic for *Cissus japonica* (fam. *Vitaceae*). (See *Bergey's Manual of Systematic Bacteriology*, vol. 1, p. 189.)
Tomato plants	*P. corrugate*	Multitrichous polar flagella. Poly-β-hydroxybutyrate is accumulated. No fluorescent pigment is produced. Yellow-green diffusible, nonfluorescent pigment is produced. Colonies are winkled, yellowish, sometimes with green center. Growth occurs at 37°C but not at 41°C. Gelatin is hydrolyzed. Among the characters that differentiate this species from *P. cepacia* and *P. gladioli* are the absence of pectate hydrolysis and rot of onion slices and the lack of utilization of D-arabinose, cellobiose, adipate, meso-tartrate, and citraconate. Isolated from tomato pith necrosis. (See *Bergey's Manual of Systematic Bacteriology*, vol. 1, p. 189.)
Rice	*P. glumae*	Two to four flagella. Fluorescent pigment produced on potato agar. Temperature range: 11–40°C. Acid produced from glucose and various other sugars. Pathogenic for rice plants (*Oryza sativa* fam. *Gramineae*). (See *Bergey's Manual of Systematic Bacteriology*, vol. 1, pp. 189–190.)
	P. fuscovaginae	Motile by one to four polar flagella. Produce a green, fluorescent, diffusible pigment. Oxidase positive. No growth at 37°C. Unable to denitrify. Acid is produced from glucose and various other carbohydrates. Arginine dihydrolase positive. Distinguished from other arginine dihydrolase positive fluorescent pseudomonads by its ability to produce a hypersensitivity reaction in tobacco plants and its inability to utilize 2-ketogluconate or inositol. Pathogenic for *Oryza sativa*, *Hordeum vulgare*, *Triticum aestivum*, *Avena arundinacea*. (For other characteristics, see Miyajimat et al., *Int. J. Syst. Bacteril.*, 33, 656–657, 1983.)
	P. plantarii	Motile by one to three polar flagella. Poly-β-hydroxybutyrate is accumulated. Colonies have a slight yellow tint and weakly produce a diffusible, reddish brown pigment under certain conditions. Fluorescent pigments not produced.

(Continued)

Source	Species	Characteristics and References
		Oxidase positive. Organic growth factors not required. Able to denitrify. Arginine dihydrolase negative. Gelatin is liquefied. Temperature range: 10°C–38°C. Tobacco hypersensitivity reaction negative. Acid is produced from glucose and various other carbohydrates. Caused seedling blight of rice. (For other characteristics, see Azegami et al., *Int. J. Syst. Bacteriol.*, 37, 144–152, 1987.)
Sugarcane	*P. rubrilineans*	Motile by a single flagellum. Poly-β-hydroxybutyrate is accumulated. No pigments are produced. Oxidase positive. Gelatin liquefaction is weak. Capable of growth at 40°C. Acid is produced from glucose and various other sugars. Caused red stripe of sugarcane. (See *Bergey's Manual of Systematic Bacteriology*, vol. 1, p. 190.)
	P. rubrisubalbicans	Slightly curved rods motile by several polar flagella. Poly-β-hydroxybutyrate is accumulated. No pigments are produced. Oxidase positive. Gelatin is not hydrolyzed. Capable of growth at 40°C. Acid is produced from glucose and various other sugars. Causes mottled stripe of sugarcane. (See *Bergey's Manual of Systematic Bacteriology*, vol. 1, p. 190.)
Carnation	*P. woodsii*	Motile by a single polar flagellum. Gelatin is not liquefied. Acid produced from glucose and various other sugars. Isolated from water-soaked lesions on carnation leaves. Pathogenic for carnation (*Dianthus caryophyllus*, fam. Caryophyllaceae). (See *Bergey's Manual of Determinative Bacteriology*, 7th ed., pp. 150–151.)

19.3   DIFFERENTIATION OF GENUS *ERWINIA* INTO GROUPS

Group	Test			
	Pigmentation	Nitrate reductase	Pectinolyse	Test tabac
Group amylovora	–	–	–	+
Group carotovora	– (+)	+	+(–)	+V
Group herbicola	+(–)	+(–)	–	– (+)*

Note: * = It may or may not produce these reaction.

19.3.1 DIFFERENTIATION OF *ERWINIA* SPECIES BY CULTURAL, PHYSIOLOGICAL, AND BIOCHEMICAL CHARACTERISTICS

Test	E. amylovora	E. ananas	E. cacticida	E. Cartovora	E. Chrysanthemi	E. cypripedii	E. mallotivora	E. nigrifluens	E. persicinus	E. psidii	E. quercina	E. rhapontici	E. rubrifaciens	E. salicis	E. stewartii	E. tracheiphila	E. uredovora
Motility	+	+	+	+	+	+	+	+	+	+	+	+	+	+	−	+	+
Anaerobic growth	W	+	+	+	+	+	+	+	+	+	+	+	+	W	+	W	+
Growth factors required	+	−		−	−	−	+	−		−	+		−	−	−	+	−
Pink diffusible pigment	−	−	−	−	−	−	−	−	+	−	−	+	+	−	−	−	−
Blue pigment	−	−	−	−	d	−	−	−	−	−	−	−	−	−	−	−	−
Yellow pigment	−	+	−	−	−	−	−	−	−	−	−	−	−	−	+	−	+
Mucoid growth	+	+	−	d	d	d	+	−	−	+	+	+	+	+	+	−	−
Symplasmata		−													−		d
Growth at 36°C	−	+	+	d	+	+	−	+	+	−	+	d	+	−	d	−	+
H₂S from cysteine	−	D	+	+	+	−	+	−	+	+	+	+	+	−		+	−
Reducing substances form sucrose	+	+	−	d	−	−	+	−		+	+	d	−	+	d	d	+
Acetoin	+	+	+	+	+	−	+	+	+	+	+	+	−	+	−	d	+
Urease	−	−	−	−	−	−	+	−	−	−	−	−	−	−	−	−	−
Pectate degradation	−	−	+	+	+	−	−	−	−	−	−	−	+	+	−	−	−
Gluconate oxidation	−	−	−	−	−	+	−	−		−	−	d	−	−	−	−	−
Gas from D-glucose	−	−	−	d	+	+	−	−	−	−	−	−	−	−	−	−	−
Casein hydrolysis	−	−		d	d	−	−	−	−	−	−	−	−	−	−	−	−
Growth in KCN broth	−	−	−	d	d	+	−	−		−	−	+	−	−	−	−	−
Cotton seed oil hydrolysis	−	−		d	d	+	−	−			−	d	−	−	−	−	−
Gelatin liquefaction	+	+	−	+	+	−	−	−	−	−	−	−	−	−	−	−	+
Phenylalanine deaminase	−	−	−	−	+	−	−	−	−	−	−	−	−	−	−	−	−
Indole	−	+	−	−	+	−	−	−	−	−	−	−	−	−	−	−	+
Nitrate reduction	−	−	+	+	+	+	−	−	+	−	−	+	−	−	−	−	+
Growth in 5 percent NaCl		+	+	+	d	+	−					+			+	−	+
Deoxyribonuclease	−	−		−	−	−	−		−	−	−	−	−	−	−	−	+
Phosphatase			d	−	+	d	−		+			d					
Lecithinase				−	+	−						−					
Sensitivity to erythromycin (15 µg/disk)			−	−	+	+			+			+					

Source: Bergey's Manual of Systematic Bacteriology, 2nd ed., vol. 2, December 14, 2007. The Proteobacter, Part B: The Gammaproteobacter, Xanthomonas, p. 323.

Note: +, 80 percent or more positive; −, 20 percent or less positive; d, 21–79 percent positive; W, weak growth; blank space, insufficient or no data.

19.3.2 Differentiation of *Erwinia* Species Based on Acid Production from Organic Compounds

Test	*E. amylovora*	*E. ananas*	*E. cacticida*	*E. Cartovora*	*E. Chrysanthemi*	*E. cypripedii*	*E. mallotivora*	*E. nigrifluens*	*E. persicinus*	*E. psidii*	*E. quercina*	*E. rhapontici*	*E. rubrifaciens*	*E. salicis*	*E. stewartii*	*E. tracheiphila*	*E. uredovora*
D-Adonitol	−	−	−	−	−	−	−	−	−		−	−	−	−	−	−	+
L-arabinose	d	+	−	+	+	+	−	+	+	+	−	+	+	−	+	−	+
Cellobiose	−	+	−	+	+	+	(+)	−	+		−	+	−	−	−	−	+
Dextrin	−	−		−	−	−	−	−			−	−	−	−	−	−	+
Dulcitol	−	−	−	−	−	−	−	−	−	+	−	d	−	−	−	−	−
Esculin	−	d	+	+	+	+	−	+	+		+	+	−	+	−	−	d
Fructose	+	+	+	+	+	+		+	+	+	+	+	+	+	+	+	+
D-Galactose	+	+	+	+	+	+	+	+	+	+	+	+	+	+	+	+	+
D-Glucose	+	+	+	+	+	+	+	+	+	+	+	+	+	+	+	+	+
Glycerol	−	+	+	d	+	d	+	+		−	+	+	d	d	−	−	+
Myo-Inositol	−	+	−	d	d	+	(+)	+	+	+	−	+	−	+	−	−	+
Inulin	−	d	−	−	d	−	−	−			−	+	−	−	d	−	+
Lactose	−	+	−	+	d	−	−	−	+		−	+	−	−	+	−	+
Maltose	−	+	−	d	−	+	−	−	+		−	+	−	−	−	−	+
D-Mannitol	−	+	+	+	+	+	−	+	+	+	+	+	+	+	+	+	+
D-Mannose	−	+	+	+	+	+	+	+	+	+	+	+	+	+	+	−	+
Melezitose	−	−	−	−	−	−	+	−			−	d	−	−	−	−	+
Melibiose	−	+	−	+	+	+	−	+	+		−	+	−	+	+	−	+
A-CH₂-D-glucoside	−	−	−	d	−	−	−	−	−	+	+	d	+	−	−	−	−
Raffinose	−	+	−	+	+	−	−	+	+		−	+	−	+	+	−	+
L-Rhamnose	−	d	+	+	+	+	−	+	+	+	−	+	−	−	−	−	+
Ribose	+	+	+	+	+	+	−	+	+	+	+	+	+	+	+	−	+
Salicin	−	+	+	+	+	+	+	+	+	+	+	+	−	+	−	−	d
D-Sorbitol	d	+	−	+	+	+	−	+	+	+	+	+	+	+	+	−	+
Starch	−	+	−	−	−	−	−				−	+	−	−	−	−	+
Sucrose	+	+	+	+	+	+	−	+	+	+	+	+	+	+	+	+	+
Trehalose	+	+	d	+	−	+	+	+	+		−	+	−	−	+	−	+
D-Xylose	−	+	d	+	+	+	+	+	−	−	−	d	−	−	+	−	+

Note: (+), delayed positive reaction. After seven days' growth at 27°C in unshaken aqueous solution of 1 percent organic compound, 1 percent peptone with bromcresol purple as an indicator, except for *E. Cacticida*, *E. Persicinus*, and *E. Psidii*. *E. tracheiphila* grows very slowly in this medium.

19.3.3   DIFFERENTIATION OF *ERWINIA* SPECIES BASED ON UTILIZATION OF SOME ORGANIC COMPOUNDS AS A SOURCE OF CARBON AND ENERGY

Species	Citrate	Formate	Lactate	Tartrate	Galacturonate	Malonate
E. amylovora	+	+	+	−	−	−
E. ananas	+	+	+	+	D	−
E. cacticida	+	+	−	b	B	+
E. carotovora	+	+	+	−	D	−
E. chrysanthemi	+	+	+	d	D	+
E. cypripedii	+	+	+	+	+	d
E. mallotivora	+	−	−	−	−	−
E. nigrifluens	−	+	+	+	−	−
E. persicinus	+	b	+	b	B	+
E. psidii	b	b	B	−	B	−
E. quercina	+	+	+	−	−	−
E. rhapontici	+	+	+	d	D	+
E. rubrifaciens	+	+	+	+	−	−
E. salicis	−	−	−	−	−	−
E. stewartii	+	+	+	+	−	−
E. tracheiphila	d	d	−	−	−	−
E. uredovora	+	+	+	+	−	−

Note: Growth symbols: +, 80 percent or more positive; −, 20 percent or less positive; d, 21–89 percent positive. In 21 days at 27°C on OY medium.

19.4   DIFFERENTIATION OF GENUS *AGROBACTERIUM* INTO GROUPS AND BIOVARS

| Characteristic | A. radiobacter | | A. rhizogenes | A. rubi | A. tumfaciens | | |
	Biovar 1	Biovar 2	Biovar 2		Biovar 1	Biovar 2	Biovar 3
Growth							
At 35°C	+	−	−	d	+	−	d
On selective medium of Scroth et al.[a]	+	−	−		+	−	−
On selective medium of New and Kerr[b]	−	+	+		−	+	−
In presence of 2 percent NaCl	+	−	−		+	−	+
3-Ketolactose produced	+	−	−	−	+	−	−
Acidic reaction produced from							
Meso-Erythritol	−	+	+	+	−	+	−
Melezitose					+	−	−
Ethanol	+	−	−	−	+	−	−

(Continued)

Characteristic	A. radiobacter		A. rhizogenes	A. rubi	A. tumfaciens		
	Biovar 1	Biovar 2	Biovar 2		Biovar 1	Biovar 2	Biovar 3
Alkaline reaction produced from							
Sodium malonate	−	+	+	+	−	+	+
Sodium L-tartrate					d	+	+
Sodium propionate					d	−	−
Simmons' citrate with 0.0005 percent yeast extract	−	+	+	−	−	+	
Reaction in litmus milk							
Alkaline	+	−	−	+	+	−	+
Acidic	−	+	+	−	−	+	
Formation of pellicle in ferric ammonium citrate solution	+	−	−	−	+	−	d
Growth factor requirements							
Biotin and/or glutamic acid	−	+	+		−	+	
L-Glutamic acid and yeast extract	−	−	−	+	−	−	
Phytopathogenicity							
Tumors produced on wounded stems of, e.g., tomato plants, *Helianthus annuis, Nicotiana tabacum* or on discs of *Daucus carota*	−	−	+	+	+	+	d
Roots produced on discs of *Daucus carota*	−	−	+	−	−	−	−

Note: 1. Symbols: See standard definitions.

2. An alkaline reaction in litmus milk is frequently accompanied by a brown discoloration; and acid reaction (pink color) is frequently accompanied by a clot formation.

3. Biovar 3 strains, phytopathogenicity can only be demonstrated on young shoots of grapevines. The majority of these isolates display a very limited host range. For such strains phytopathogenicity can only be demonstrated on young shoots of grapevines (d) Represent variability in pathogenicity among host.

[a]Scroth et al., *Phytopathology*, 55, 645–647, 1965.

[b]New and Kerr, *J. Appl. Bacteriol.*, 34, 233–236, 1971.

20 Differentiation of Bacterial Group into Species

The bacterial plant pathogens in every genus are specific in their host range and on this basis are further divided into group or species. Among the group or species these are further divided into pathovars based on the plant species in which they initiate the disease and cause symptoms, for example, *Xanthomonas* attacking a strawberry plant is specific to its host and is classified as *X. fragariae*. Similarly, *X. albilineans* are specific to sugarcane. *Xanthomonas* attacking the host of *cruciferae* family is classified as *X. campestris* and further into pathovar/race on the basis of the host they infect in *cruciferae* family. The *Xanthomonas* of *axonopodis* group has a wide host range and is further divided into pathovars based on the host they infect like *X. a.* pv. *punicae* infecting pomegranate, *X. a.* pv. *malvacearum* infecting cotton, *X. a.* pv. *sesame* infecting sesamum, and so on.

The genus *Xanthomonas* is subdivided into eight groups/species on the basis of various biochemical tests/characteristics, and each species is further divided into pathovars on the basis of the host they infect. The genus *Pseudomonas* is subdivided into eight groups/species and further into pathovars on the basis of host they infect. The genus *Erwinia* is divided into four groups/species. The genus *Pantoea* is divided into two species. The genus *Clavibacter* (*corynebacterium*) is divided into four species. The genus *agrobacterium* is divided into four groups/species and each group has a number of biovars in it, depending on the biochemical tests.

Thus, biochemical tests play an important tool in the differentiation of the bacterial isolate into separate distinct species or biovar. The differentiation of different genus into group/species/biovar is illustrated in the following tables.

20.1 DIFFERENTIATION OF GENUS *XANTHOMONAS* INTO SPECIES BASED ON MORPHOLOGICAL AND BIOCHEMICAL CHARACTERISTICS

Characteristic	X. albilineans	X. axonopodis	X. campestris	X. citri	X. fragariae	X. maltophilia	X. phaseoli	X. populi
Number of flagella	1	1	1	1	1	>1	1	1
Reduction of NO_3^- to NO_2^-	–	–	–	–	–	+	–	–
Lysine decarboxylase	–	–	–		–	+		–
Methionine or cystine required for growth	+		D	–		+	–	
Mucoid growth on nutrient agar + 5 percent glucose	–	–	+		+			+
Xanthomonadins produced	+	+	+	+	+	–	+	
Hydrolysis of								
Gelatin	d	–	D		+	+		–
Esculin	+	+	+		–			
Starch	–	+	D	+	+	–	+	
Milk proteolysis	–	–	+		–	+	+	Slow
H_2S from peptone	–	+	+		–			–

(Continued)

131

Characteristic	X. albilineans	X. axonopodis	X. campestris	X. citri	X. fragariae	X. maltophilia	X. phaseoli	X. populi
Pectinase activity in culture			D	+			−	
Maximum growth temperature, °C	37	35–37	35–39	38	33		38	27.5
Maximum NaCl tolerance, %	0.5	1.0	2.0–5.0		0.5–1.0			0.4–0.6
Acid production within 21 days on dye's medium C from								
Arabinose	−	−	+		−			−
Mannose	+	−	+	+	+		−	+
Galactose	d	−	+		−			+
Trehalose	−	+	+		−			+
Cellobiose	−	−	+		−			−
Fructose	−	−	+		+			+
Opportunistic pathogen of humans	−	−	+	−	−	+	−	−

Source: Bergey's Manual of Systematic Bacteriology.

Notes: Symbols: see standard definitions. *Xanthomonas citri* and *X. phaseoli* were not included in *Bergey's Manual of Systematic Bacteriology*, vol. 1. The two species were revived in 1989 by D.W. Gabriel et al. (*Int. J. Syst. Bacteriol.*, 39, 14–22).

20.2 DIFFERENTIATION OF GENUS *PSEUDOMONAS* INTO SPECIES ON THE BASIS OF BIOCHEMICAL TESTS

	Test							
Species	Gelatinolyse (1)	Esculine	Sorbiotol	Inositol	Erythritol	DL Lactate	D(−) Tartrate	L(+) Tartrate
P. syringae	+	+	+	+	+	+	−	−
						−	+	
P. morsprunorum	−	−	+	+	D	−	−	+
P. tomato	D	+	+	+	−	−	+	−
P. savastanoi	−	−	D	D	−	−	−	+
P. pisi	−	D	+	+	−	−	−	−
P. glycinea	−	−	−	−	D	−	−	−
P. phaseolicola	D	−	−	−	−	−	−	−
P. persicae	−	−	+	−	+	−	−	−

20.3 DIFFERENTIATION OF GENUS *ERWINIA* INTO SPECIES ON THE BASIS OF BIOCHEMICAL TESTS

	Test						
Species	Pectinolyse	S.R.S.	A. methyl glucoside	Pigment rose	Gelatinase	Indole	Malonate
Carotovora	+	−	−	−	+	−	
Atroseptica	+	+	+	−	−	−	
Chrysanthemi	+	d	−	−	+	+	
Rhapontici	−	+	D	+	−	−	

20.4 DIFFERENTIATION OF GENUS *PANTOEA* INTO SPECIES ON THE BASIS OF BIOCHEMICAL TESTS

Test	P. agglomerans	P. dispersa
Phenylalanine deaminase	[+]	−
Malonate utilization	+	−
L-Arabinose, acid	+	d
Glycerol, acid	−	[−]
Myo-Inositol, acid	−	d
Lactose, acid	[−]	−
Salicin, acid	+	−
Mucate, acid	−	d
Tartrate, Jordans	−	d
Esculin hydrolysis	+	−
Nitrate reduction	+	d

20.5 DIFFERENTIATION OF GENUS *CLAVIBACTER (CORYNEBACTERIUM)* INTO SPECIES ON THE BASIS OF TESTS

Species	Test				
	Flagella/mobility	Growth	Pigment	Urease	Esculine
C. fascians	−	Medium	Orange	+	−
C. sepedonicum	−	Slow	−	−	+
C. flaccumfaciens	+	Rapid	Yellow	−	+
C. michiganense	−	Medium	Yellow	−	+

20.6 DIFFERENTIATION OF *AGROBACTERIUM TUMEFACIENS* INTO BIOTYPE ON THE BASIS OF DIFFERENT TESTS

	Biotype 1	Biotype 2	Biotype 3
3 ceto lactose	+	−	−
Growth on 2 percent NaCl	+	−	+
Maximum temperature	37°C	29°C	33°C
Acidification of			
Erythritol	−	+	−
Ethanol	+	−	−
Melezitose	+	−	−
Alkalinization of			
Malonate	−	+	+
L(+) tartrate	−	+	+
Propionate	+	−	−

21 Identification of Races of Bacterial Plant Pathogen

The variation in the reaction (susceptible or resistant) of a bacterial isolate depends on the cultivar of the host plant. The host crop plant with different genetical makeup reacts differently to the same bacterial isolate to produce susceptible or resistant reaction, and this forms the basis for variation in the given bacterial species or pathovars. The isolates of the pathovars, on the basis of their reaction to different genotypes/cultivars/varieties of the host crop, are further subdivided into races.

These races are specific in attacking the particular genotype of the crop while not to the other genotype. On the basis of differential reaction of the host cultivars the races of the given pathogen is determined.

21.1 IDENTIFICATION OF RACES OF *X. CAMPESTRIS* PV. *CAMPESTRIS* (Xcc)

Kamoun et al. (1992) separated the isolates of *X. campestris* pv. *campestris* into five different races (0 to 4) based on the response of certain cultivars of turnip (*B. Rapa*) and a cultivar of mustard (*B. Juncea*).

Vicente et al. (1998) suggested that race 1 could be subdivided into three races (tentatively, 1a, 1b, and 1c) on the basis of their reaction on several accessions of *B. Oleracea* and one of the *B. Carianata*. In 2001 they studied 164 isolates of *Xcc* and other *X. campestris* pathovars known to infect cruciferous hosts by inoculating onto a differential series of *Brassica* spp. to determine both pathogenicity to *Brassica* and race. They grouped 144 isolates of *X. campestris* pv. *campestris* into six races (1–6), with race 1 (62 percent) and 4 (32 percent) being predominant.

A set of six differential crucifer hosts, as given below, is used for the characterization of races of *X.c. campestris*:

1. Just Right Turnip F1
2. Seven top turnip
3. Indian mustard (*B. Juncea*) Florida broadleaf
4. Cauliflower (*B. Oleracea* var.) Miracle F1
5. Cabbage wirosa F1
6. PI 199947 (*B. Carinats*)

Material Required

Crucifer host differential, culture of *X.c.* pv. *campestris*, hypodermic syringe, distilled water, scrapping needle, tags, and so on.

Procedure

Plant the seeds of six differential crucifer hosts separately in individual pots or in a field. When the plants are at least one month old, their leaves are used for inoculation of bacterial suspension.

Prepare the bacterial suspension of *Xanthomonas campestris* pv. *campestris* at 0.1 OD and inoculate in the dorsal side of leaves of the above host differential by hypodermic syringe inoculation or swab inoculation method. Tag the plant leaves properly for the respective bacterial culture.

TABLE 21.1
The Scheme of Race Identification[a] for *X.c.* pv. *Campestris*

		Race								
Sr. No.	Differential Cultivares	1	2	3	4	5	6	7	8	9
1.	Wirosa F1 (*B. Oleracea*)	+	+	+	+	+	+	+	+	+
2.	Just Right Hybrid Turnip (*B. Rapa*)	+	+	+	−	+	+	+	+	−
3.	Seven top turnip (*B. Rapa*)	+	−	+	−	+	+	+	−	−
4.	PI 199947 (*B. Carinata*)	−	+	−	−(+)	+	+	+	−	−
5.	Florida Broadleaf Mustard (*B. Juncea*)	−	+	−	−	(+)	+	−	−	−
6.	Miracle F1 (*B. Oleracea*)	+	−	−	+	−	+	+	−	−

Note: (+) = Weekly pathogenic. On the basis of differential reaction of these hosts to bacterial isolates, three races are identified in India. These races are race 1, race 4, and race 6.

[a] As proposed by Vicente et al., 2001; Fargier and Maneau, 2007.

Maintain the plant under humid condition for 46 hours and then at field temp (28 ± 2°C) for development of black leg symptom or vein blackening symptoms. Record the observation at 8–10 days of inoculation as + (disease symptoms) or − (no disease symptoms) to determine the race of the pathogen (Table 21.1).

21.2 IDENTIFICATION OF RACES OF *X. AXONOPODIS* PV. *MALVACERAUM*

Material Required

Cotton differential plant, culture of *Xanthomonas axonopodis* pv. *malvacearum*, hypodermic syringe, distilled water, scrapping needle, tags, and so on.

Name of cotton differentials/CVs: Acala-44, stonevil 2B S9, Stonevil-20, mebane B-1, 1-10B, 20-3, 101-102 B.

Procedure

Plant the seeds of the above differentials either in a pot or in a field and raise the cotton plant. When the plants are at least one month old, their leaves are used for inoculation of bacterial suspension.

Prepare the bacterial suspension of *Xanthomonas axonopodis* pv. *malvacearum* at 0.1 OD and inoculate in the dorsal side of leaves of above cotton differentials by hypodermic syringe inoculation. Tag the plant leaves properly for the respective bacterial culture.

Maintain the plant under humid condition for 48 hours and then keep at field temperature (28 ± 2°C) for the development of water soaking reaction (+) or hypersensitive reaction (HR).

The hypersensitive browning reaction is usually developed by 48 hours while water soaking susceptible reaction is developed by the third day onward. Record the reaction on the fourth day to determine the race number of the bacterial culture (Table 21.2).

Subdivision of *Xanthomonas axonopodis* pv. *malvacearum* races 18 on the basis of additional host differential was proposed by Bhosale and Borkar (2011). This race was divided further into two biotypes by using additional cotton differential (Table 21.3).

Name of additional cotton differential: Gregg and DPX4.

TABLE 21.2

Identification Scheme for Races of *Xanthomonas Axonopodis* pv. *Malvacearum*

Cotton Differentials and Their Reaction to *X.c.* pv. *Malvacearum*

1	2	3	4	5	6	7	
Acala 44	Stoneville 2BS9	Stonevil 20	Mebane B-1	1-10B	20-3	101-102B	Race No.
+	HR	HR	HR	HR	HR	HR	1
+	+	HR	HR	HR	HR	HR	2
+	HR	+	HR	HR	HR	HR	3
+	HR	HR	+	HR	HR	HR	4
+	HR	HR	HR	+	HR	HR	5
+	HR	HR	HR	HR	+	HR	6
+	+	+	HR	HR	HR	HR	7
+	+	HR	+	HR	HR	HR	8
+	+	HR	HR	+	HR	HR	9
+	+	HR	HR	HR	+	HR	10
+	HR	+	+	HR	HR	HR	11
+	HR	+	HR	+	HR	HR	12
+	HR	+	HR	HR	+	HR	13
+	HR	HR	+	+	HR	HR	14
+	HR	HR	+	HR	+	HR	15
+	HR	HR	HR	+	+	HR	16
+	+	+	+	HR	HR	HR	17
+	+	+	HR	+	HR	HR	18
+	+	HR	HR	+	HR	HR	19
+	+	HR	+	+	HR	HR	20
+	+	HR	+	HR	+	HR	21
+	+	HR	HR	+	+	HR	22
+	HR	+	+	+	HR	HR	23
+	HR	+	+	HR	+	HR	24
+	HR	+	HR	+	+	HR	25
+	HR	HR	+	+	+	HR	26
+	+	+	+	+	HR	HR	27
+	+	+	+	HR	+	HR	28
+	+	+	HR	+	+	HR	29
+	+	HR	+	+	+	HR	30
+	HR	+	+	+	+	HR	31
+	+	+	+	+	+	HR	32

Source: Verma and Singh, 1975. *Indian Phytopath.* 28:459–463.

21.3 IDENTIFICATION OF RACES OF *X. AXONOPODIS* PV. *VITICOLA*

Material Required

Grapevine differential plant, culture of grapevine bacterium, hypodermic syringe, distilled water, scrapping needle, tag, and so on.

Name of grapevine differentials: Thompson seedless/Haithi; Tas-A Ganesh/Bhokari; Perlette/Arkashyam; Beautiseedless/Kalisahebi Arkawati/Dogris America, and Chenin Blanc/Karolina Blackrose.

TABLE 21.3
Subdivision of Race 18 into Biotype

Xam Race	Cotton Differential and Their Reaction to Xam		Biotype
	Gregg	DPX4	
Xam race no. 18	+	+	B
Xam race no. 18	+	−	A

Procedure

Prepare a bacterial suspension of grapevine bacterium at 0.1 OD and inoculate in the dorsal side of leaves of above grapevine differential plants by hypodermic syringe inoculation. Tag the plant leaves properly for the respective bacterial culture. Inoculation should be done during rainy season or humid season so as to maintain the natural humidity for development of disease reaction. The water-soaking reaction develops within 72–96 hours on susceptible grapevine cvs. Record the reaction as + (water soaking disease reaction) or − (no reaction or resistant reaction) to determine the race number of the bacterial culture (Table 21.4).

21.4   IDENTIFICATION OF RACES OF *X. ORYZAE* PV. *ORYZAE*

Characterization of races was done by using 10 differential rice varieties. The differentials are

1. 1RRB-1   (Resistance gene xa-1)
2. 1RRB-3   (Resistance gene xa-3)
3. 1RRB-4   (Resistance gene xa-4)
4. 1RRB-5   (Resistance gene xa-5)
5. 1RRB-7   (Resistance gene xa-7)
6. 1RRB-10   (Resistance gene xa-10)
7. 1RRB-11   (Resistance gene xa-11)
8. 1RRB-13   (Resistance gene xa-13)
9. 1RRB-14   (Resistance gene xa-14)
10. 1RRB-21   (Resistance gene xa-21)
11. 1R24   (No resistance gene)

Ten near isogenic lines with single resistance gene as specified above (Huang et al., 1997) were developed as international differential varieties of BCB isolates, and 1R 24 used as the susceptible check.

Material Required

Eleven differential rice cvs; bacterial culture of *X.o.* pv. *oryzae*, scissor, tags, and so on.

Procedure

Sow the seed of these varieties in a plastic pot and after 20 days, transplant the seedlings into pots containing natural paddy soil. Grow rice plants under greenhouse condition. The bacterial inoculation is done when rice plants are 40 days old. Prepare the bacterial suspension for inoculation using 48-hour-old cultures of each isolate in 20 ml sterile water and adjusted to 10^8 CFU/ml. Inoculate the fully expanded leaves with bacterial inoculums by the leaf-cutting method (Kauffman et al., 1973).

Measure the lesion length from the cut leaf tip in centimeters (cm) at 18 days after inoculation and categorize the disease reaction according to lesion length.

TABLE 21.4

Identification Scheme for Classification of Races of *Xanthomonas Axonopodis* pv. *Viticola*

Reaction of Bacterial Isolates on Grapevine Differentials

1	2	3	4	5	6	
Thompson Seedless/Haithi	Tas-A-Ganes/Bhokari	Perlette/Arkashyam	Beautx Seedless/Kalisahebi	Arkawati/Dogris America	Chenin Black/Karolina Blakrose	Race Number
+	−	−	−	−	−	1
+	+	−	−	−	−	2
+	−	+	−	−	−	3
+	−	−	+	−	−	4
+	−	−	−	+	−	5
+	−	−	−	−	+	6
+	+	+	−	−	−	7
+	+	−	+	−	−	8
+	+	−	−	−	−	9
+	+	−	−	−	+	10
+	−	−	−	+	+	11
+	−	+	−	+	−	12
+	−	+	+	−	−	13
+	−	−	+	+	−	14
+	−	+	−	−	+	15
+	−	−	+	−	+	16
+	+	+	+	−	−	17
+	+	+	−	+	−	18
+	+	+	−	−	+	19
+	+	−	+	+	−	20
+	+	−	+	−	+	21
+	+	−	−	+	+	22
+	−	+	+	+	−	23
+	−	+	+	−	+	24
+	−	+	−	+	+	25
+	−	−	+	+	+	26
+	+	+	+	+	−	27
+	+	+	+	−	+	28
+	+	+	−	+	+	29
+	+	−	+	+	+	30
+	−	+	+	+	+	31
+	+	+	+	+	+	32

Source: Borkar, S. G., *Indian J. Plant Pathol.*, 20(1&2), 67–69, 2002.

The lesion length of 0 to 6 cm is classified as resistance (R) and more than 6 cm as susceptible (s) (Sanchez et al., 2000).

Six races or pathotypes are identified based on the differential reaction of these hosts (Table 21.5). Most races producing the susceptible reaction produce the same lesion lengths as on susceptible check variety 1R 24.

TABLE 21.5
A Race Differentiation Scheme for *X. Oryzae* pv. *Oryzae*

Sr. No.	Differential Rice Variety	Race					
		A	B	C	D	E	F
1.	1RRB 1	S	S	S	S	S	S
2.	1RRB 3	S	S	S	S	S	S
3.	1RRB 4	S	S	S	S	S	S
4.	1RRB 5	S	R	R	S	R	R
5.	1RRB 7	S	S	R	R	S	S
6.	1RRB 10	S	S	S	S	S	S
7.	1RRB 11	S	S	S	S	S	S
8.	1RRB 13	R	S	S	S	S	S
9.	1RRB 14	S	S	S	S	S	S
10.	1RRB 21	S	R	S	S	R	S

Source: Hoang Dinh Dinh et al., 2010. *Omonrice* 17:147–151.

21.5 IDENTIFICATION OF RACES OF *RALSTONIA SOLANCEARUM*

The five races of *Ralstonia solanacearum* can be differentiated on the Solanaceae host differential and the respective host plant where the wilting symptoms are induced.

Material Required

Solanacearum host differential, bacterial culture of *Ralstonia solanacearum*, scissors, tags, and so on.

Procedure

Raise the differential host plant from seed (tomato, chili, brinjal, tobacco). When these are ready for transplanting, uproot these. Clip off their roots with scissors and dip in bacterial suspension (0.1 OD or 10^t CFU/ml) for 5 minutes and transplant to pots.

TABLE 21.6
Race Identification Scheme for *Ralstonia Solanacearum*

Differential Hosts	Race No.
Tomato	Race 1 produces wilt symptoms on these plants
Chili	
Brinjal	
Tobacco	
Pepper	
Triploid	Race 2 produces wilt symptoms on these plants
Banana	
Heliconia sp.	
Potato	Race 3 produces wilt symptoms on these plants
Tomato	
Ginger	Race 4 produces wilt symptoms on ginger
Mulberry	Race 5 produces wilt on mulberry

Note: Race 3 is weakly virulent on host of race 1.

TABLE 21.7
Determination of Biovars of *Ralstonia solanacearum* Race

Biovar	Utilization of Carbohydrate
III	Oxidizes both disaccharide and hexose alcohols
II	Oxidizes only disaccharide
I	Oxidizes hexose alcohols only
IV	Oxidizes only alcohols

Injure the potato sprout tuber, ginger sprout rhizome, banana sprout rhizome, and its relatives with pinprick needles and dip in bacterial suspension for 5 minutes. Injure the root of mulberry and pepper cutting and dip in bacterial suspension for 5 minutes.

Plant these bacterial inoculums treated transplants/rhizomes/cutting in the pot soil and drench the pot soil with bacterial inoculums. Maintain the differential plants under glasshouse humid condition for development of wilt disease symptoms. Observe the wilting of plants from 20 days onward and up to 45 days to determine the race number of *Ralstonia solanacearum* (Table 21.6).

Differentiation of *Ralstonia* race in biovars: The differentiation of races of *Ralstonia solanacearum* into biovars is based on the utilization of carbohydrates.

Material Required

Media/broth of respective carbohydrate, bacterial culture of *Ralstonia*, inoculating needles, and so on.

Procedure

Prepare the media/broth of respective carbohydrate, sterilize, and inoculate with bacterial culture. Incubate the inoculated tubes at $28 \pm 2°C$ for 96 hours and record the utilization of carbohydrate. Different biovars oxidize/utilize different carbohydrates. Based on the carbohydrate utilization, these are termed as biovar 1 to 4 (Table 21.7).

22 Studies on Bacterial Cell Wall-Related Biological Compounds

The bacterial cell wall provides a rigid structure and shape to the bacterial cell. The bacterial cell wall also helps to differentiate bacteria into Gram-positive or Gram-negative groups based on the cell wall composition. The cell wall also possesses the antigenic properties. The cell wall-related compounds produced by phytopathogenic bacteria play an important role in the induction of disease water-soaking (WS) reaction in the host plant (Borkar, 1989). Therefore, the cell wall and cell wall-related biological compounds are important for study Borkar and Vermna (1991a).

22.1 ISOLATION OF BACTERIAL CELL WALL (PEPTIDOGLYCAN) (SCHLEIFER AND KANDLER, 1972)

The bacterial cell wall is critical for the determination of cell shape during growth and division, and maintains the mechanical integrity of cells in the face of turgor pressures several atmospheres in magnitude. Across the diverse shapes and sizes of the bacterial kingdom, the cell wall is composed of peptidoglycan, a macromolecular network of sugar strands cross-linked by short peptides.

The preparation of the peptidoglycan starts either from intact bacteria or from the previously isolated cell wall. The isolation of cell wall is performed by disintegration of the cell using ultra-sonication, repeated freezing and thawing, strong shear forces, or grinding with quartz sand. The cell wall can be separated from other cell components by differential centrifugation.

The standard preparation procedures for peptidoglycans of Gram-positive and Gram-negative bacteria are summarized below.

Gram-Positive Bacteria

Material Required:

- Formamide (150–180°C for 30 minutes)
- HNO_3, 0.8 M (37°C for 15 minutes)
- Trichloroacetic acid, 5–25 percent (90°C for 10 minutes, followed by O°C for several days)
- H_2SO_4, 0.1 N (60°C for 24 hours)
- NaOH, 0.1 N (100°C for 1 hour, followed by 35°C for 8 hours)
- Deoxycholate, 1 percent (O°C for 16 hours)
- N,N, Dimethylhydrazine, 2 percent (80°C for 2 hours)

Gram-Negative Bacteria

Material Required:

- Phenol-water (45:55) at 68°C for 5–35 minutes
- Sodium dodecyl sulfate, 4 percent at 100°C

22.1.1 Isolation and Preparation of Bacterial Cell Wall for Compositional Analysis by Ultra Performance Liquid Chromatography (UPLC)

The method is described by Desmarais et al. (2014). High performance liquid chromatography (HPLC) is a powerful analytical method for quantifying differences in the chemical composition of the walls of bacteria grown under a variety of environments and genetic conditions, but its throughput is often limited. A straightforward procedure for isolation and preparation of bacterial cell wall for biological analysis of peptidoglycan is ultra performance liquid chromatography (UPLC).

The method is used to provide information such as the identification of the muropeptide components and their concentrations, the average length of glycan strands, and the fraction of material involved in cross-link between strands. The protocol is as follows:

Protocol

1. Grow bacterial cultures in 2.5 ml of media overnight: Back dilute cultures 1:100 into 250 ml of fresh media and grow to 0.7–0.8 OD at 600 nm. Prepare a solution of 6 percent sodium dodecyl sulfate (SDS) in water. *Caution*: SDS powder is hazardous; avoid inhaling SDS powder, and wear a mask over nose and mouth.

2. Day 1: Lysing bacterial culture is performed over the course of one day and overnight: While diluted cultures are growing, put water to boiling in a 1 L beaker. When on a hot plate and magnetic stirrer the water starts boiling, prepare an aliquot of 6 ml of 6 percent SDS in to 50 ml polypropylene tubes and add one small stir bar to each tube. Secure the tube lid to finger tight, place the tubes in boiling water, and turn on stirring to 500 rpm on the hot plate.

   Harvest 250 ml cultures at 5000 × g for 10 minutes at room temperature and resuspend pellets in 3 ml of media or 1× phosphate buffered saline. Slowly pipette cell suspension into 50 ml tubes with 6 percent boiling SDS to lyse the cell (final concentration 4 percent SDS) while the tubes are submerged in the boiling water bath, and reclose the lid to finger tight.

   Note: Cell suspension must be quickly transferred in to boiling SDS once resuspended. Abrupt environmental changes should be avoided, as this could cause erroneous alteration of cell wall structure.

   Cover boiling water beaker and allow cells to boil for 3 hours, checking the water level periodically and refilling the water beaker when necessary. After 3 hours, turn off the heating element of the hot plate, but continue to stir overnight at 500 rpm.

3. Day 2: Enzymatic digestion is performed over the course of 1 day: If SDS precipitates in the 50 ml tubes during overnight, get the water to boil for an additional 1–2 hours. Turn on the heat block to 60°C. Prepare a 1 mg/ml stock of pronase E in 10 ml Tris-HCl (pH 7.2) + 0.06% w/v NaCl and activate Pronase E at 60°C for at least 30 minutes. Use an ultracentrifuge set at 400000 × g to spin samples for 20 minutes at room temperature in order to pellet the large PG polymers and thereby purify them from other cellular components. Remove the supernatant carefully and then resuspend each pellet at room temperature in ultrapure water.

   Note: Resuspension volume depends upon the volume of the ultracentrifuge tube used; use a volume that fills the tubes at least halfway but does not exceed the maximum volume of the tubes.

   Repeat centrifugation/washing until the water does not form bubbles during resuspension, indicating that the SDS has been fully removed (typically three washes), and now stop washing the pellet. If a white precipitate forms, this indicates that the sacculi are clumping together. Clumping is not catastrophic, but clumps bind very strongly to plastic and glassware, causing large sample loss. In this case, proceed with the protocol using the clumped sacculi sample. On the last centrifugation/washing step, resuspend the sample in 900 μl of 100 mM tris-HCl (pH 7.2) + 0.06% w/v NaCl and transfer to 2 ml tubes previously poked with holes in the tops with a small needle. Add 100 µl of 1 mg/ml activated pronase E (100 µg/ml final concentration) to each sample and incubate at 60°C for 2 hours. Get a different heat block to 100°C. Stop the pronase E digestion by adding 200 µl of 6% SDS to each sample and boil the samples in the 100°C heat block for 30 minutes. Get a different heat block to 37°C and make a 1 mg/ml stock of muramidase (mutanolysin) in 50 mm phosphate buffer (pH 4.9). As in the previous step, use an ultracentrifuge set at 400000 × g to spin samples for 20 minutes at room temperature and wash with room temperature ultrapure water until SDS is fully removed (typically three washes). The resuspension volume depends upon the volume of the ultra-centrifuge tube used; use a volume that fills the tubes at least halfway but does not exceed the maximum volume of the tube. On the last centrifugation/washing step, resuspend samples in 200 µl of 50 mm sodium phosphate buffer (pH 4.98). This volume can be adjusted according to the amount of peptidoglycan (PG) in sample, and may be species dependent. If there are previously published reports of HPLC analysis for the species of interest, this volume can be estimated based on these quantitations (a compilation of HPCL PG studies can be found in the supplemental information). For other species, the sacculus preparation can be executed up to this step and then different amounts of resuspension volumes can be added to replicate samples in order to determine the minimal volume that allows the PG to remain in solution. If the sample contains more peptidoglycan, increase the resuspension volume; if the sample has little peptidoglycan, reduce the resuspension volume to a minimum of 50 µl. Transfer samples to 1.5 ml tubes and add 1 mg/ml muramidase to give a final concentration of 40 µg/ml. Incubate 6–8 hours or overnight at 37°C.

4. Day 3: Preparation of sample for UPLIC is performed on the last day: Turn on a heat block to 100°C. Boil the sample without SDS for 5 minutes to stop the muramidase digestion. Cen-trifuge samples for 10 minutes at 16000 × g at room temp, then transfer the supernatant (muropeptides are now soluble) to 13 mm × 100 mm glass tubes. Try to recover as much supernatant as possible, getting very close to the pellet without disturbing it. Adjust the pH by adding 500 mm borate buffer (pH 9) to the sample for a final concentration of 100 mm borate buffer. Borate buffer is compatible with the reducing agent sodium borohydride. Add one to two grains of sodium borohydride to reduce each sample and let the reaction proceed for at least 30 minutes at room temperature.

   Caution: Sodium borohydride is highly reactive and dangerous to handle. Avoid contact with skin (wear gloves) and eyes (wear safety glasses).

   Adjust sample to pH 3–4 (the muropeptide isoelectric point is 3.5) with 50 percent w/v orthophosphonic acid using 20 µl increments until pH 6, as measured with pH indicator paper, then using 2 µl increments.

   Caution: Orthophosphoric acid is corrosive and dangerous to handle. Avoid contact with skin and eyes.

   The sample should bubble in response to addition of orthophosphoric acid; when the sample stops bubbling, this typically indicates that a pH of 6 has been reached. If no bubbling occurs in the sample, this may indicate that the amount of sodium borohydrite added was too small. In this case, stop lowering pH, carefully add one or two grains of sodium borohydrite, let it react for 5–10 minutes, then resume pH adjustment. Filter sample through 0.22 µm syringe filter directly into UPLC vial. If a precipitate has formed, heat the tube with several passes through a flame before filtering. If the sample will not be injected in to the UPLC instrument within a day, freeze at −20°C overnight. Samples can be stored for up to a year at −80°C. The sample can be thawed by passing through a flame several times. Inject 10 µl of each sample onto a UPLC instrument equipped with a C18 1.7 µm reversed phase column and an absorbance detector set to monitor 202–208 hours. Samples are injected sequentially, but auto samples capabilities allow up to 96 samples to be processed in a batch. Use 50 mM sodium phosphate (pH 4.35) + 0.4% v/v sodium azide for solvent A, and 75 mM sodium phosphate (pH 4.95) + 15% v/v methanol for solvent B.

Note: Sodium azide is added to compensate for the 205 nm absorption methanol to avoid baseline drift. Get the flow to 0.25 ml/minute and use a linear gradient over 25 minutes to achieve 100 percent solvent B and sequential elution of muropeptides within 30 minutes. If mass spectrometry will be used to characterize muropetide after UPLC, collect and dry the fractions using a centrifugal evaporator.

22.2   ISOLATION OF CELL WALL–LESS (L-FORM) BACTERIA

Bacterial morphology is determined by the cell wall. As L-form has no cell wall, its morphology is different from that of the strain of bacteria from which it is derived. Typical L-form cells are spheres or spheroids.

Although L-forms can be developed from Gram-positive and Gram-negative bacteria, in Gram staining it always tests Gram-negative, due to the lack of a cell wall. These are multiplied by budding, that is, extension of thin protrusion from the cells surface and then these protrusion pinch off to form new cells.

Protocol

L-forms can be generated in the laboratory for many bacterial species that usually have cell walls. This is done by inhibiting peptidoglycan synthesis with antibiotics or treating the cells with lysozyme, an enzyme that digests cell wall.

Prepare the nutrient agar plates of various concentrations of penicillin. Streak the bacterial inoculums and incubate at $28 \pm 2°C$ temp for 5–7 days. Observe the plates for development of cell wall–less bacterial colonies.

Observation

The colonies of cell wall–less bacteria look like the poached egg omelets.
   Note: No cell wall–less bacteria is yet reported as plant pathogen.

22.3   ISOLATION OF BACTERIAL EXOPOLYSACCHARIDE (EPS)

Protocol

Grow the bacterial culture in exopolysaccharide production medium (K_2HPO_4, 1.2 g; KH_2PO_4, 0.8 g; $MgSO_4$. $7H_2O$, 0.2 g; peptone, 5 g; sucrose, 5 g; distilled water, 1 L; pH, 6–8) for five days at 30–32°C on a rotary shaker. Remove the bacterial cell by centrifugation at 15000 rpm for 15 minutes and collect the cultural filtrate. Add N-cetylpyridinium chloride (monohydrate) at 2 gm per L of culture filtrate and allow to stand for precipitation of exopolysaccharide. Add one or two pellets of potassium hydroxide to hasten the precipitation. Remove the precipitate by centrifugation and dissolve in the required quantity of 10 percent NaCl depending upon the volume of precipitate. Reprecipitate the EPS with two volumes of ethanol. At this stage the EPS appears as a thick gel that floats at the top and does not settle down easily. Remove the EPS carefully with the help of a bent glass rod.

Dissolve again in distilled water and reprecipitate with ethanol. Finally wash with petroleum ether, dry under vacuum, and store overnight over calcium chloride in a desiccator. Make the powder of EPS in sterile pestle and mortar and store in freezer at 4°C for further use.

22.3.1   TESTING OF EPS BY IODINE TEST

Material Required

Polysaccharide suspension, dilute HCl, iodine solution.

Procedure

Take 2 ml of polysaccharide suspension (1 mg per ml polysaccharide). Acidify with two drops of dilute HCl (0.1N). Add two drops of iodine solution and observe for the color development.

Observation

Presence of starch in EPS gives a blue color, while presence of glycogen in EPS gives a red-brown color (the major content of EPS is glycogen).

22.3.2 ESTIMATION OF EPS

Measure the quantity of EPS in mg/L of bacterial cell filtrate for individual bacterial culture, race, or pathovar.

22.3.3 ESTIMATION OF SUGAR IN EPS

Estimate the total sugars in EPS by anthrone method (Dubois et al., 1951).

Solution and reagents

Anthrone reagent, 2 g/L con H_2SO_4, standard sugar glucose 100 µg/ml.

Procedure

Dissolve the EPS (10 mg/ml) in distilled water. Add 4 ml of anthrone reagent to 1 ml of test EPS solution and mix rapidly. Place the tubes in boiling water bath for 10 minutes with a marble on top to prevent water loss by evaporation. Cool and read the extinction at 620 nm against reagent blank. Estimate the sugar concentration from a standard curve prepared with known amount of glucose.

22.3.4 ESTIMATION OF PROTEIN IN EPS

Procedure I

Material Required

EPS, ultrapure water, spectrophotometer.

Procedure

Prepare 1/10, 1/20, 1/40, and so on dilution of 1 gm EPS in ultrapure water and read the extinction at 280 nm. Calculate the amount of protein in the sample by employing the following formula:

$$\text{Mg protein}/\text{ml} = \frac{0.5 \times \text{OD} \times \text{dilution}}{0.67}$$

or

$$\text{Mg protein}/\text{ml} = \frac{1 \times X \times \text{dilution}}{1.3}$$

where X = Optical density of sample
(*Note*: 1 mg albumin protein/ml gives 1.3 OD.)

Procedure II

Material Required

EPS, ultrapure water, reagent, spectrophotometer.

Reagent A = Nak-tartarate, 0.2 gm/10 ml
B = $CuSO_4.5H_2O$
C = Na_2CO_3, 10 gm/500 ml 0.1 N NaOH solution
D = 1N Folin reagent
E = 98 ml of C + 1 ml of B + 1 ml of A

Procedure

To 1 ml test solution add 5 ml of solution E, shake the tubes, and keep at room temp for 10 minutes. Add 0.5 ml of 1 N Folin reagent in each tube and shake well. Incubate the tubes at 30°C for 30 minutes and read the extinction at 750 nm. Calculate the protein from the standard curve of albumin (use 0–100 μg albumin per ml for preparation of standard curve).

22.4   ISOLATION OF BACTERIAL LIPID

There has been widespread interest for analyzing lipid profiles of bacteria for the taxonomical identification as variations in lipid profile within the pathovars do exit. Over the past decade, a significant portion of mass spectrometry studies on bacteria have focused on the peptides and proteins within the cell. In whole cell, matrix-assisted laser desorption/ionization (MALDI) analysis yield characteristic peptide fingerprint pattern that are specific to a given bacterium allowing differentiation of bacteria to the genus, species, and even strain.

The method recommended by Lewis et al. (2000) for the extraction of lipids from unicellular organisms such as bacteria is as follows.

Protocol I

Grow the bacteria in nutrient broth for 3 days. Harvest the bacterial cells by high speed centrifugation for 15 minutes. Discard the supernatant and resuspend the pellet in 100 ml 1.0% NaCl (w/v) and recentrifuge. Discard the supernatant and collect the cell pellet to be frozen overnight at –30°C. Freeze-dry the frozen biomass for 15 hours and store in a sealed glass container at –30°C. To freeze-dried cells (about 100 mg) add ¼ ml solvent in sequence: Chloroform, methanol, and water to achieve a final chloform/methanol/water ratio of 1:2:0.8 (V/V/V). Shake the samples for 15 seconds immediately following the addition of each solvent and allow to stand for about 18 hours, with occasional shaking by hand. Carry out the phase separator of biomass–solvent mixture in a separating funnel by addition of chloroform and water to obtain a final chloroform/methanol/water ratio of 1:1:0.9 (v/v/v). A known portion of each total lipid extract recovered from the lower chloroform phase can be used for further analysis.

Note: If only fatty acid determination is required, a direct saponification using KOH in ethanol is effective. Similarly, a hexane–isopropanol solvent system is found to be effective for the bacteria *Pseudomonas* (Guckert et al., 1988).

Protocol II

Grow the bacteria in nutrient broth for 3 days and harvest the bacterial cells by centrifugation at 4°C. Wash the pellet with ice-cold 75 percent ethanol twice. For extraction of the lipid in the bacterial cell, the cells are mixed with 30 ml of 1:1:1 (v/v/v) mixture of dichloromethane, ethanol, and water. Shake the mixture for 1–2 minutes and allow to stand overnight. The result is a biphasic system, with the bottom organic layer containing the lipids. This layer is then transferred by pipette to another container and the solvent evaporated under nitrogen until approximately 1 ml remains. 1 µl of the lipid extract is then mixed with 1 µl of 1 M 2.5, dihydroxy benzoic acid (DBH; in 90 percent methanol with 0.1 percent formic acid) and spotted onto a MALDI target.

22.4.1 ESTIMATION OF BACTERIAL LIPID

Mass spectra can be obtained on an ultra-flex II MALDI-TOF/TOF mass spectrometer equipped with a smartbeam solid state laser (Bruker Daltonics Bremen, Germany) operated in positive ion retraction mode. The laser power is adjusted to a point slightly above the ionization threshold of the sample and fired at a rate of 10 Hz with 1000–1500 laser shots accumulated per scan. Five to eight replicate samples are to be obtained for each bacterial species.

Three to five mass spectra are to be obtained and average for each individual sample using 1000–1500 laser shots blasted over the entire spot on the MALDI target. To test the variability in the spectra, the relative intensity of a given peak is calculated as a percentage of the sum of the 10–15 highest intensity peaks in the spectrum. For sample replicates, the average relative intensity of a given peak varies less than 5 percent.

The results of mass spectra of the extract analyzed by matrix-assisted laser desorption/ionization time of light (MALDI-TOF) mass spectrometry and TOD tandem mass spectroscopy show a series of peaks corresponding to sodiated phospholipids—primarily phosphatidyl ethanolamine (PE) and phosphatidylglycerol (Ph). The relative amount of phospholipids and fatty acid compositions inferred from the spectra can be compound with previously reported values from GC/MS studies.

22.5 DETERMINATION OF GLYCOPROTEIN SECRETION FACTOR

Extracellular glycoprotein (GP) secreted by plant pathogenic bacteria is capable of inducing WS symptoms in leaf spot diseases, and the quantity of GP plays an important role in the induction of WS symptoms. The quantitative as well as qualitative secretion of GP during multiplication of bacterial plant pathogens seems to be important to induce WS reaction.

The GP secretion factor as defined by Borkar (1989a) is a quantity of GP (ml) produced by 1×10^7 initial colony-forming unit (CFU) of bacterial clone during their growth period in an hour of incubation and is denoted in ml.

$$\text{GP secretion factor} = \frac{\text{Amount of GP secreted (ml)} - \text{Previous amount of secreted GP (ml)}}{\text{Number of hours of secretion}}$$

Protocol

For production of GP plate 1×10^7 CFU of *Xanthomonas* on LPGA medium (liver extract, 7 g; peptone, 7 g; agar, 20 g; water, 1 pH 7.0) and incubate at 26°C for different period, for instance, 1–7 days. Harvest the cultivated bacterial growth and suspend in 500 ml of distilled water and shake well. The bacterial suspension is centrifuged at 19000 rpm for overnight in ultracentrifuge at 4°C and the supernatant is filtered through a Millipore (0.44 µ) filter. The filtrate is precipitated with ammonium sulfate. The precipitated GP is collected, measured, and dialyzed for 2–3 days at 4°C. Determine the GP secretion factor by employing the above formula.

Sugar and protein content of GP secreted at different time interval is to be estimated. Sugar content is estimated by anthrone method while protein content by Folin–Lowry method.

22.6   TESTING OF EPS/GP FOR INDUCTION OF WS REACTION

Material Required

EPS of test bacterial species, distilled sterile water, hypodermic syringe, test host plant, tags, and so on.

Protocol

The protocol given by Borkar and Verma (1989a) must be followed. Dissolve 10 mg EPS in 1 ml distilled sterile water and make serial dilutions up to 10^{-10}. Inoculate measured quantity of EPS dilutions from the above serial dilutions into the dorsal side of leaves of test host plant with the help of hypodermic syringe. Incubate the plants in humid chambers and observe for the development and persistence of the WS reaction by the EPS up to 3 days.

Observation

Estimate the quantity of EPS required for the induction of WS reaction and the quantity of EPS required for persistent WS.

23 Isolation of Bacterial Metabolites

Bacteria secrete and release certain metabolites, which help them to derive nutrients from their surrounding environments, to play a role in disintegration of host tissues and symptoms developments, or to survive in a competitive habitat. These metabolites can be enzymes, toxins, bacteriocins, and so forth.

23.1 EXTRACTION OF TOTAL CULTURAL FILTRATE

Protocol

Exponential phase culture of virulent strain of test plant pathogenic bacterium is grown in a polysaccharide-production medium (K_2HPO_4, 1.2 g; KH_2PO_4, 0.8 g; $MgSO_4.7H_2O$, 0.2 g; peptone, 5 g; sucrose, 5 g; distilled water, 1 L; pH 6.8 for 5 d at 30°C on a rotary shaker. Bacterial cells are harvested by centrifugation (250 Hz, 2 h, 4°C) and the supernatant thus obtained is termed *total cultural filtrate* (TCF).

Solvent extraction of TCF: Reduce TCF (200 ml) to 50 ml in a rotary evaporator in vacuo at 60°C and extract 5 ml of TCF per 10 ml solvent individually with diethyl ether, butanol, and chloroform. Store the different solvent phases aseptically at 4°C till further use. Evaporate the solvent extracts in vacuo and dissolve in 5 ml sterilized water before using on the test plant Borkar and Verma (1989b).

23.2 ISOLATION OF BACTERIAL TOXINS

23.2.1 NON-HOST-SPECIFIC TOXINS

23.2.1.1 Tabtoxin

It is produced by *Pseudomonas syringae* pv. *tabaci*, a pathogen of wildfire disease of tobacco. The toxin is involved in the production of wildfire disease symptoms on tobacco leaves, exhibiting necrotic spots surrounded by a yellow halo. Identical symptoms of the disease may be induced by cultural filtrate of the bacteria or purified toxin. Similar effects can be observed on a relatively wide range of hosts, thus making the toxin non-host-specific. Tabtoxin, chemically, is a dipeptide composed of the amino acids threonine and tabtoxinine. In the cell, tabtoxin is cleaved releasing the tabtoxinine moiety which is an active toxin in the inhibition of the enzyme glutamine synthetase and this is the primary mode of action of the toxin.

Protocol

Grow the bacterium in nutrient broth for 4–5 days on rotary shaker at 27 ± 2°C temp. Centrifuge at 10000 × g for 10 minutes to collect the supernatant. Pass the supernatant through cation exchange resin so as to absorb the toxin on resin. Recover the toxin from resin by elution with 5 percent ammonia. Purify the toxin by ion-exchange chromatography.

Note: Tabtoxin is difficult to work with because it is chemically unstable. This is because of labile B-lactum grouping in one of the amino acid residue. Tabtoxin slowly loses biological activity in aqueous solution, with the formation of inactive isomer, isotabtoxin. The isomerization or toxin inactivation is minimal in the pH range 3–5 and at low temperatures, but is enhanced at higher pH values (7.7) and higher temperatures.

23.2.1.2 Phaseolotoxin

It is produced by *Pseudomonas syringae* pv. *phaseolicola*, a halo blight disease pathogen of beans. The toxin is involved to induce halo blight symptoms on bean. Chemically, the toxin is a tripeptide of ornithine–alanine–arginine with a phosphosulfinyl group. Within cells the toxin is enzymatically cleaved, releasing phosphosulfinylornithine, which is the toxic moiety; cellular effects are a result of the inactivation of the enzyme ornithine carbamoly transferase.

Protocol

Grow the bacterium in nutrient broth for 4–5 days on rotary shaker at $27 \pm 2°C$ temp. Centrifuge at $10000 \times g$ for 10 minutes to collect the supernatant. The three-step purification is routinely used for phaseolotoxin.

Adsorption on to charcoal: Pass the cultural supernatant through charcoal column, washed with water, and recover the phaseolotoxin from it by extraction with methanol-chloroform-0.5 M aqueous ammonia (3:1:1). *Carry out ion-exchange column chromatography* on GAE-sephadex, utilizing gradient elution with ammonium bicarbonate and *partition column chromatography* on LHZO sephadex using methanol –0.05 M aqueous ammonia (1:1).

A major requirement while manipulating phaseolotoxin is control of pH at neutral or slightly alkaline values, so as to maintain chemical stability of the toxin. At pH<3, phaseolotoxin is very labile. Solutions of the ammonium salt of phaseolotoxin are very labile. Solutions of the ammonium salt of phaseolotoxin, obtained from this purification sequence, are chemically stable under refrigeration (water or methanol-water 1:1).

23.3 ISOLATION OF BACTERIAL ENZYMES

Plant cell walls are primarily polysaccharide in composition. A simple but major pathogenic mechanism in plants involves degradation of the cell wall by a battery of enzymes secreted by pathogens. Most of the derivative enzymes are glycoside hydrolyses, which degrade the cellulose and pectate matrices by the addition of water to break the glycosidic bonds. The pectate network is also degraded by polysaccharide lyases, which cleave the glycosidic bonds via a B-elimination mechanism.

23.3.1 Isolation of Cellulases

Cellulose is a long-chain polymer of glucose linked by means of glycosidic bonds with β 1, 4 linkages, in which number one (#1) carbon on a glucose bonds is linked to the #4 carbon on a second glucose. It is a major component of the cell wall in plants. Breakdown of cellulose is brought about by cellulases, a group of enzymes hydrolyzing β1-4 glycosidic bonds or its derivatives. Hydrolysis of cellulose ultimately yields glucose, which is an important energy source for pathogenic microorganisms, and due to cellulose hydrolysis cell walls lose their strength and collapse.

This experiment deals with the production of extracellular cellulase *in vitro* and *in vivo* by a pathogen and testing of activity of its various components (endoglucanase or exoglucanase and cellobiase).

A diverse group of fungi and a few bacteria produce cellulase in a nutrient medium broth containing cellulose.

23.3.1.1 Extraction of Cellulase from Pathogen (*In Vitro*)

Material Required

Test bacterial culture, Czapek's liquid medium (containing 2–3 percent carboxyl methyl cellulose or cellulose powder) of pH 6.5, centrifuge, 500-ml Erlenmeyer flasks, Büchner funnel, Whatman no. 1 filter papers, dialysis tubes, spirit lamp/Bunsen burner.

Protocol

Pour 100 ml of Czapek's broth in each 500-ml flask and plug with a cotton plug. Autoclave these at 15 lb psi for 15 minutes. Aseptically inoculate the flasks, with test bacterial cultures. Incubate the inoculated flasks at room temperature (25°C) for 7 days. Harvest the bacterial cells by centrifugation at 10000 × g and collect the supernatant. Pour the supernatant into the dialysis tubes and dialyze the enzyme against distilled water at 2–4°C for 8 hours.

Testing for Presence of Enzyme Cellulase

Prepare the microtome section of leaf tissues of available plant. Put a few sections in the isolated enzyme solution while a few sections are placed in distilled water for 8–10 hours. Make microscopic observations for the breakage of cell wall structure in the specimen or follow the procedure as described in Section 3.3.1 to observe the degradation of cellulose in the cell wall. The degradation of cell wall denotes the presence of cellulase enzyme in the preparation.

The clear dialyzed solution (or nondialyzed solution) as the enzyme source can be used for various tests. It can be stored at 2–4°C when not in use, after adding a few drops of toluene.

23.3.1.2 Extraction of Cellulase Enzyme from Diseased Plants (*In Vivo*)

Cellulases are produced by several phytopathogenic fungi and bacteria. In living tissues, these play an important role in softening or disintegration of the cell wall material, thus facilitating penetration and spread of pathogen. The extraction of cellulases is done from an infected host plant.

Material Required

Bacterial infected disease samples, waring blender, pestle and mortar, centrifuge, cheesecloth, funnel, 100-ml Erlenmeyer flask, dialysis tubes, scalpel/razor blade.

Protocol

Take 25 g of infected tissue of bacterial disease where cell wall structures are disintegrated due to bacterial infection. Cut the tissue into pieces each of 1–2 cm dimension. Transfer it to a waring blender. Add 100 ml distilled water or acetic acid–acetate buffer at pH 5.2. Blend for 5–10 minutes. Filter through two layers of cheesecloth. Centrifuge at 2000 g at 35°C for 20 minutes. Take the supernatant into the dialysis tubes and dialyze it against distilled water at 2–4°C for 8 hours. (Dialysis for a longer period may be avoided because it decreases the activity of cellulase enzyme.)

Testing for Presence of Enzyme Cellulase

Testing for the presence of enzyme cellulose can be carried out as described earlier or as in Section 3.3.1. The clear dialyzed solution in the dialysis tube (or crude enzyme in the flask) is the enzyme which can be used for various tests. It may be stored at 2–4°C, adding a few drops of toluene when not in use.

Precautions

For best results, work at low temperature (2–4°C). Dialysis may be avoided because some enzymes get absorbed into tubing during dialysis and lessen the cellulase activity. The chilled buffer should be used for extraction of enzyme.

23.3.1.3 Measurement of Endoglucanase

Endo β-1-4–glucanase (or CMCase) that is produced extracellular by microbes attacks at random on the β-1-4–glycosidic bonds of cellulose molecule and results in the release of reducing sugars. These are absorbed by the microbe's cell and used for the production of energy for their metabolic activities.

The assaying of CMCase is done by the estimation of reducing sugars spectrophotometrically by the use of dinitrosalicyclic acid reagent (DNS).

Material Required

Enzyme source (culture filtrate or diseased tissue extract), test tubes (18 × 150 mm), 1-ml and 5-ml pipettes, hot water baths (at 37°C and 100°C), sodium acetate–acetic acid buffer, pH 5.2. 1 percent carboxymethyl cellulose (CMC; prepared in acetate buffer at pH 5.2), spectrophotometer (spectronic 20), 250-ml beaker, dinitrosalicyclic acid (DNS) reagent, 40 percent Rochelle salt solution.

Protocol

Prepare DNS reagent by dissolving 1 g of DNS, 200 mg of crystalline phenol, and 50 mg of sodium sulfate in 100 ml of NaOH (1 percent). Prepare a 40 percent solution of potassium sodium tartarate (Rochelle salt). Pipette 1 ml of CMC followed by 1 ml of buffer and 2 ml enzyme into a test tube. Keep the tube containing the reaction mixture in the water bath at 37°C for 30–120 minutes. Pipette 1 ml of the reaction mixture into a test tube after 30 minutes incubation. Add 1 ml of DNS reagent to the above tube. Keep the mixture in a boiling water bath for 5 minutes or till the color develops. Add 1 ml of Rochelle salt (40 percent) while the mixture is still warm. Cool the tube under tap water.

Measure the absorbance of the solution at 575 nm on the spectrophotometer. Repeat the same processes of withdrawing 1 ml aliquot of reaction mixture at an interval of 60, 90, and 120 minutes to carry out the test.

Observations

Calculate the amount of reducing sugars released in the reaction mixture after intervals of time using standard prepared from glucose. Express the enzyme activity as the amount of glucose released/ml of the enzyme extract/unit time.

23.3.2 Pectinases or Pectolytic Enzymes of Pathogens

Pectic substances composed mainly of galacturonic acid units (linked in an α-1, 4 fashion), are found in plant's primary cell wall and middle lamella. Decomposition of these is brought about by the microbial enzymes which are divided into three major categories: Hydrolytic (PG,PMG), transeliminative (PTE, PATE), and pectin esterase (PME). These enzymes are secreted by the infecting parasite or activated in the tissue during infection and are thought to play an important role in pathogenesis/disease development. Pathogens secrete enzymes either inductively, that is, only when its substrate is present called induced (or inducible) enzymes, or noninductively, that is, regardless of the presence of its substrate, called constitutive enzymes both in culture media and inoculated plants.

Some of the pectolytic enzymes (PG, PME, PTE, PATE) can be individually assayed by employing appropriate substrates.

23.3.2.1 Extraction of Pectolytic Enzymes from Diseased Plants (*In Vivo*)

Pectolytic enzymes play an important role in pathogenesis/disease development and are produced in infected tissues of plants. These can be extracted either from the plant parts showing symptoms of a particular disease or from the deliberately infected plant parts. The extraction of crude pectic enzymes is done from *Erwinia*-inoculated tubers/rhizomes of colocacea.

Material Required

Erwinia-inoculated tubers of colocacea, healthy tubers of colocacea, blender/pestle and mortar, 5-ml pipette, beaker (250 ml), NaCl (0.15 M), toluene, cheesecloth, cellophane tubes, centrifuge.

Protocol

Take 20 g of *Erwinia*-infected (naturally or artificially inoculated) colocacea tubers. Transfer the material to a waring blender and add 100 ml 0.15 M NaCl. Blend for 2 minutes. Filter through two layers of cheesecloth. Transfer the filtrate to centrifuge tubes. Run the centrifuge at 2000 g for 30 minutes. Take the supernatant in cellophane tubes and dialyze it against distilled water at 4°C for 24 hours, changing the water at every 8 hours.

Testing for Presence of Enzyme

Prepare the microtome section of healthy tissues of colocacea tuber and put them in isolated enzyme solution while few sections in distilled water for 8–10 hours. Remove the microtome sections and stain them as described in Section 3.3.1 for degradation of pectate substance and observe under a microscope.

The clear dialyzed solution (or the nondialyzed solution) is the crude pectic enzyme which can be used for various tests. Add a few drops of toluene, when not in use, and store at 4°C.

23.3.2.2 Extraction of Pectolytic Enzymes from a Pathogen (*In Vitro*)

The bacterial plant pathogens, for example, *Erwinia*, produce pectic enzymes in culture media supplemented with pectic substances (i.e., induced enzymes).

The extraction of pectolytic enzymes is done from inoculated Czapek's broth supplemented with pectin.

Material Required

Czapek's liquid medium (broth) containing 1–2 percent pectin (pH 6.5), 500-ml Erlenmeyer flasks, Büchner funnel, 5-ml pipette, cellophane dialyzing tube, centrifuge, refrigerator.

Protocol

Take 100 ml of Czapek's broth each in 500 ml conical flasks. Aseptically inoculate the flasks with *Erwinia* culture and incubate the inoculated flasks at 26°C for 7 days. Filter through Whatman No. 1 filter paper. Centrifuge the filtrate at 10000 rpm for 30 minutes to remove the bacteria in pallet form. Take supernatant into the cellophane tubes and dialyze against distilled water at 4°C (refrigerator) for 24 hours, changing the water every 8 hours.

Testing for Presence of Pecotolytic Enzymes

Follow the procedure as described earlier. The clear dialyzed filtrate (or nondialyzed filtrate) is the crude pectic enzyme. Add a few drops of toluene and store at 4°C when not in use.

23.3.2.3 Assaying of Polygalacturonase

Polygalacturonase (PG) split or hydrolyze the α-1, 4-glycosidic bonds of pectic substances (pectic acids), releasing galacturonic acids of different sizes. Hydrolysis of pectic acids by PG can be at random (endo-PG) or terminal (exo-PG) linkages.

PG (endo) can be assayed by a physical method, based upon the loss in viscosity of sodium polypectate or pectic acid, or chemically, measuring the amount of galacturonic acids by titration. Exo-PG is determined by descending paper chromatography.

The assay of Endo-polygalacturonase (Endo-PG) is determined by the loss in viscosity of sodium polypectate and estimation of reducing groups by titration method.

PG by Estimation of Reducing Groups

Material Required

Enzyme extract/filtrate, 1 percent sodium polypectate (prepared in sodium acetate–acetic acid buffer, pH 5.2), 1 M Na_2CO_3 (anhydrous) solution, 0.1 N sodium thiosulfate, 2M H_2SO_4, 0.1 N iodine solution, 100-ml Erlenmeyer flasks, 100-ml glass stoppered Erlenmeyer flasks, 5-ml pipette, 25-ml burette, water bath (30°C).

Protocol

Take 20 ml of sodium polypectate (1 percent) prepared in acetate buffer in an Erlenmeyer flask. Add 20 ml of enzyme into the flask to make a final concentration of the substrate to 0.5 percent. Incubate the mixture at 30°C for 2 hours. Pipette 5 ml aliquots from the incubated reaction mixture after 30 minute into a glass Stoppard flask. Add 0.9 ml of 1 M Na_2CO_3 and 5 ml 0.1 N iodine solution to the reaction mixture. Acidify the reaction mixture after 20 minutes with 2 ml 2 M H_2SO_4. Titrate the residual iodine with 0.1 N sodium thiosulfate. Repeat the same process at definite intervals of 30 minutes on the incubated mixture for 2 hours to estimate the amount of reducing sugars released with prolonging the incubation period.

Observation

Record the titration results after intervals of time. Calculate the mille equivalent reducing groups liberated from sodium polypectate from a standard curve prepared from glucose because both galacturonic acid monohydrate and glucose have the same value.

Express PG activity in mille equivalents of reducing groups from sodium polypectate/min/ml of the enzyme extract/filtrate or mg protein.

23.3.2.4 Assaying of Pectin Transliminases (PTE, PATE)

Pectin transeliminases (pectin-transliminase or pectin lyase [PTE] and polygalacturonate transeliminase [PGT or PATE]) bring about nonhydrolytic breakdown of pectic substances by removal of H^+ atom at C-5 position of the galacturonic acid chain resulting in the formation of a double bond between C-4 and C-5 and elimination of a single unsaturated monouronic acid and a homologous series of oligouronides. Transeliminases act in an alkaline pH (7.0–7.2) when compared with PG and PMG which are functional at a pH range of 4.4–5.6. The enzyme extract or culture filtrate should be dialyzed at pH 8.7 in 0.01 M Tris-HCl buffer.

Pectin Transeliminase (Pectin Lyase, PTE): Pectin transeliminase enzyme brings about nonhydrolytic breakdown of pectin releasing unsaturated galacturonic acids. The activity of this enzyme can be estimated by loss in viscosity of pectin by TBA.

TBA (Thiobarbituric Acid) Method

Material Required

Crude enzyme (extract or culture filtrate), 1 percent pectin (as prepared in viscosity method), 0.01 M thiobarbituric acid, HCl (0.5 N, 1 N), 9 percent $ZnSO_4$, $7H_2O$, 0.5 N NaOH, distilled water, pipettes (5 ml, 100 ml), 100-ml beaker, measuring cylinder (100 ml), test tubes, water bath (100°C), spectrophotometer/calorimeter.

Protocol

Prepare a reaction mixture of substrate, buffer, and enzyme in the ratio of 4:1:2 in a test tube. Prepare another mixture of substrate, buffer, and boiled enzyme (control).Incubate both the tubes at 30°C for 2 hours. Add 1 ml each of $ZnSO_4$, $7H_2O$, and 0.5 N NaOH to both the tubes to stop enzyme action and to precipitate enzyme protein and excess substrate. Pipette 5 ml of reaction mixture to tube containing 5 ml of TBA reagent (3 ml of TBA, 1.5 ml 1 N HCl, and 0.5 ml distilled water). Place the tube in a boiling water bath for 40 minutes to permit pink color development. Allow the tube to cool to room temperature. Measure the absorbance of the solution against a control between 450 and 575 nm on a spectrophotometer.

Observation

Observe the tube held in the boiling water bath for pink color development. Maximum absorbance of the solution will be at 547 nm; light absorbance at this nm indicates the presence of PTE activity.

Polygalacturonase transeliminase (PGTE): Polygalacturonase transeliminase (PGTE) or pectic acid transeliminase (PATE) brings about nonhydrolytic breakdown (i.e., transeliminative split of pectic acid releasing unsaturated galacturonic acids). PGTE activity is tested by various tests (viscosity and TBA methods) as used for PTE, using sodium-polypectate or pectic acid as a substrate in place of pectin.

23.3.3 ASSAYING PECTOLYTIC ACTIVITY OF *ERWINIA CAROTOVORA*

Material Required

Waring blender, 200-ml flask, aluminum foil, sterile Petri plates, sodium polypectate medium, culture of *Erwinia carotovora* var. *zeae*.

Protocol

Add 4.5 ml (1 N) NaOH, 3.0 ml (10 percent) $CaCl_2 2H_2O$, 1.5 g Difco agar, 1.0 g $NaNO_3$, and 0.05 g Difco yeast extract in 300 ml boiling water and blend at high speed in hot water sterilized blender. Add 15 g sodium polypectate and blend slowly adding 200 ml boiling water. Place in 2 L flask and cap with aluminum foil. Autoclave at 15 pounds of pressure for 25 minutes. Allow pressure to drop slowly to avoid bubbles. Pour plates as soon as possible. The medium cannot be again melted in water bath. Place dilute bacterial suspension on poured solidified medium and incubate at 25°C.

Observation

Pectolytic organisms form deep, circular depressions in the polypectate medium.

23.3.4 DETERMINATION OF PECTIC ENZYME PRODUCTION BY *XANTHOMONAS MALVACEARUM*

Material Required

Viscous meter; pipettes; stopwatch; 1 percent pectin solution in phosphate buffer, pH 5.0; 0.5 percent solution of pectinol 100 D in water; culture filtrate of 36-hour-old culture of *X. malvacearum* grown on nutrient broth + 0.5 percent pectic or pectic acid.

Protocol

Mix 1 ml of water thoroughly to 10 ml of pectin solution, add 6 ml of this mixture to the viscous meter, and determine the time with stopwatch for this preparation to flow through the viscous meter at 5 and 20 minutes after the water is added. Determine also the flow time for a comparable sample of water. Boil a small sample of pectinol 100 D solution gently for 5 minutes and add to the pectin solution. Mix thoroughly 1 ml of pectinol 100 D (not previously heated) with 10 ml of pectin solution and add 6 ml of this mixture to viscous meter. The time for the sample to run through viscous meter is recorded at 0, 5, 10, and 20 minutes after the enzyme sample is added to the pectin solution.

Repeat the second and third procedures with the pectin solution using culture filtrate of *X. malvacearum* instead of pectinol 100 D. Plot curve showing the change in viscosity of pectin solution with time and estimate the activity of the bacterial enzyme preparation relative to that of commercial preparation. Percentage reduction in viscosity can be calculated by the following formula:

$$(T_o - T_t)/(T_o - T_w) \times 100$$

where T_o is the flow time immediately after the addition of enzyme extract, T_t is the flow time after specified time interval, and T_w is the flow time for water.

Note: Pectinol 100 D contains a mixture of the enzyme polygalacturonase. Enzyme preparations are inactivated by boiling. An active enzyme preparation will reduce measurably the viscosity of a pectin solution in a short time. The enzymes produced by phytopathogenic bacteria are considered to play a role in the breakdown of host tissue.

23.4 ISOLATION OF BACTERIOCIN

Bacteriocin are defined as proteinaceous antibacterial substances produced by bacterial species, which are inhibitory to the isolates of the same species or closely related species. The chemical composition of these proteinaceous inhibitory molecules distinguish them from other classic anti-biotics. Investigation of bacteriocin production by plant pathogens have found application in typing of bacterial species and as biological control agent. Even though bacteriocins are generally produced in appreciable amounts in ordinary media, the use of inducing agent can improve their production.

23.4.1 Bacteriocin Production

Protocol I

The protocol described by lavermicocca et al. (1999) is as follows.

The following media can be used for bacteriocin production: NBY (Smidt and Vidaver, 1986); Woolley's medium (Woolley-Peptone supplemented with Bacto-Peptone [Difco]), $15 \, g \, l^{-1)}$; or potato-dextrose broth–casamino acids (PDB–CA), that is, PDB (Difco) supplemented with Bacto Casamino acids (Difco) $4 \, g \, l^{-1}$.

Add *Pseudomonas syringae ciccaronei* bacterial suspension (200 μl; A_{600} nm, 0·3) to 250-ml Erlenmeyer flasks containing 100 ml media. Shake the cultures ($100 \, rev \, min^{-1}$) at 26°C. Every 24 hours, for a period of 7 days, aliquots (1 ml) are taken from the cultures, centrifuged (9000 g, 10 minutes, 4°C) and filter-sterilized (0·45 μm cellulose acetate, Millipore Corp., Bedford, Massachusetts).

Quantitative Assay

In order to quantify bacteriocin production in the culture filtrates, an agar spot assay is used. Deter-mine the bacteriocin titres by a quantitative serial dilution test. Spot twofold serial dilutions of *Ps. syringae ciccaronei* culture filtrate (10 μl) onto the surface of KB agar plates seeded with the indicator strain *Ps. Syringae* subsp. *savastanoi* PVBa204. After 24 hours of incubation at 26°C, inhibition of growth of the indicator strain is expressed in arbitrary units (AU) of activity, that is, the amount of bacteriocin in an endpoint dilution will completely inhibit the growth of the test strain in the area of application of a 10 μl droplet.

Protocol II: Production and Purification of Bacteriocin

Grow *Pseudomonas syringae ciccaronei* strain NCPPB2355 in 1 L Erlenmeyer flasks containing 400 ml WP. Use bacterial suspension (A_{600}nm, 0·3; 800 μl per flask) as inoculums. After 4 days of incubation in the shaken culture ($100 \, rev \, min^{-1}$, 26°C), remove the cells by centrifugation (9000 g, 10 minutes, 4°C) and the supernatant fluid filter-sterilized. The antimicrobial activity of the prepa-ration is determined against the indicator strain *Ps. Syringae* subsp. *savastanoi* PVBa204, using the agar spot assay. Precipitate the culture filtrate (1800 ml) with ammonium sulfate (66 percent w/v saturation) and store overnight at 4°C. Remove the precipitate and the surface pellicles by centri-fugation (12000 g, 20 minutes, 4°C), resuspend in Tris-phosphate buffer (0·01 mol l^{-1}, pH 6·5; 1/10 of the initial volume), and dialyze against the same buffer for 48 hours at 4°C with several changes (dialysis tube, porosity 24, cut-off 17 kDa; Union Carbide Corporation, Danbury, Connecticut).

The dialyzed precipitate, designated as crude bacteriocin, is sequentially ultra-filtered through a series of Spectra/Pore membranes (Spectrum Medical Industries, Laguna Hills, California) of decreasing pore sizes (300, 100, 50, and 20 kDa). Record the volumes and the antimicrobial activities of the retentates and filtrates. The retentate recovered from the 100 kDa ultra filtration is lyophilized, and the residue (1.20 g) is dissolved in 6 ml ultrapure Milli-Q water and apply to a Sephadex G-150 fine column (Pharmacia, Uppsala, Sweden; 4.5×40 cm; flow rate 2.5 ml min^{-1}). Collect the column fractions (7 ml each) in homogeneous groups according to the u.v. diagram obtained by monitoring at 267 nm, the maximum for absorption of ultra-filtered crude bacteriocin. Lyophilize fractions and taste for protein content and antimicrobial activity against the indicator strain.

Electrophoresis (Analytical gel electrophoresis): Recover the nondenaturing discontinuous PAGE of the most active fraction from gel filtration (fraction IV) performed according to the method of Gross and DeVay (1977) with minor modifications, in a 7.5 percent (w/v) discontinuous gel with a Mini-Protean II electrophoresis system (BioRad Laboratories, Hercules, California). Use bovine serum albumin (BSA; Sigma) as a protein marker. Run samples at a constant voltage of 200V 1 mm^{-1} gel for 1 hour. After electrophoresis, cut the gel in half, leaving identical marker and sample lanes on each half. One half is stained with Coomassie brilliant blue R-250 (Sigma) and the other is rinsed in deionized water for 20 minutes with frequent changes, laid on KB agar plates, and overlaid with soft KB-agar (0.7 percent w/v) seeded with the indicator strain (Bhunia et al., 1987). Incubate the plate overnight at 26°C. Following growth of the indicator strain, the location of inhibition halo(s) on the unstained gel is correlated with the protein band(s) on the stained gel.

Preparative PAGE: Gel preparation and electrophoresis conditions to be the same as those described for analytical PAGE. Resuspend a 5 mg sample of fraction IV in 200 μl distilled water and apply to the gel. After electrophoresis, the stained band with the *Rf* corresponding to the location of the inhibition halo in the unstained lane is cut out of the gel using a razor blade. Recover the protein band with Micropure 0.22 μm separator located in a Microcon-30 centrifugal micro concentrator (Amicon, Beverly, Massachusetts) according to the manufacturer's instructions.

SDS-PAGE: Analyze the protein recovered from the preparative gel by SDS-PAGE performed as described by Laemmli (1970). Protein standards and their molecular weights include the following: Thymoglobulin, 330 kDa; ferritin (half unit), 220 kDa; albumin, 67 kDa; catalyses, 60 kDa; lactate dehydrogenase, 36 kDa; ferritin, 18.5 kDa (Pharmacia).

Protein estimation and antimicrobial activity: Determine the concentration of protein in the samples either by the method of Bradford (1976), or using a Bio-Rad (Bradford) Low Protein Assay Kit. Use bovine serum albumin fraction V (Merck) as protein standard. Assay the antimicrobial activity of bacteriocin preparations in the agar-spot assay against *Ps. Syringae* subsp. *savastanoi* strain PVBa204.

23.4.2 CHEMICAL AND PHYSICAL STABILITY OF BACTERIOCIN

Treat aliquots (1 ml) of crude bacteriocin (3200 AU ml^{-1}) with the following enzymes (Sigma): Proteinase K, protease, trypsin, α–chymotrypsin, ficin, lipase, and α–amylase. Perform the assays at a final concentration of 1 mg ml^{-1} at pH 6.5. Hold the samples with and without enzymes at the appropriate temperature (depending on the enzymes) for 1 hour.

To test stability at various temperatures and pH values, crude bacteriocin aliquots (1 ml, 3200 AU ml^{-1}, pH 6.5) is to be adjusted to pH values ranging from 2.0 to 10.0 with 1 mol l^{-1} HCl or NaOH; and keep for 1 hour at 25 and 50°C, and for 15 minutes at 60 and 100°C, then rapidly cool and bring back to the initial pH value.

To test the effect of organic solvents on the stability, freeze-dried crude bacteriocin aliquots (1 ml; 3200 AU ml^{-1}) are to be treated with various organic solvents, such as formaldehyde, (10 percent v/v), chloroform (10 percent v/v), acetone 10 percent v/v), isopropanol (10 percent v/v), ethyl alcohol (25 percent v/v), acetonitrile (70 percent v/v), and hexane (25 percent); keep at room temperature for 1 hour and the solvents are then evaporated under vacuum. The dried samples are reconstituted with sterile distilled water.

After each treatment, samples are to be assayed for antimicrobial activity against the indicator strain *Ps. Syringae* subsp. *savastanoi* strain PVBa204.

Mode of Action

To study the mode of action of the bacteriocin, the viability and lysis of sensitive cells of *Ps. Syringae* subsp. *savastanoi* strain PVBa229 is examined. Dilute an overnight culture of the strain to one-tenth into fresh Woolley's medium and aliquots of *Ps. syringae ciccaronei* culture filtrate, crude bacteriocin, or fraction IV from Sephadex G150 column (1200 AU ml^{-1}, final concentration) is added to the culture which is incubated at 26°C. The optical density (A$_{600}$) of the cultures is measured at intervals (Beckman DU-65 spectrophotometer) and the number of viable bacterial cells is determined on KB agar plates.

23.4.3 Induction of Bacteriocin Production by Mitomycin C

Grow *Pseudomonas syringae ciccaronei* in WP and NBY. Induce the cultures for bacteriocin production with mitomycin C (final concentration 0·1 µg ml^{-1}) when the cell concentration reaches about 3×10^8 cells ml^{-1}. After induction, cultures are shaken at 26°C for 4 hours and then stored at 4°C overnight. Remove cell debris by centrifugation (9000 g, 10 minutes, 4°C) and filter sterilize the supernatant fluids and assays for antimicrobial activity.

23.4.4 Detection of Antimicrobial Activity in Bacteriocin

To detect the antagonistic activity of *Ps. syringae ciccaronei*, an agar spot deferred method assay is used (Vidaver et al., 1972). A bacterial suspension (5 µl; A$_{600}$ nm, 0·3) of the strain is spot-seeded on the surface of KB agar and incubate the plates at 26°C for 3d. The colonies are killed with chloroform vapors and overlaid the plates with 3 ml soft agar (agar 0·7 percent w/v) seeded with 100 µl of a bacterial suspension (A$_{600}$ nm, 0·15) of the strains available. After 24 hours of incubation at 26°C, inhibition is scored positive if the width of the clear zone around *Ps. syringae ciccaronei* colonies was 1·0 mm or larger. Perform assays twice in triplicate.

To exclude bacteriophage activity, agar plugs are picked out from the area of inhibition and tested for the presence of bacteriophages by standard procedures (Pugsley and Oudega, 1987).

To evaluate the antimicrobial activity of *Ps. syringae ciccaronei* culture filtrate, grow the bacterium in WP medium for 4 days. Spot the aliquots (10 µl) of twofold serial dilutions of the culture filtrate onto KB agar plates and taste in the agar spot assay against the indicator strains.

23.5   ISOLATION OF SIDEROPHORE

Siderophores (Greek: "iron carrier") are small, high-affinity iron chelating compounds secreted by microorganisms such as bacteria. Siderophores are among the strongest soluble Fe^{3+} binding agents known.

Iron is essential for almost all life processes, such as respiration and DNA synthesis. Despite being one of the most abundant elements in the Earth's crust, the bioavailability of iron in many environments such as soil or sea is limited by the very low solubility of the Fe^{3+} ion. This is the predominant state of iron in aqueous, nonacidic, oxygenated environments. It accumulates in common mineral phase such as iron oxide and hydroxide; hence, it cannot be readily utilized by organisms or plants. Some plants like grasses, including barley, wheat, and oat, and certain microbes release siderophores to scavenge iron from these mineral phases by formation of soluble Fe^{3+} complexes that can be taken up by active transport mechanisms.

Siderophores are also important for some pathogenic bacteria for their acquisition of iron. Siderophores enable bacteria to take up iron under conditions of limited availability of the element in the environment. They are responsible for the dissolution, chelation, and transport of iron (III) into the cell.

Bacteria frequently infect plants by gaining entry to the tissues via the stomata. Having entered the plant, they spread and multiply in the intercellular spaces. With bacterial vascular disease, the infection is spread within the plants through the xylem. Once within the plant, the bacteria need to be able to scavenge iron from two main iron-transporting legends, nicotianamic and citrate. To do this they produce siderophores. The enterobacterium *Erwinia chrysanthemi* produces two siderophores, chrysobactin and achromobactin.

Pseudomonas sp. is known to produce a siderophore, known as pyoverdins. Fluorescent pseudomonas have been recognized as a biocontrol agent against certain soilborne plant pathogens. They produce yellow green pigment (pyoverdines) which fluoresce under UV light and function as siderophores. They derive the iron required for their growth and pathogenesis.

23.5.1 Siderophore Detection Assay

The method described by Mastan et al. (2014) is as follows: Siderophore production is studied using succinate medium (SM; Meyer and Abdullah, 1978) consisting of the following components: Succinic acid (4 g), K_2HPO_4 (6 g), KH_2PO_4 (3 g), $(NH_4)_2SO_4$ (1 g), $MgSO_4$ (0.2 g), and pH (7.0). In a 250 ml flask containing succinate medium 0.1 ml of inoculums is added and incubated on orbital shaking incubator for 48 hours at 28°C. For the detection of siderophores, each *Pseudomonas* isolate is grown in synthetic medium containing 0.5 m of iron, and incubated for 24 hours on rotary shaker at room temperature. The assays used to detect siderophores are the Chrome Azurol S assay and Atkin's assay.

Chrome Azurol S (CAS) Agar medium Assay (Schwyn and Neilands, 1987): The CAS plates are used to check the culture supernatant for the presence of siderophores. Culture supernatant is added to the wells made on the CAS agar plates (mannitol, 10.0 g; sodium glutamate, 2.0 g; K_2HPO_4, 0.5 g; $MgSO_4.7H_2O$, 0.2 g; NaCl, 0.1 g; distilled water, 1000 ml; pH 6.8–7.2) and incubated at room temperature for 24 hours. Formation of a yellow- to orange-colored zone around the well indicates siderophore production.

Note: All glassware used to store the stock solution of the medium are treated with concentrated HNO_3. The containers were dipped with concentrated HNO_3 and left overnight. After 24 hours, the acid is removed and the glassware are rinsed thoroughly with double-distilled water. CAS plates are prepared in three separate steps.

Preparation of CAS indicator solution: Initially 60.5 mg of chrome azurol S is dissolved in 50 ml of double-distilled H_2O. 10 ml of Fe III solution (27 mg FeCl, $6H_2O$ and 83.3 ml concentrated HCl in 100 ml double-distilled H_2O) is added along with 72.9 mg hexadecyl trimethyl ammonium bromide (HDTMA) dissolved in 40 ml double-distilled water. The HDTMA solution is added slowly while stirring, resulting in a dark blue solution (100 ml total volume) which is then autoclaved.

Preparation of basal agar medium: In a 250 ml flask, 3 g of 3-(N-Morpholino) propane sulfonic acid (MOPS) (0.1M), 0.05 g NaCl, 0.03 g KH_2PO_4, 0.01 g NH_4Cl, and 0.05 g L-asparagines are dissolved in 83 ml of double-distilled water. The pH of the solution is adjusted to 6.8 ml using 6 M NaOH. The total volume is brought to 88 ml using double-distilled water and 1.5 g agar is added to the solution while stirring and heating until melted. The solution is then autoclaved.

Preparation of CAS agar plates: The autoclaved basal agar medium is cooled to 50°C in a water bath. The CAS indicator solution is also cooled to 50°C, along with a 50 percent solution of glucose. Once cooled, 2 ml of the 50 percent glucose solution is added to the basal agar medium with constant stirring, followed by 10 ml of the CAS indicator solution, which is added carefully and slowly along the walls of the flask with constant stirring. Once mixed thoroughly the resulting solution (100 ml) is poured into sterile plates. Under minimal iron conditions, siderophores are produced and released into the culture medium.

To isolate and collect siderophores, grow *Pseudomonas* isolates in iron restricted (0.5 ml added iron) synthetic medium and synthetic medium with high concentration of iron (20 ml). After 24 hours of growth, the culture is centrifuged and the cell free supernatant is separated and collected by centrifugation for 10 minutes at 13500 rpm. Supernatant is applied to CAS plates by using cork borer to make a well on the plate. Culture supernatant is added to the well (60 ml), and plates are incubated at room temperature and observed for color change to develop. If siderophores are present, an orange halo is visible. A halo is formed by the supernatant of cultures grown in iron restricted media while cultures grown under high iron conditions did not create any color change. In addition to using supernatant from culture grown in high iron medium as a control, uninoculated medium is also added to a separate well to ensure that the medium alone does not cause a color change.

$$\% \text{ siderophore units} + A_r - A_s \times 100\, A_r$$

where A_r = absorbance of reference at 630 nm (CAS reagent) and A_s = absorbance of sample at 630 nm.

Estimation of siderophores: To study the effect of iron concentration and various carbon sources on siderophore production, cultures are grown for 40 hours at 25°C with shaking (200 rpm) in 500 ml Erlenmeyer flasks containing 125 ml medium, with the pH adjusted to 7. Four basal media are employed with $FeCl_3$ added in increasing amounts (5, 10, 50, 100, 150, 200, 250, and 300 g/ml). The media contain the following components (Meyer and Abdallah, 1978):

Asparagines medium: Asparagines 5 g/L, $MgSO_4$ 0.1g/L, and K_2HPO_4 0.5 g/L, King, sB: Glycerine-10 g/L, Proteose-peptone 20 g/L, and $MgSO_4$–1.5 g/L.

Glycerol medium: Glycerol –10 g/L, $(NH_4)2SO_4$–1 g/L, $MgSO_4.7H_2O$ – 1g/L, K_2HPO_4 – 4g/l.

Succinate medium: KH_2PO_4– 6 g/L, K_2HPO_4–3 g/L, $(NH_4)2SO_4$– 1 g/L, $MgSO_4.7H_2O$ – 0.2 g/L, sodium succinate – 0.2 g/L.

23.5.2 Effect of Iron Concentration, Sugars, Organic and Amino Acids, and Nitrogen on Siderophores Production

1. *Effect of iron concentration*: In order to determine the threshold level of iron at which siderophore biosynthesis is repressed in *Pseudomonas* under study; the cultures are grown in SM, externally supplemented with 1–100 µM of iron ($FeCl_3.6H_2O$). Following the incubation at 29°C and 120 rpm, growth and siderophore content is estimated.

   Optimization for the production of siderophores: Medium SM is prepared with different pH in the range of 2, 7, 10, and 14 and separately inoculated with cultures to check the effect of varying pH on growth and siderophores production.

2. *Influence of sugars, organic acids, and amino acids*: In order to examine the effect of different sugars, organic acids and amino acids on growth and siderophores production: in the first set, each 100 mL of SM is externally supplemented with 1 g/L each of glucose, dextrose, sucrose, maltose, and mannitol. Second set of SM is individually supplemented with 4.0 g/L each of citric acid and malic acid. The third set of SM is separately fortified with 1 g/L each of proline, histidine, tyrosine, threonine, cystein, and alanine. Each set is separately inoculated with cultures and incubated. Following the 24-hour incubation at 29°C, each set is subjected for growth and siderophore quantification.

3. *Influence of nitrogen sources*: In this experiment, ammonium sulfates in SM is replaced separately by different concentrations of urea (commercial grade) in the range of 0.1–1.0 g/L, and sodium nitrate, soy flour at the rate of 1.0 g/l. Growth and siderophores production in this media is compared with that of SM containing ammonium sulfate.

4. *Influence of other metal ions*: For detecting the influence of different heavy metals on growth and siderophore production, the cultures are separately grown in SM. 100 ml of SM is supplemented with 10 µM of different heavy metals, such as mercury ($HgCl_2$), magnesium chloride ($MgCl_2$), cobalt chloride ($COCl_2$), and molybdenum chloride ($MoCl_2$). Following the incubation at 29°C and 120 rpm, growth and siderophore content is estimated.

23.5.3 CHARACTERIZATION OF SIDEROPHORES

Hydroxomate type of siderophores is determined by hydrolyzing 1 ml supernatant of overnight grown culture with 1 ml of 6N H_2SO_4 in a boiling water bath for 6 h or 130°C for 30 minutes. Further this hydrolyzed sample is buffered by adding 3 ml of sodium acetate solution. To this 0.5 ml iodine is added and allowed to react for 35 minutes. After completion of reaction the excess iodine is destroyed with 1 ml of sodium arsenate solution. Finally 1 ml alpha–napthlamine solution is added and allowed to develop color. Wine-red color formation indicates production of hydroxamate type of siderophore (Gillan et al., 1981). While catecholate type of siderophore is determined by taking 1 ml of supernatant in a screw-capped tube. To this 1 ml of nitrite molybdate reagent with 1 ml NaOH solution is added. Finally 1 ml of 0.5 N HCl is added and allowed to develop color. Yellow color formation indicates production of catecholate type siderophore (Arnow, 1937).

24 Role of Bacterial Component in Induction of Disease Reaction or Pathogenesis

The term pathogenesis means the chain of events leading to the disease due to a series of changes in the structure or function of a cell/tissue/organ being used by a microbial, chemical, or physical agent. The pathogenesis of a disease is the mechanism by which an etiological factor causes the disease. There are several chemical weapons secreted by pathogens that are utilized by the pathogen to kill their target cell to derive their nutrient for their growth, multiplication, and survival, which ultimately leads to the disease condition in the host.

24.1 EPS IN INDUCTION OF DISEASE REACTION

Water-soaking is the first symptom of bacterial infection. The EPS induces a water-soaking reaction in the host plant, which creates humidity in the host tissue for bacterial multiplication.

Material Required

EPS, host cotton plant, water blank, hypodermic syringe, label, and so on.

Procedure

Dissolve different quantities of EPS in 5 ml distilled sterile water. Infiltrate 0.2 ml of this EPS suspension into the cotton leaves by hypodermic syringe. Maintain the plant at 27 ± 2°C with 80 percent humidity under glasshouse condition. Record the persistence of WS reaction in the host plant leaves.

Observation

The minimum quantity of EPS required for the persistent water soaking at inoculation/infection site seems to be 10 mg/ml (Table 24.1).

TABLE 24.1
Effect of EPS Concentration on Persistence of WS in Cotton

	Persistence of WS (in min) in Leaves of		
	Susceptible CV		Resistance CV
Concentration of EPS/ml (mg)	Acala-44	1-10 B	101-102 B
Water (control)	8	8	8
5	10	10	10
10	31 hr	31 hr	15
20	31 hr	31 hr	15

Source: S.G. Borkar and J.P. Verma. 1989. *Cot. Fib. Trop.* XLIV(2): 149–153.

24.2 BACTERIAL ENZYMES IN INDUCTION OF DISEASE REACTION

Cutinase, cellulase, pectinase, and lignase are often secreted by the pathogenic organism in a specific pathogen–host combination.

In a pathogen–host combination, the first surface with which the organism comes in contact is the cuticle and cell wall of the plant. The cuticle is comprised of a complex wax, cutin, which impregnates the cellulose wall. The cell wall is comprised of cellulose, which makes up the structural framework of the wall, along with the matrix molecules hemicelluloses, glycoprotein, pectin, and lignin. Therefore, penetration into living parenchymatous tissues and degradation of middle lamella is achieved by the secretion and action of one or more enzymes which degrades these chemical substances of the host cell wall.

Cutinase degrades the cutin on the cuticle layer to presoften the tissue for mechanical penetration as a first step of tissue degradation. At least one bacterial species produces cutinase. Evidences indicate that cutinase are continuously produced, albeit in low concentration, with degradation product often inducing an even higher level of cutinase secretion. In some pathogens the cutinase production may be linked to virulence.

Pectic substance comprise the middle lamella and also form an amorphous gel between the cellulose micro fibrils in the primary cell wall. Pectin degrading substances often termed as pectinase or pectolytic enzymes which include pectin methyl esterase (PME), polygalacturonse (PG), and pectin lyases or transeliminase. Pectin methyl esterase remove small groups such as methyl groups (CH$_3$), often altering solubility and thus affecting the rate of chain splitting by polygalacturonase and pectin lyase. Polygalacturonase split chains by adding a molecule of water, while pectin lyses split chains by removing a water molecule from a linkage. Pectin degrading enzymes are involved in a wide range of plant diseases, particularly soft rot caused by *Erwinia carotovora*, the action of pectin enzymes leads to tissue maceration, and these enzymes are known as macerating enzymes.

Cellulose is the major framework molecule of the plant cell wall existing as micro fibrils with matrix molecules (glycoprotein, hemicelluloses, pectin, lignin), filling the space between the micro fibrils and cellulose chains. Cellulase have been shown to be produce by bacteria. Cellulotytic enzymes play a role in softening and disintegration of cell wall. Cellulotytic enzymes are involved in the inversion and spread of the pathogen and also are instrumental in the collapse of cells and tissues.

24.2.1 STUDIES WITH PECTOLYTIC ENZYMES OF *ERWINIA*

Material Required

A solution of pectolytic enzyme, *Erwinia caratovora* bacterial culture, potato tuber, distilled sterile water, desiccators, and so on.

Procedure

Wash the potato tuber in running tap water and sterilize with 0.1 percent mercuric chloride solution. Cut the potato tuber in two equal parts and make a cavity in the center of the tuber with cork borer. In one cut tuber part, add solution of pectolytic enzyme; in another cut tuber part, add distilled sterile water as a control. In another cut potato tuber, add cultural suspension of *Erwinia caratovora*. Incubate all three sliced tubers in desiccators for up to 48 hours and observe the development of reaction.

Observation

The tuber part with pectolytic enzymes start disintegration of the cells, their softening and rot, while the tuber part with sterile water does not show softening and rot. Similarly, the tuber part with *Erwinia* bacteria also shows the softening and disintegration of the cells indicating the tissues maceration by the pectolytic enzyme.

24.3   BACTERIAL TOXINS IN INDUCTION OF DISEASE REACTION

Toxins may act directly on living host cells, damaging or even killing the host and are implicated in plant diseases. Some toxins are active on a wide range of plant species (non-host-specific toxins) or on a specific host species (host specific toxin).

The non-host-specific toxins produced by bacterial plant pathogen are tabtoxin (produced by *Pseudomons syringae* pv. *tabaci*) and phaseolotoxin (produced by *P. syringae* pv. *phaseolicola*). The tabtoxin induce the symptoms identical to those of induced by the pathogen *Pseudomons syringae* pv. *tabaci*, that is, necrotic spots surrounded by a yellow halo. Similar effects may be observed on a variety of hosts, making it a non-host-specific toxin. Tabtoxin is chemically dipeptide, composed of amino acid threonine and tabtoxinine. In the cell, tab toxin is cleaved, releasing the tabtoxinine moiety which is the active toxin. The inhibition of enzyme glutamine synthetase is a primary mode of action of the toxin. Similarly, phaseolotoxin produced the symptoms similar to that of the pathogen *Pseudomons syringae* pv. *phaseolicola*. Chemically it is tripeptide of ornithine alanine arginine with phosphosulfinyl group. Within the cell, the toxin is enzymatically cleaved, releasing phosphosulfinyl ornithine, which is a toxic moiety. In cells it affects the inactivation of the enzyme ornithine carbamoly transferase.

24.3.1   STUDIES WITH PHASEOLOTOXIN OF PSEUDOMONAS

Material Required

A solution of phaseolotoxin, susceptible bean host, culture of *Pseudomonas syringae* pv. *phaseolicola*, distilled sterile water, infiltration syringe.

Procedure

Take a susceptible bean host plant to *Pseudomons syringae* pv. *phaseolicola* with other crops. Prepare bacterial suspension of *Pseudomons syringae* pv. *phaseolicola* (10^8 cfu ml) and inoculate by hypodermic syringe in the dorsal side of the leaves of the plant and tag them appropriately. Prepare the solution of commercial phaseolotoxin and infiltrate in the leaves of the host and tag them appropriately. Incubate the plant at $27 \pm 2°C$ temp to observe the symptoms up to 5–7 days.

Observation

Observe the symptoms produced by phaseolotoxin on the bean and other hosts within 12 to 24 hours and their subsequent changes.

Observe the symptoms produced by bacterial pathogen *Pseudomons syringae* pv. *phaseolicola* on the bean and other hosts by 5–7 days.

Compare the symptoms produced by phaseolotoxin and the bacterium *Pseudomons syringae* pv. *phaseolicola* on the bean and other hosts.

25 Enumeration of Bacterial Population

Microbiological analysis of plant, soil, and air samples requires quantitative determination, that is, total population of microorganisms in these substrates. The density of cells of microorganisms can be measured in the laboratory by several methods, either by direct or indirect counts. In the direct microscopic count, a known volume of liquid is added to the slide and the number of microorganisms are counted by examining the slide with the bright-field microscope. For direct microscopic counts, Neubauer or Petroff–Hausser counting chamber, breed smears, or an electronic cell counter (such as Coulter counter) are used. Various methods for indirect counts are determining cell mass (dry weight determination) or cellular constituents (DNA and protein), oxygen uptake, carbon dioxide production, turbid metric measurements for increase in cell number (spectrophotometric or colorimetric analysis), membrane–filter count and the serial dilution–agar plate (plate count) method. Excepting the last two methods, which are used to determine the viable cells, the major disadvantage common to all these is that the total count includes both the dead as well as living cells. The two most common indirect methods are plate count and turbidity measurements.

25.1 ENUMERATION (COUNTING) OF BACTERIA BY PLATE COUNT OR SERIAL DILUTION AGAR PLATE TECHNIQUE

The plate count technique is one of the most routinely used procedures because of the enumeration of viable cells by this method. This method is based on the principle that when material containing bacteria is cultured, every viable bacterium develops into a visible colony on a nutrient agar medium. The number of colonies, therefore, are the same as the number of bacteria contained in the sample. In this procedure a small measured volume (or weight) is mixed with a large volume of sterile water or saline called the diluents or dilution blank. Dilutions are usually made in multiples of ten. A single dilution is calculated as follows:

$$\text{Dilution} = \frac{\text{Volume of the sample}}{\text{Total volume of the sample and the diluents}}$$

Serial dilutions are later prepared by transferring a known volume of the dilution to a second dilution blank and so on. Once diluted, the specified volume of the dilution sample (1 ml or 0.1 ml) from various dilutions is added to sterile Petri plates (in triplicate for each dilution) to which molten and cooled (45–50°C) suitable agar medium is added. The colonies are counted on a Quebec colony counter. The number of organisms developed on the plates after an incubation period of 24–48 hours per ml is obtained by multiplying the number of colonies obtained per plate by the dilution factor, which is the reciprocal of the dilution. To facilitate calculations, the dilution is written in exponential notation. For example, 1:1000 dilution would be written as 10^{-3}.

$$\text{Number of cells/ml} = \frac{\text{Number of colonies}}{\text{Amount plated} \times \text{dilution}}$$

Material Required

Sample or bacterial suspension, 9 ml dilution blanks (7), sterile Petri plates (12), sterile 1 ml pipettes (7), nutrient agar medium (200 ml), colony counter.

Procedure

Label the dilution blanks as 10^{-1}, 10^{-2}, 10^{-4}, 10^{-5}, 10^{-6}, and 10^{-7}. Prepare the initial dilution by adding 1 ml or 1 g of the sample into a 9 ml dilution blank labelled 10^{-1}, thus diluting the original sample 10 times ($\frac{1}{1} \times \frac{1}{10} = \frac{1}{10}$ and is written 1:10 or 10^{-1}). Mix the contents by rolling the tube back and forth between your hands to obtain uniform distribution of organisms (cells). From the first dilution, transfer 1 ml of the suspension while in motion, to the dilution blank 10^{-2} with a sterile and fresh 1 ml pipette diluting the original specimen/suspension to 100 times ($\frac{1}{10} \times \frac{1}{10} = \frac{1}{100}$ or 10^{-2}).

From the 10^{-2} suspension, transfer 1 ml of suspension to 10^{-3} dilution blank with a fresh sterile pipette, thus diluting the original sample to 1000 times (1:1000 or 10^{-3}). Repeat this procedure till the original sample has been diluted 10000000 (10^{-7}) times, using a fresh sterile pipette every time.

From the appropriate dilutions (10^{-1} to 10^{-7}) transfer 1 ml or 0.1 ml of suspension while in motion with the respective pipettes to sterile Petri dishes. Three Petri dishes are to be used for each dilution (if 0.1 ml is plated; the dilution is increased 10 times). Add approximately 15 ml of the nutrient medium, melted and cooled to 45°C, to each Petri plate containing the diluted sample. Mix the contents of each plate by rotating gently to distribute the cells throughout the medium. Allow the plates to solidify. Incubate these plates in an inverted position for 24–48 hours at 37°C.

Observations

Observe all the plates for the appearance of bacterial colonies. Count the number of colonies in the plates that have colonies in the 30–300 range by placing each plate one by one on the platform of a Quebec colony counter.

Calculate the number of bacteria per ml of the original suspension/sample as follows:

$$\text{Bacteria per ml/gram of the sample}$$

$$= \frac{\text{Number of bacterial colonies (average of three replicates)}}{\text{Amount plated} \times \text{dilution}}$$

For example, if 60 colonies were counted on a $1:10^5$ dilution, then

$$\text{Number of cells/ml} = \frac{60 \text{ colonies}}{1 \text{ ml} \times 10^{-5}} = 6000000$$

$$= 6 \times 10^6 \text{bacterial/ml or gram of sample}$$

25.2 COUNTING OF BACTERIAL POPULATION BY THE USE OF SPECTROPHOTOMETER

Bacterial population or amount of growth can be determined by measuring turbidity or optical density (i.e., cloudiness of a suspension) of a broth culture. The more turbid a suspension, the less light will be transmitted through it. Since turbidity is directly proportional to the number of cells, this property is used as an indicator of bacterial concentration in a sample. Turbidity is quantified with the spectrophotometer that measures the amount of light transmitted or absorbed directly through a sample. It transmits a beam of light at a single wavelength through a liquid culture. The cells suspended in the culture interrupt the passage of light, allowing less light to reach the photoelectric cell and the amount of light energy transmitted through the suspension is measured as percentage of transmission or % T (i.e., the amount of light getting through the suspension) on the spectrophotometer as zero percent to 100 percent. The density of a cell suspension is expressed as absorbance or optical density (a value derived from the percentage of transmission) which is directly proportional to the cells' concentration. Absorbance is a logarithmic value and is used to plot bacterial growth on a graph.

Material Required

Test tubes each containing 5 ml of nutrient broth (13), 24-hour nutrient broth culture of bacteria, sterile 1 ml serological pipette, Bausch & Lomb Spectronic 20 spectrophotometer, Bunsen burner.

Procedure

Using a sterile 1 ml pipette, inoculate six labeled tubes (as 0, 2, 4, 6, 8, 16, and 24) of nutrient broth each with 0.1 ml of bacterial culture. The remaining seven uninoculated broth tubes serve as blanks for adjusting the spectrophotometer. Incubate all the 13 tubes at 30°C for 24 hours.

Observe the inoculated cultures following incubation at an interval of 0, 2, 4, 6, 8, 12, and 24 hours for the amount of growth which is determined by measuring the turbidity (i.e., percent transmission) of inoculated broth by using spectrophotometer. The steps for the measurement of % T are as follows:

Turn the spectrophotometer on by rotating the zero control knob clockwise. Allow the instrument to warm up for 15 minutes. Set wavelength at 620 nm (millimicrons). Adjust the zero control to set percent transmittance to "0" percent (OD to 2) by bringing the knob on the left. Wipe clean an uninoculated tube of broth, with tissue paper to remove all liquids and fingerprints, which will serve as the blank. Place the blank into the sample holder and close the cover. To standardize the instrument set the percent transmission to 100 by turning the knob to the right. Thoroughly mix the contents of an inoculated tube (labeled "O"). Wipe the test tube clean and keep for some seconds, allowing the medium to settle. Remove the blank from the sample holder. Place the inoculated tube into the sample holder, close the holder cover, and read the percent transmittance from the scale. Remove the broth culture from the sample holder. Take readings for percent transmittance for all the inoculated tubes after the desired period of growth and calculate the absorbance (or optical density) for all the inoculated tubes by applying the following formula:

$$\text{Absorbance} = -\log \frac{\% \, T}{100}$$

Observation

Plot the readings in terms of absorbance versus the time at which the readings are taken. It will be observed that the turbidity of some inoculated tubes increases with the incubation period, the OD will be found to increase while % T will be found to decrease, indicating more growth of the cell population in the culture with increased incubation period.

Calculate the number of cells per ml of the suspension by multiplying the appropriate factor. Record the data of cell numbers at intervals of 2, 3, 4, 5, 6, 7, 8, 9, and 10 days or more. Plot the log10 of cell numbers against the days after inoculation for the preparation of a growth curve. A typical growth curve thus obtained may show a lag phase, an exponential phase, stationary phase, and a death phase.

Note: The number of bacterial cells at a given optical density may vary because of the extracellular material produced by the bacterial cells of different species.

It is better to employ plate count method for each optical density measured concentration of a given species to calculate the number of cells/ml at a given optical density to plot a curve and this can be routinely used afterward for population estimation of that species. Generally, 0.1 OD means 10^7 cfu/ml of bacterial cells of *Xanthomonas axonopodis* pv. *malvacearum* (Borkar, 1984).

26 Determination of Ice-Nucleation in Plant Pathogenic Bacteria

Plant pathogenic bacteria particularly of the *Pseudomonas syringae* group have been reported to possess an ice-nucleation property (Gross et al., 1983). Ice-nucleation is the ability of bacteria to convert water molecules or dew molecules into the ice at a temp above 2°C. Generally these bacteria convert water molecules into ice at a temp of 3–4°C. These bacteria are generally prevalent in temperate regions of the world. They are present on the plant parts and increases the damage caused by frost by their ice-nucleation activity.

26.1 ASSESSMENT OF ICE NUCLEATION PROPERTY OF *PSEUDOMONAS SYRINGAE*

Material Required

Bacterial culture of *P. syringae*, distilled water, freezer.

Procedure

Prepare the suspension of *Pseudomonas syringae* bacteria (0.1 OD) and dispense in 5 tubes with 5 ml in each tube. Prepare the suspension of other bacterial plant pathogen (*Xanthomonas*) (0.1 OD) and dispense in 5 tubes with 5 ml in each tube. This will serve as the control. Incubate the tubes at 2, 4, 6, 8, and 10°C temperature for an hour. Note the formation of ice in the tubes.

Observation

Ice-nucleation bacteria convert the water into ice at 4°C and below.

27 Transmission of Plant Pathogenic Bacteria through Vector

Vectors play an important role in the transmission of plant pathogenic bacteria from plant to plant in the field and from field to field over a long distance. The classical example is the fire blight of apple and pear bacterium, which are transmitted through insect vectors over a long distance. Cucurbit wilt pathogen *Erwinia* is transmitted through spotted beetle in the field. These bacterial pathogens also survive in their vector and are transmitted when these vector visit the healthy plants to feed on them.

27.1 TRANSMISSION OF BACTERIA (*XANTHOMONAS*) THROUGH LEPIDOPTERON LARVAE (SPOTTED BOLLWORMS)

The transmission of *Xanthomonas axonopodis* pv. *malvacearum* in cotton was demonstrated by Borkar and Verma (1981) through spotted bollworm and red cotton bugs. The insects acquire the bacterium from infected leaves/bacterial oozing under high rainy humid conditions and transmit it to healthy leaves/plants while feeding/crawling on them.

Material Required

Spotted bollworm larvae, bacterial culture, infected cotton leaves with bacterium, and so on.

Procedure

Feeding of insects to acquire the bacterial pathogen: The fasting (3–4 hours in empty Petri plates) bollworm caterpillars are allowed to crawl on 48–70-hour-old highly virulent *Xanthomonas malvacearum* cultures in Petri plates. Similarly, another set of bollworm is fed on diseased leaf bits on moist filter paper in Petri plates or on plants with diseased leaves. These caterpillars are termed, after feeding and acquiring the *Xanthomonas malvacearum*, as *loaded*.

Method of inoculation and detection of Xanthomonas malvacearum: The loaded caterpillars (CP) are released on healthy plants put in glass chambers to provide moist conditions, and observed for symptom production.

Alternatively, the insect macerate (1 Cp/2 ml water) is infiltrated with the help of a hypodermic syringe onto the lower surface of cotton cv. Acala-44, which is highly susceptible to *Xanthomonas malvacearum*. The plants are observed for disease reactions.

The presence of bacteria (*Xanthomonas malvacearum*) in the insect can also be determined by specific bacteriophages and direct isolation.

Observation

Observe the leaves for disease water-soaking symptoms where loaded caterpillars are released and allowed to feed. Similarly observe the leaves where caterpillar macerate is inoculated for the development of a disease water-soaking reaction.

It is reported (Borkar and Verma, 1980) in both the cases that the bollworm transmit the disease and the disease symptoms develops within 5–7 days.

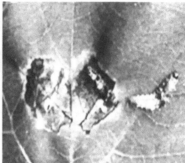

FIGURE 27.1 Crawling and feeding of spotted bollworm on infected cotton leaves with *Xanthomonas axonopodis* pv. *malvacearum*. (Courtesy of Dr. S. G. Borkar and Dr. J. P. Verma, Division of Plant Pathology, IARI, New Delhi.)

The excreta of the bollworm tested positive for the presence of *Xanthomonas axonopodis* pv. *malvacearum* in bacteriophage typing and direct isolation, indicating that the bollworm and its excreta serve as transmitting agent of this bacterial pathogen. (See Figure 27.1.)

27.2 TRANSMISSION OF BACTERIA (*XANTHOMONAS*) THROUGH INSECT/BEETLE (RED COTTON BUG)

Material Required

Red cotton bug, infected disease leaves, healthy plant, incubation humid chambers, insect cages, and so on.

Procedure

Collect the red cotton bugs from the fields and maintain them on cotton plants in cages. Fasting of the bugs is done by keeping the insects in empty Petri plates for 8–12 hours before placing them on a pure culture of *Xanthomonas campestris* pv. *malvacearum* in Petri plates or on diseased leaf bits for feeding/movement during which the bugs acquire the bacterium. These are termed *loaded* bugs, which are to be released on healthy plants kept under polythene cover to avoid the escape of the bugs and provide humid conditions for disease development.

Observation

To ascertain the presence of bacterium in the cotton bugs, the direct isolation from the bug and leaf infiltration techniques (LIT) for bugs' macerate were used (Borkar et al., 1980). In LIT, the bugs were macerated and the suspension was infiltrated with the help of an hypodermic syringe on to the lower surface of cotton cv., Acala-44, which is highly susceptible to *Xanthomonas campestris* pv. *malvacearum*. The plant is observed for disease reaction. It was reported (Borkar and Verma, 1980) in both the cases that the bacterium was recovered from the insect, thereby indicating it as a transmission vector. (See Figure 27.2.)

27.3 TRANSMISSION OF *ERWINIA AMYLOVORA* THROUGH BEES

In apple and pear orchards, the fire blight pathogen *Erwinia amylovora* is transmitted through pollinating insects, particularly bees from flower to flower in search of nectar. This can be demonstrated by following methods.

FIGURE 27.2 Crawling and feeding of beetle on bacterial blight infected cotton leaves. (Courtesy of Dr. S. G. Borkar and Dr. J. P. Verma, Division of Plant Pathology, IARI, New Delhi.)

Material Required

Pollinating honeybees from infected gardens, healthy honeybees, healthy apple/pear plants, muslin cloth net, nutrient agar plates, and so on.

Procedure

1. Collect the pollinating insects or honeybees from the fire blight–infected apple or pear orchards with the help of an insect collecting net.
   a. Allow the few insects/bees to crawl on the nutrient agar plates for a minute and incubate the plates at 28°C in incubator for 3 days to observe the development of bacterial colonies of *Erwinia*.
   b. Another set of collected insects is to be transferred on the bloom of healthy flowering pear plants, covered with muslin cloth, and tied firmly to avoid the visit of other insects. Observe the cover areas for the development of disease symptoms within a week period.
2. Collect the pollinating insects or honeybees from the healthy apple or pear orchards with the help of insect collecting net.
   a. Allow the few insect/bees to crawl on the nutrient agar plates with the *Erwinia amylovora* culture for a minute and then transfer these bees to the bloom of healthy flowering pear plants. Cover with a muslin cloth and tie firmly to avoid the visit of other insects. Observe the cover areas for the development of disease symptoms within a week period.

Observation

Formation of *Erwinia* colonies in the plates where honeybees collected from infected orchards crawl is a positive test that the honeybees carry the fire blight pathogen. Similarly, the development of fire blight symptoms on healthy blossoms through these bees indicate the transmission of this pathogen by the insect.

27.4 TRANSMISSION OF *ERWINIA* THROUGH SPOTTED BEETLES

Bacterial wilt of cucurbits are transmitted by beetles, which is a serious disease in cucurbit cultivation. The transmission of bacterial pathogen by beetle can be demonstrated by the following methods.

Material Required

Beetles from a wilt-infected cucurbit field, healthy cucurbit plants, beetles from healthy cucurbit field, muslin cloth net, nutrient agar plates, and so on.

Procedure

1. Collect the beetles from the wilt-infected cucurbit field with the help of an insect collecting net.
   a. Allow the few beetles to crawl on the nutrient agar plates for a minute and incubate the plates at 28°C in incubator for three days to observe the development of bacterial colonies of *Erwinia*.
   b. Another set of beetles collected is to be transferred to the healthy plant of cucurbit, covered with muslin cloth, and tied firmly to avoid the visit of other insects. Observe the cucurbit plant for the development of wilt disease symptoms within a fortnight period.
2. Collect the beetles from the healthy cucurbit field with the help of an insect collecting net. Allow the few beetles to crawl on the nutrient agar plates with the *Erwinia trichiphila* culture for a minute and transfer these beetles to healthy cucurbit plants. Cover with muslin cloth and tie firmly to avoid the visit of other insects. Observe the cucurbit plants for the development of wilt disease symptoms within a fortnight period.

Observation

Formation of *Erwinia* colonies in the plates where the beetles from infected cucurbit cultivation crawl indicate positive results for carrying of the bacterium by the beetle. Development of disease symptoms in healthy cucurbits through the infected beetles indicate transmission of the pathogen by the insect.

27.5   TRANSMISSION OF *ERWINIA* THROUGH NEMATODES

Rhizome rot or collar rot pathogen of banana, that is, *Erwinia chrysanthemi*, is transmitted by nematodes (Nagarale, 2000), which is a serious disease in banana cultivation. The transmission of bacterial pathogen by nematodes can be demonstrated by the following methods.

Material Required

Nematodes collected from collar rot–infected rhizome of bananas from an infected field, healthy banana rhizomes, nematodes collecting apparatus, nutrient agar plates, and so on.

Procedure

1. Collect the nematodes from the collar rot–infected rhizomes with the help of a nematode-collecting apparatus.
   a. Allow the few nematodes to crawl on the nutrient agar plates for a minute and incubate the plates at 28°C in an incubator for three days to observe the development of bacterial colonies of *Erwinia*.
   b. Another set of nematodes collected is transferred to the healthy rhizome of banana and planted in healthy soil in a pot. Keep the inoculated plants in the glass house under suitable climatic conditions for development of collar rot symptoms. Observe the banana plant for the development of collar rot disease symptoms within a fortnight period.
2. Collect the nematodes from the healthy banana field with the help of a nematode-collecting apparatus.
   a. Allow the few nematodes to crawl on the nutrient agar plates with the *Erwinia chrysanthemi* culture for a minute and then transfer these nematodes to a healthy rhizome

of banana and plant it in healthy soil in a pot. Keep the inoculated plants in the glass house under suitable climatic conditions for development of collar rot symptoms. Observe the banana plant for the development of collar rot disease symptoms within a fortnight period.

Observation

Formation of *Erwinia* colonies in the nutrient agar plates, where the nematodes from infected banana rhizome crawled, indicate the positive results for carrying the bacterium by the nematode. Development of disease symptoms in healthy banana rhizomes through infected nematode indicate the transmission of the pathogen by the insect.

28 Determination of Perpetuation of Bacterial Plant Pathogen

Perpetuation of bacterial plant pathogen can be referred to as survival of bacteria in a given substrate or material over a long period of time.

28.1 IN SOIL

The plant pathogenic bacteria of the genus *Agrobacterium*, *Ralstonia*, and *Erwinia* are known to present and survive in soil, as these are soilborne bacterial plant pathogens.

Material Required

Samples of soil exhibiting wilt/crown rot/soft rot symptoms, distilled sterile water, nutrient agar plates.

Procedure

Collect the soil sample, label properly, store at ambient condition, and analyze the same at different intervals. Dissolve 1 gm soil sample in 25 ml distilled sterile water with agitation on magnetic shaker for 10 minutes. Filter the sample solution through Whatman filter paper no. 24 to remove the soil material and collect the filtrate. Make the serial dilution up to 10^{-6}. Plate 0.2 ml of each dilution on separate nutrient agar plate, spread with glass rod, and incubate the plates in an incubator at $27 \pm 2°C$.

Observation

Observe the bacterial colonies on each plate, note their morphological characteristics, and compare them with the original culture of the bacterial plant pathogen exhibiting the disease symptoms.

For determining perpetuation of the bacteria in the soil, the same procedure or experimentation has to be carried out at different intervals (in days/in month/in years) and determine the perpetuation. The time period at which the last bacterial population was detected from soil is considered as perpetuation period of the bacterial pathogen in the soil. (See Figure 28.1.)

28.2 IN PLANTING MATERIAL

Most of the plant pathogenic bacteria survive in the infected plant material, particularly plant debris fallen on the ground/buried in soil.

Material Required

Infected fallen plant leaves/debris/root/leftovers, distilled sterile water, nutrient agar plates.

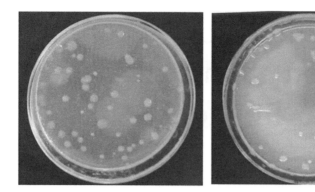

FIGURE 28.1 Survival of *Xanthomonas campestris* pv. *campestris* in moist and dry soil. (Courtesy of Dr. S. G. Borkar and Maria D'Souza, Department of Plant Pathology, Mahatma Phule Krishi Vidyapeeth, Rahuri.)

Procedure

Collect the plant debris/fallen leaves/leftovers, label them properly, and store at ambient condition. Crush 1 g of sample material in 20 ml of distilled sterile water with a pestle and mortar and allow to settle for 5 minutes. Pipette out the suspension from the crushed material and filter through Whatman filter paper no. 24 to remove the suspended crushed material. Prepare the serial dilution up to 10^{-6} from the filtrate. Plate 0.2 ml of each dilution on separate nutrient agar plate, spread with glass rod, and incubate the plates in incubator at $27 \pm 2°C$.

Observation

Observe the bacterial colonies on each plate, note their morphological characters, and compare with the original culture of the bacterial plant pathogen exhibiting disease symptoms.

Note: The colonies of plant pathogenic bacteria generally appears after 2 to 3 days of incubation.

For determining perpetuation of the bacteria in the planting material, the same procedure or experimentation has to be carried out at different intervals (in days/in month/in years) and the perpetuation determined. The time period at which the last bacterial population was detected from collected and stored sample is considered as perpetuation period of the bacteria in the infected planting material.

28.3   IN SEED MATERIAL

Some bacterial plant pathogen survive and perpetuate in seeds as external or internal seedborne plant pathogens and transmit the disease from one field to another, from one season to next, and from one locality to far-off places. Therefore, detection of perpetuation of bacterial plant pathogen in seed lots is important.

Material Required

Cotton seed infected with black arm pathogen, bean seed and sesamum seed infected with bacterial leaf spot pathogen, distilled sterile water, nutrient agar plate.

Procedure

Collect the seed material from previously infected crop with bacterial pathogen, label them properly, and store at ambient condition. Crush one gram of seed sample in 20 ml of distilled sterile water in pestle and mortar and allow to settle for 5 minutes. Pipette out the suspension from the crushed material and filter through Whatman filter paper no. 24 to remove the suspended crushed material. Plate 0.2 ml of suspension on nutrient agar plate, spread with glass rod, and incubate the plates in incubator at 27 ± 2°C.

Observation

Observe the bacterial colonies on each plate, note their morphological characters, and compare with the original culture of the bacterial plant pathogen exhibiting disease symptoms in the field from where the seeds lot is collected.

Note: The colonies of plant pathogenic bacteria generally appears after 2 to 3 days of incubation.

For determining perpetuation of the bacteria in the seed material, the same procedure or experimentation has to be carried out at different intervals (in days/in month/in years) and the perpetuation determined. The time period at which the last bacterial population was detected from collected and stored seed sample, is considered as perpetuation period of the bacteria in the seed lot material.

28.4   IN VECTOR

Some of the plant pathogenic bacteria are transmitted by insect vectors. On these insect vectors the bacteria survive or perpetuate externally or internally.

Material Required

Insect vectors and their instars, distilled sterile water, nutrient agar plates.

Procedure (For External Perpetuation)

Collect the insects from the diseased field and allow them to crawl on the nutrient agar plate. Incubate the plates at 27 ± 2°C.

Collect the insects from the diseased field and store them in the laboratory at ambient conditions. At different period intervals, take a few insect specimens and suspend them in distilled sterile water (2 ml) and agitate on a magnetic stirrer. Remove the insect and filter the suspension through sterile Whatman filter paper no. 24 in a sterile test tube. Plate 0.2 ml of suspension on nutrient agar plate and incubate at 27 ± 2°C.

Observation

Observe the plate for colonies of plant pathogenic bacteria, plated at different time intervals to determine the bacterial perpetuation in insects.

Procedure (For Internal Perpetuation)

Collect the insects from the diseased field and allow them to molt into different stages/instar. Take the different molts at appropriate time and crush them with a pestle and mortar in distilled sterile water. Filter the suspension through sterile Whatman filter paper no. 24 in a sterile test tube to remove insect crush. Plate 0.2 ml of suspension on nutrient agar plate and incubate at 27 ± 2°C.

Collect the insect from the diseased field and collect their excreta. Store the excreta in laboratory under ambient conditions for a long period. Dissolve the excreta in distilled sterile water at various interval periods and agitate on a magnetic stirrer. Filter the suspension through sterile Whatman filter paper no. 24 in sterile tube. Plate 0.2 ml of suspension on the nutrient agar plate and incubate at 27 ± 2°C.

Observation

Observe the bacterial colonies in each experimentation, note their morphological characters, and compare with the original culture of the bacterial plant pathogen exhibiting disease symptoms.

On the basis of colonies appearing in individual experimentation determine the external/internal transmission of the bacterial pathogen.

If the bacteria survive through different molts of the insect, or on/in insect, in their excreta for a different period, then it is assumed that the perpetuation of the bacteria occurs in the insect.

28.5   ON IMPLEMENT

Farm implements and pruning shears act as transmission tools for plant pathogenic bacteria. During the process of cultivation or pruning, these gets contaminated with the bacteria, which may survive or perpetuate.

Material Required

Farm implement, pruning shears, distilled sterile water, nutrient agar plates.

Procedure

Swab the farm implement or pruning shear used in bacterial-infected plant orchard or field with wet sterilized cotton. Suspend the cotton in sufficient quantity of sterile distilled water and allow to stand for 10 minutes. Plate 0.2 ml of these suspension on a nutrient agar plate, spread with glass rod, and incubate the plate in incubator at $27 \pm 2°C$.

Observation

Observe the bacterial colonies in the plate, note their morphological characteristics, and compare with the original culture of the bacterial plant pathogen exhibiting the disease symptoms.

For determining the perpetuation of the bacteria on the implement, the same procedure or experimentation has to be carried out at different interval (in days) and the perpetuation determined. The time period at which the last bacterial colonies were detected from implement swabbing is considered as perpetuation period of the bacterial pathogen on implement.

29 Determination of Plant Resistance

During the plant cultivar–bacterial pathogen interaction, in a favorable environmental condition for disease development, if the plant host cv shows no reaction or hypersensitive reaction without the induction of disease symptoms, this type of reaction is known as *plant resistance* to the given pathogen. The plant cv may have either vertical or horizontal resistance. The plant variety may show resistance to one race of a given pathogen (known as *vertical resistance*) and may be susceptible to other races of the same pathogen. The plant variety that shows resistance to several races or all the known races of the given pathogen is known to possess *horizontal resistance.*

29.1 DETERMINATION OF VERTICAL RESISTANCE

Material Required

Available germplasm of a crop, a known bacterial race of the pathogen, infiltration syringe, water, tags, and so on.

Procedure

Grow the available germplasm of a crop in the polyhouse/glasshouse/field with proper management. When the crop plant is 30–45 days old, tag the leaves of individual germplasm plant. Prepare the bacterial suspension (0.1 OD) of the known race and infiltrate it into the dorsal side of the tagged leaves. The infiltration should be carried out when proper environmental conditions are present in the field for disease development. If the experiment is carried out in the polyhouse/glasshouse, maintain the humidity (above 87 percent) and temperature ($27 \pm 2°C$) for the proper disease development. Observe the water-soaking disease reaction or resistant hypersensitive reaction at the second, fourth, and eighth day of inoculation.

Note: For wilt diseases, use the root dip method or soil drench method as described earlier.

Observation

Development of either hypersensitive browning reaction with no further spread or no reaction indicate the presence of vertical resistance in the crop cultivar against the specific race/isolate of the bacteria. (See Figure 29.1.)

29.2 DETERMINATION OF HORIZONTAL RESISTANCE

Material Required

Available germplasm of a crop, available races of the bacterial pathogen, infiltration syringe, water, tag, and so on.

FIGURE 29.1 (A) Race 18 produced disease water-soaking reaction. (B) Race 4 produced immune reaction indicating the presence of vertical resistance for race 4 in the grapevine plant. (Courtesy of Dr. S. G. Borkar and Swapanali Kadam, Department of Plant Pathology, Mahatma Phule Krishi Vidyapeeth, Rahuri.)

Procedure

Grow the available germplasm of a crop in the polyhouse/glasshouse/field with proper management. When the crop plant is 30–45 days old, tag the leaves of individual germplasm plant with the bacterial race number to be inoculated. Prepare the bacterial suspension (0.1 OD) of the individual race aseptically. Care should be taken that no bacterial cells of another race are mixed with each other. Infiltrate the suspension of the individual race into the tagged leave for the same. Thus, all the races should be infiltrated into their proper tagged leaves. The infiltration should be carried out when proper environmental condition are present in the field for disease development. If the experiment is carried out in the polyhouse/glasshouse, maintain the humidity (above 87 percent) and temperature ($27 \pm 2°C$) for the proper disease development. Observe the water soaking disease reaction or resistant hypersensitive reaction at the second, fourth, and eighth day of inoculation.

Note: For wilt diseases, use the root dip method or soil drench method as described earlier.

Observation

Record the hypersensitive reaction or water-soaking reaction of individual races on the same plant. If the plant shows resistance to move than one or several races, it is presume to possess horizontal resistance against the described races.

29.3 DETERMINATION OF HORIZONTAL RESISTANCE BY USING SINGLE RACE

The concept of using single race to determine horizontal resistance is based on the use of a more virulent race capable of attacking most of the resistant genes in a plant cv. The concept was given and demonstrated by Borkar (1990) for cotton–Xam interaction determining horizontal resistance.

Material Required

Available germplasm of a crop, a most virulent race (race 32 of Xam for cotton), infiltration syringe, water, tags.

Procedure

Grow the available germplasm of a crop in the polyhouse/glasshouse/field with proper management. When the plant is of 30–45 days old, tag the leaves of individual germplasm plant. Prepare the bacterial suspension (0.1 OD) of the most virulent race (known to attack more than one resistant gene) and infiltrate it into the dorsal side of the tagged leaves. Infiltration should be carried out when the proper environmental conditions are present in the field for disease development. If the experiment is carried out in the polyhouse/glasshouse, maintain the humidity (above 87 percent) and temperature ($27 \pm 2°C$) for proper disease development. Observe the water soaking disease reaction or resistant hypersensitive reaction at the second, fourth, and eighth day of inoculation.

Note: For wilt disease, use the root dip method or soil drench method as described earlier.

Observation

Development of hypersensitive, browning reaction with no further spread or no reaction indicate the presence of horizontal resistance in a given cv/variety to the bacterial pathogen.

30 Identification of Bacteria by Using the Immunodiagnostic Technique

The substances, which have now been identified as modified globulins, are formed not only as the result of an infection, but also as the consequence of the administration of certain dangerous substances of high molecular weight such as toxins of bacteria, animal, plants, or of dead bacteria.

Apart from the initial interest in resistance (immunity) to diseases, other research in immunology found that immunity caused by bacteria and toxins is only a part of a more general principle: The same mechanism is triggered when animals are inoculated with materials, such as cells or proteins derived from foreign species. In this case, antibodies appear in the blood causing agglomeration, cell destruction, or precipitation of those proteins.

All immune-antibodies are specific. This means, literally, that they react only to one antigen (the one used for immunization). Antigens may be proteins, cells from the blood of a different species, bacteria or viruses.

Modern biological sciences have developed efficient and precise biochemical and immunological techniques for the diagnosis of various important diseases, thus replacing more conventional methods.

Immunodiagonostic techniques are used in the identification of plant pathogenic bacteria in the laboratory or directly under field condition to identify the pathogen. These are generally used in the identification of plant pathogenic bacteria from the imported seed material or nursery plant material. These are specific in the identification of plant pathogen where specialized antibodies or antiserum are used. This chapter deals with general terminology in immunology; preparation of antigen and antiserum; and the techniques employed in the identification of plant pathogenic bacteria.

General Terminology and Meaning

Leucocytes: These are white blood cells of the immune system that are involved in protecting the body against infectious disease and foreign invaders. Leucocytes are found throughout the body, including in the blood and lymphatic systems. There are different types of leucocytes, among which the most important are

Monocytes: They are the largest type of leucocyte, generally 15–18 μm, and can differentiate into macrophages or dendritic cells. As a part of vertebrates' innate immune system, monocytes also influence the process of adaptive immunity. Monocytes comprise 2 to 10 percent of all leucocytes in the human body and serve multiple role in immune system.

Lymphocytes: They are one of the subtypes of white blood cells in a vertebrate's immune system. Lymphocytes include natural killer cells (NK cells, which function in cell mediated cytotoxic innate immunity), T cells (for cell mediated cytotoxic adaptive immunity), and B cells (for humoral, antibody-driven adoptive immunity). They are the main types of cell found in the lymph and are therefore known as lymphocytes. They are 8–9 μm and comprise 30–35 percent of all leucocytes.

Eosinophils: Also known as *eosinophiles* or *acidophils*, they are a variety of white blood cells and one of the immune system components responsible for combating multicellular parasites and certain infections in vertebrates. They also control mechanism associated with allergy and asthma. They are acid-loving and stain with eosin dye. In a normal individual they make up about 1–6 percent of white blood cells and are about 12–17 μm in size.

Antigens: Substances capable of inducing an immunological response (humoral, cellular, or mixed T and B cells), when introduced in animals are called antigens or immunogens. Most antigens are macromolecular proteins, but also may be immunogens, polysaccharides, synthetic polypeptides, as well as other synthetic polymers.

Immunogenic molecules have the following characteristics: The molecules should be foreign to the host. Molecules with a molecular weight higher than 10,000 are weak immunogens; proteins with a molecular weight higher than 1,00,000 are stronger immunogens. The molecule must have a certain degree of complexity to be antigenic. Immunogenicity increases with structural complexity. Moreover, aromatic amino acids contribute to larger degree than residues of non-aromatic amino acids in immunogenicity. The capacity to respond to an antigen varies with the animal species, and even with its genetic constitution. The production of immunoglobulin requires the linkage of the antigen to the surface of the lymphocyte. Combination sites on the surface of the lymphocyte that consist of molecules similar to antibodies are called *antigen receptors*. Only restricted portions of the antigenic molecules are related to antibody combination sites. These areas are called antigenic determinants, and they determine antigen–antibody reactions. An *antigenic determinant* can comprise only 6 to 7 amino acids of a protein's total number. In other words, a virus, for example, induces the production of a mixture of antibodies that react specifically to various antigenic determinants present in the particle. Thus, we can define the antisera containing a mixture of antibodies as *polyclonal antisera*. The number of antigenic determinants in a single molecule varies with the size of the molecule and its complexity.

Hapten: This is a small, chemically defined molecule that cannot induce the production of specific antibodies against itself. However, when it is covalently linked to a larger molecule, it can act as an antigenic determinant and induce antibody synthesis.

Adjuvant: Adjuvants are chemical substances used to enhance immunological response. They not only stimulate the formation of antibodies, but also localize them at the site of injection as deposits from which they are slowly released during the period of antibody synthesis, either through adsorption to solid particle or through their incorporation in to an oily emulsion. The most frequently used adjuvants are calcium phosphate, inorganic gels, aluminum hydroxide, bentonite, mentholated serum albumin, and incomplete Freund adjuvant. This is the most commonly used adjuvant with viral antigens and is made of 9 parts mineral oil (lanolin) and 1 part detergent; it allows the formation of a stable emulsion. If the mixture also contains dead *Mycobacterium* sp. cells, it is called a complete Freund adjuvant.

Antibodies: Antibodies are immunoglobulin produced by an organism as a response to the invasion of foreign compounds such as proteins, glycosides, or nucleic acid polymers. The antibody molecule associates noncovalently to the foreign substance starting a process of eliminating the foreign substance from the body.

The antibody molecule is basically formed by four polypeptide chains forming the basic unit: Two identical heavy (H) chains (MW 53,000 to 75,000) and two light (L) chains (MW 23,000) united by disulfur bonds. There are two regions in the antibody molecule. These include the *common fraction* (Fc) and the *variable fraction* (Fab).

Fc: Constituted of a portion of heavy chain.
Fab: This fraction chain (MW 50,000) determines the specificity of the antibodies and is constituted by the other portion of the heavy chains (MW 1,00,000).

Immunoglobulins made up of more than one basic monomeric unit are called *polymers*. Electrophoretic and ultracentrifugation studies have allowed the identification of five groups of immunoglobulin (Table 30.1):

TABLE 30.1
Properties of Human Immunoglobulin

	LgG	IgA	IgM	IgD	IgE
Molecular weight	1,50,000	1,60,000	9,00,000	1,80,000	1,90,000
Sedimentation coefficient	6–7	7	19	7–8	8
Concentration in serum (mg %)	1,000	200	120	3	0.05
Placenta transfer	Yes	No	No	No	No
Bacterial lysis	+	+++	+	?	?
Antiviral activity	+	+++	+	?	?

Source: Harper, S. J., in *Radioiodination: Theory, Practice, and Biomedical Application*, ed. M. K. Dewanjee, Hagerstown, Maryland, 1980.

1. *IgG.* This is the main fraction of antibodies and comprises 80 percent of immunoglobulin having molecular weights of between 1,50,000 and 1,60,000; it contains 2–4 percent carbohydrates and shows the lowest electrophoretic mobility among all immunoglobulin. It is the only immunoglobulin capable of crossing the placenta.
2. *IgA.* This immunoglobulin has a molecular weight of between 1,80,000 and 4,00,000. It has a higher carbohydrate content (5–10%) than IgG, IgA is found in high concentrations in blood, in secretions such as colostrums, saliva, tears, and bronchial and digestive tube secretions. IgA cannot cross the placenta.
3. *IgM.* This immunoglobulin has the most proteins and its amino acid sequence has not yet been determined. It contains 576 amino acids and has a molecular weight of 9,50,000. It is the first antibody synthesized by a newborn animal or human being. The cells producing IgM are divided into two daughter cells, which produce IgG. The IgMs can promote phagocytosis of microorganisms by macrophages and polymorph nuclear leukocytes. IgMs do not cross the placenta.
4. *IgD.* No antibody activity is attributed to IgDs; its role in the general scheme of immunoglobulin is still unknown.
5. *IgE.* This appears in serum at very low concentrations; it has a molecular weight of 1,90,000. Approximately half of the patients with an allergic diseases show high IgE concentrations.

Monoclonal antibody: An antibody produced by a single clone of cells or cell line and consisting of identical antibody molecules.

Monoclonal antibodies are specific antibodies to only one antigenic determinant. Monoclonal antibodies are produced by cloned cells (hybridomes) resulting from the union of a B lymphocyte with a carcinogenic myeloma cell. The lymphocyte confers the monoclonal antibody the capability to produce antibodies, and the myeloma cell gives it the ability to reproduce itself indefinitely. Because all cell clones from a hybridome come from a single B lymphocyte, they produce specific antibodies to only one antigenic determinant.

Monoclonal antibodies (mAb or muAb) are monospecific antibodies that are made by identical immune cell that are all clones of a unique parent cell, in contrast to polyclonal antibodies which are made from several different immune cells.

The antigen–antibody union is noncovalent and irreversible under normal laboratory conditions, and it includes hydrogen bonds, Van der Waals forces, and hydrophobic and coulombic interactions. The resulting chemical complementarily is similar to that between a key and lock.

The union area between the antigen and the antibody is only 1 percent of the total globulin surface and its specificity depends on the sequence of amino acids present in that region. The degree of association between the antigen and the antibody depends on the characteristics of each molecule and is determined by the combined effects of the interactions between them. If affinity and avidity are high, the molecules will unite more rapidly and dissociation will be very slow. Antisera with strong precipitation reactions show more avidity than those reacting weakly at a similar degree of dilution.

Antiserum affinity depends on, the temperature at which the reaction takes place. Higher temperatures cause stronger reactions. However, temperatures over 40°C result in protein denaturation. Salt concentration also affect the union. Low salt content favors the union of reacting substances.

The pH can cause changes in antigen–antibody affinity and the concentration of reacting substances for the formation of precipitates.

Immunization and production of antibodies: Animals commonly used in the production of antisera for research purposes include rabbits, guinea pigs, mice, goats, sheep, monkeys, horse, and hens. Selection of species depends on availability and volume of antiserum required. The mechanism regulating antibody synthesis and the reactions produced in the cells of the immune system in the presence of viral particles are extremely complex, and some characteristics are still unknown.

The immunization process can be summarized as follows: Some of the pathogens, for instance, bacteria/viruses entering the animal, are ingested by the macrophage cells. Some of the many millions of the helper T lymphocytes normally flowing in the blood system are activated to identify the new enemy, in this case a pathogen particle. The T cell is activated by its union with the macrophage that captures the particle. The union takes place when the T lymphocyte recognizes in the macrophage one of its own markers and another corresponding to the antigen. The activated helper T lymphocyte multiplies itself and thus stimulates the multiplication of killer T lymphocytes and of B lymphocytes.

Once the B lymphocytes have multiplied (approximately a million different lines), the T lymphocytes signal them to start the production of a great number of antibodies that will join the blood stream.

When a bacteria or virus infects an animal or human being, some particles penetrate and start multiplying in the infected cells. Killer T lymphocytes cause chemical damage to the membranes of the infected cells, thus interrupting virus replication.

Some of the synthesized antibodies neutralize the remaining virus particles. When the infection stops, the suppressor T lymphocyte stops all activity of the immune system, thus preventing uncontrolled reactions.

T lymphocytes and memory B lymphocytes store information on the viral particles and remain in the blood and lymphatic systems. Thus they are ready to mobilize and start a reaction if the same viral particles invade the animal again.

Immunization can be intravenous, intramuscular, subcutaneous, intraperitoneal, intradermal, intraarticular, or intranodular. This increases the stimulating effect upon the immune response.

Antibody production increases during the first days following the first immunization to reach a maximum concentration that can be maintained for a few days. If a new dose of antigen is injected, the synthesis of antibodies increases rapidly to a higher concentration than initially produced by the first dose of antigen. Antibodies are found in the blood serum of immunized animals. This is called *serum*.

To produce antiserum, allow a blood sample to stand to separate serum from other blood components. The serum is centrifuged (7840 g × 15 minutes) to remove debris that have not separated. Serum may be stored frozen (–20°C to –70°C) and lyophilized, or as a liquid at 4°C mixed with an antiseptic. Adequately stored, the antisera can remain active for many years.

Immunodiagnostic techniques are employed in detection of specific bacterial pathogens or closely related species of bacterial plant pathogens. These techniques help in the identification of the bacterial genus or species and also to detect the latent infection of bacterial plant pathogen in the healthy-looking planting material. Immunodiagnostic techniques are helpful in detection of phytobacterial pathogens of plant quarantine importance. Preparation of antibodies through antigens and their uses are essential part of immunodiagnostic technique.

30.1   PREPARATION OF BACTERIAL ANTIGENS OR ANTIGEN GLYCOPROTEIN FROM BACTERIAL PATHOGENS

The method demonstrate the preparation of antigen glycoprotein for bacterial pathogen *Xanthomonas corylina* infecting a noisetier plant.

Procedure

Grow a culture of *Xanthomonas corylina* on a liver–peptone–glucose–agar medium (liver extract, 5 g; peptone, 5 g; glucose, 5 g; agar, 20 g; distilled water, 1 L; pH 7) for 54 hours at 26°C. Harvest the growth from six plates by adding sterilized water which is diluted to 1 L and centrifuge at 19000 rpm overnight in ultracentrifuge at 4°C. Filter the supernatant through a Millipore membrane (0.45 μm pore size) and precipitate with ammonium sulfate. Dialyze the precipitate in running tap water overnight and subsequently in distilled water on a magnetic stirrer for three days at 4°C by changing water frequently. The removal of ammonium sulfate from the glycoprotein is tested by sulfate test. Detoxify the glycoprotein (GP) by treating it at 90°C hot water for 15 minutes. Protein is estimated from GP by UV absorption at 280 nm by employing following formula:

$$\text{Mg protein/ml} = \frac{0.5 \times \text{OD} \times \text{dilution}}{0.67}$$

Prepare 1 mg GP/ml as an antigen for inoculation in rabbits.

30.2   IMMUNIZATION AND PREPARATION OF ANTIBODIES

This method demonstrates the production and extraction of immunoglobulin for *Xanthomonas corylina* bacterium.

Procedure

Total immunoglobulin: Use two rabbits for each immunization schedule. The rabbits selected for the immunization schedule should be of the same age (4 months old) and weight. Give the injection of GP in rabbit (0.25 mg on first day followed by 0.5 mg on day 2, 3, 4, and 5, respectively, and 1 mg thereafter at an interval of a week) and collect the blood serum at different time intervals, for instance, 2, 3, 9, and 14 weeks of immunization by the vacuum puncture method. Store the collected blood serum at room temperature for 2 hours followed by at 4°C for 24 hours. Centrifuge the blood serum in a hemolyzed tube at 4000 rpm for 15 minutes and collect the serum. Estimate the immunoglobulin content of antiserum by UV absorbance at 280 nm.

Immunoglobulin "G": Immunoglobulin "G" is extracted from total immunoglobulin by the method of Bar–Joseph and Garnsey by using adsorption of IgG to protein-A sepharose followed by desorption at low pH. Estimate the immunoglobulin "G" content of antiserum by UV absorbance at 280 nm.

30.2.1   EFFECT OF IMMUNIZATION PERIOD ON RECOVERY OF IMMUNOGLOBULIN

Follow the process of immunization as described above. Collect the serum after each immunization schedule and estimate the immunoglobulin content of antiserum by UV absorbance at 250 nm.

Observation

Immunoglobulin content of antiserum is dependent on the immunization period and found to increase up to a period of 9 weeks immunization. The percent increase of immunoglobulin in the antiserum for

X. corylina was 1.24, 12.36, and 8.66 at 3, 9, and 14 weeks immunized serum over 2 weeks immunized serum. The immunoglobulin increases at 11.12 percent in antiserum during 3 to 9 weeks immunization; whereas it decreases at 3.70 percent during 9 to 14 weeks immunization. Difference in immunoglobulin content (mg)/ml of antiserum during 2 to 3, 3 to 9, and 9 to 14 weeks of immunization was 0.44, 3.99, and 1.31 percent, respectively. Thus an increase in the production of immunoglobulin was more during 2 to 9 weeks immunization whereas it decreased during 9 to 14 weeks immunization.

The immunoglobulin "G" content of antiserum is found to increase constantly up to 14 weeks of immunization. Immunized serum obtained after 2, 3, 9, and 14 weeks of immunization contained 10.33, 12.99, 20.49, and 23.94 percent IgG, respectively. On the other hand, IgM and other protein content of antiserum varied from 89.67 to 76.06 percent and was invariably related to the immunization period. IgM and other protein content in immunized serum decreased as immunization period increased. Percent increase of IgG in antiserum was 27.32 to 151.91 whereas the percent decrease of IgM and other proteins in the antiserum was 0.37 to 7.83 during 14 weeks of immunization (percent increase/decrease values are based over 2 weeks immunization). Immune serum obtained too soon after injection had predominance of low avidity IgM antibodies whereas later bleeds yielded proportionally richer and more useful Ig. IgG content in the serum increased constantly with increasing immunization period, though the quantum of increase was dependent on immunization period. Difference in IgG content (mg)/ml of antiserum during 2 to 3, 3 to 9, and 9 to 14 weeks of immunization were 1.0, 3.50, and 1.06, respectively. Fluctuation in IgM content of serum was also found to be dependent on immunization period (Borkar, 1989b).

30.2.2 EFFECT OF EXTRACTION METHOD ON RECOVERY AND AFFINITY OF IMMUNOGLOBULIN

Extraction of immunoglobulin from antiserum is mostly done to obtain concentrated form (as in case of salt-precipitation method) or to separate immunoglobulin of different molecular weight (as in case of ion-exchange chromatography). Immunoglobulin being sensitive protein molecules may be affected by chemicals during the process of extraction.

Procedure

Extraction of immunoglobulin: Antiserum (18 ml) obtained from 2, 3, 9, and 14 weeks of immunization of the rabbit (containing 35 to 39 mg protein/ml antiserum), which was dependent on the immunization) is to be used separately for extraction of immunoglobulin by standard salt precipitation and ion-exchange chromatography methods.

1. Standard salt precipitation of immunoglobulin from antiserum.
   Preparation of saturated ammonium sulfate salt: Dissolve 766.80 g $(NH_4)_2$ SO_4 in 1 L ultrapure water with stirring on magnetic stirrer. When only a little amount of salt is left to dissolve, heat the solution at 50°C for 15 minutes to dissolve the complete salt. Adjust the pH of the solution to 7.0 with the help of ammonia or H_2SO_4 as required. Store the saturated salt solution in bottle at room temperature.

Procedure

Take 5 ml antiserum in a beaker and add drop by drop 5 ml saturated ammonium sulfate while stirring on a magnetic stirrer. Allow the suspension to precipitate for 30 minutes on magnetic stirrer while stirring. Centrifuge the precipitated solution at 10000 rpm for 20 minutes. Discard the supernatant. Dissolve the precipitate in 1 ml ultrapure water and add 1 ml $(NH_4)_2$ SO_4 solution drop by drop while stirring and allow to stir for 30 minutes. Centrifuge the solution at 10000 rpm for 20 minutes and discard the supernatant. Dissolve the precipitate in 1 ml ultrapure water and add 1 ml $(NH_4)_2$ SO_4 solution drop by drop while stirring and allow to stir for 15 minutes. Centrifuge the suspension at 10000 rpm for 20 minutes and discard the supernatant. Dissolve the precipitate in 0.5 ml ultrapure water. Wash the centrifuge tube with another 0.5 ml water and put in dialysis sack. Carry out the dialysis for 45 minutes against the ultrapure water and then against the physiological buffer ($Na_2HPO_4.12H_2O$, 32.926 g; K_2HPO_4, 1.088 g; Nacl, 80 g; ultrapure water, 1 L. Dissolve the ingredient in small volume of water and make the quantity to 1 L. Store this in the freezer as stock buffer and use whenever required. Dilute stock buffer at the rate of 1:10 with ultrapure water and adjust the pH 7.2 whenever used. To avoid contamination of buffer add nitrium azide 0.1 percent in buffer or theomersel at 0.0005 concentration) for 48 hours at 4°C on the magnetic stirrer. Test the presence of ammonium sulfate. Centrifuge the dialyzer suspension at 10000 rpm for 20 minutes. Discard the pallet and filter supernatant. Measure OD at 280 and calculate the protein/ml and store in the freezer for further use.

2. Separation of immunoglobulin by ion-exchange chromatography.
   Buffers for ion-exchange chromatography:
   Buffer I: Tris Hcl, 0.025M; NaCl, 0.035 M
   Add 8.8 ml of 0.1 N HCl in 25 ml tris (0.2 M) to make tris HCl 0.025 M. Make vol. 200 ml.
   NaCl 0.035 M (i.e. 2.04 g/L)
   pH 8.8
   Buffer I is used for elution of IgG from serum.
   Buffer II: Tris HCl, 0.1M; NaCl, 1 M
   Tris solution 0.4 M (67.76 g in 1400 ml water).
   Mix 50 ml tris solution with 50 ml HCl and 100 ml water to obtain tris HCl of 0.1 M
   pH 8.8 buffer. It is used for elution of other molecules fixed in the column.

Procedure

Take 90 ml of DEAE tris acrylM and put in vacuum flask. Remove the air bubbles from this suspension with the help of vacuum pump. It takes around 15–20 minutes. Pour the DEAE suspension in the column (whose one end is closed with a stopper) until the other end of column and put in the other stopper. Connect the column to the flow adjuster with the help of small plastic tubes. Adjust the flow of buffer present in DEAE to settle down the DEAE gel. Put the Buffer I in buffer container and attached it with the help of small plastic tubes to the gel column via flow regulator to regulate the flow of buffer in the column gel (100 ml/hr). Run the equipment and allow the buffer to flow down the gel for overnight to adjust the pH of the gel 8.8 pH. Take the pH of buffer that flows out from gel and it must be an accurate 8.8 pH.

Remove the serum blood from dialysis sack and measure the pH. It must be an accurate 8.8 pH. Put the absorbance monitor on before 1–2 hours of utilization. Put lab current on 360 DC MA. Adjust the charted speed on graph paper. It is around 1.5 CM/HR. Put the absorbance range at 1.0. Adjust the baseline of the graph paper with a needle with the help of baseline adjustment as well as with an absorbance calibrator at 0.44 as the absorbance standard. When equipment has to run, put the absorbance standard at 0. For protein filter at 280 nm must be used.

Put the serum sample in a tube and joint the tube with column and buffer flow run the instrument. Put the sample collector on and adjust the speed on sample collector (80 drops/tube). Observe the movement of serum in gel and when the needle start working on graph paper to make graph and mark the tube in which the sample is going to be collected until the fall of graph curve up to the baseline.

Mix the samples collected, adjust the pH 7.0, and store in a freeze. Stop the buffer flow controller. Remove Buffer I and put Buffer II in the buffer container. Start the equipment (buffer flow controller) when needle start making graph mark the tube until the graph falls up to the base line. Stop the equipment. Take out Buffer II from the equipment through the Buffer flow and put Buffer I in the equipment. Run the equipment to wash the DEAE gel overnight to utilize further. Collect the other immunoglobulin sample and mixed them and put in freezer. Take out graph paper and mark the speed of paper, absorbance range, and amount of serum used.

Estimation of protein (IgG) after ion-exchange chromatograph:

Measure the optical density at 280 nm

$$\text{Protein IgG content} = \frac{1 \times X \times \text{dilution}}{1.6}$$

Where X = OD of sample (1 mg igG protein/ml give 1.6 O.D)

Affinity of immunoglobulin separated by different method to antigen: This is estimated by indirect method of immunofluorescence. The affinity of normal, slat-precipitated and ion-exchanged immunoglobulin to the antigen is determined on the basis of end point dilution.

In immunological assay the binding ability of the immunoglobulin molecules to the antigen depends on the recognition of the molecules at the binding site and their affinity. This character can be influenced by the method used for extraction of immunoglobulin.

Extraction of immunoglobulin from antiserum by salt-precipitation and ion-exchange chromatography is further affected by the age of immunoglobulin, that is, the period of immunization. The longer the immunization period, the better the percent recovery of eluted immunoglobulin protein.

30.2.3 ESTIMATION OF EFFECTIVE IMMUNOGLOBULIN

Effective immunoglobulin content of different antiserum samples collected during different immunization period and its affinity towards antigen *Xanthomonas corylina* is determined by using immunofluroscence (IF) technique.

Procedure

Estimate the effective immunoglobulin content of antiserum by indirect method of immunofluroscence. Put 10 µl of *Xanthomonas corylina* suspension (log phase 10^8 cfu/ml) in each well of IF slide and dry at 26°C. This is fixed with alcohol for 10 minutes and the slide is dried with blotter paper. 10 µl of antiserum of different dilution (1/25 to 1/6400) is added in each well separately and allowe to react for 30 minutes. Flood the excess of antiserum twice with phosphate buffer (Nacl, 8 g; Na_2HPO_4, 2.7 g; NaH_2PO_4, 2.4 g; distilled water, 1L; pH 7.2) for 5 minutes. Add 10 µl of fluorescein conjugated goat and rabbit IgG or IgM (whichever has to be estimated) at 1/300 dilution in each well and allowed to react for 30 minutes. Remove the excess of fluorescein conjugate by washing the slide with phosphate buffer thrice for 3 minutes and drying with blotting paper. Add a drop of glycerin buffer ($Na_2HPO_4.12H_2O$, 3.2 g; $NaH_2PO_4.2H_2O$, 0.15 g; distilled water, 100 ml; glycerin distilled, 50 ml) on the slide and observed under oil immersion in immunofluorescence light. Effective immunoglobulin content in antiserum is estimated on the basis of dilution end point of the antiserum to detect *Xanthomonas corylina* under IF.

30.2.4 ESTIMATION OF AFFINITY OF IMMUNOGLOBULIN

Procedure

To determine the affinity of immunoglobulin produced during different immunization period, prepare equal concentrations of immunoglobulin (equal amount of protein content/ml antiserum) from all the serum (serum obtained from 2, 3, 9, and 14 weeks immunization) and used during IF technique. Prepare the slide for observation under IF as described earlier. Affinity of immunoglobulin, produced during different immunization period, to *Xanthomonas corylina* is determined on the basis of end point dilution of immunoglobulin.

A key successful diagnosis of a bacterial strain using immunodiagnostic methods is a thorough knowledge of the specific antibody used in the test. Certain plant pathogens are more heterogeneous than others with respect to detectable antigens displayed in the natural pathogen population. A few bacterial pathogens can be reliably identified with polyclonal antibodies, but in most cases, the greater specificity of monoclonal antibodies is required for a reliable diagnosis. One must first consult relevant literature to determine the characteristics of the antibody as well as the serological composition of the bacterial population to be detected or identified.

30.3 PRODUCTION AND CHARACTERIZATION OF MONOCLONAL AND POLYCLONAL ANTIBODIES (EVALUATION OF SPECIFICITY)

Procedure

The procedure described is for *Erwinia carotovora* subsp. *atroseptica*. Immunize BALB/c mice by intraperitoneal injection of living whole cells of *Erwinia carotovora* subsp. *atroseptica* 1001, serogroup I. Hybridization of spleen cells is conducted by standard procedures. Screening for the presence of antibodies is performed by an ELISA-I biotin-streptavidin system against the homologous strain. A reaction is considered positive when the optical density reading is three times that obtained with a hybridoma culture supernatant of *Xylophilus ampelinus*, used as a control. Specific antibody-secreting hybridomas are cloned under conditions of limited dilution, using feeder layers. Cloning is repeated three times, and established hybrids are grown in hypoxanthine-thymidine medium. Nab isotypes are determined in culture supernatant by a commercial kit based on immunodiffusion (The Binding Site, Ltd. Birmingham, England). Ascetic fluid is produced, and MAbs are purified by affinity chromatography with protein A (Beckman).

Biotin conjugation: Prepare a solution containing 1 mg of purified MAb 4G4 per ml of 0.1 M sodium borate buffer (pH 8.8) to biotinylated antibodies with N-hydroxysuccinamide biotin (Sigma). Add the biotin in an amount equal to one-fifth of the weight of the immunoglobulins used. The N-hydroxysuccinamide biotin is previously dissolved in a volume of dimethyl sulfoxide equal to the volume of the immunoglobulins divided by 10. Incubate the solution for 4 hours at room temperature, and add 20 µl of 1 M NH_4 Cl per mg of immunoglobulins and incubate the solution for 10 minutes at room temperature. The uncoupled biotin is removed by 24 hours of dialysis against 0.1M phosphate buffered saline (PBS; pH 7.2 to 7.4).

Production of PAbs: Prepare antiserum from a rabbit (Californian × New Zealander) by immunization with cells of *E. Carotovora* subsp. *atroseptica* 1001, dialyzed against glutaral dehyde for 3 hours followed by dialysis against PBS for 24 hours at 4°C. Immunization and antiserum production is done as described by Alarcon et al. (1995) and purify the immunoglobulin's from the antiserum.

Specificity of MAbs and PAbs: The specificities of the antibodies are evaluated in the ELISA-I biotin-streptavidin system. Polystyrene micro plates (Polisorp; Nunc) are coated with suspensions of 10^8 CFU of different bacterial species per ml in sodium carbonate buffer (0.05 M, pH 9.6). Specificity is also tested against 162 isolates of potato saprophytic flora. MAb 4F6 is used when required, for comparison.

Various xanthomonads, such as *X.campestris* pv. *vesicatoria* (Xcv), *X. axonopodis* pv. *citri* (Xac) and *X. oryzae* pv. *oryzae* (Xoo) are tested for cross-reactivity at 10^4 and 10^2 cells ml^{-1} with different doses of antigen (10–50 µl $well^{-1}$). Culture filtrates (100 µl) of all these organisms at equivalent cell concentration of 10^8 cells ml^{-1} are also examined by ELISA.

30.3.1 VALIDITY OF pAB-XCC IN DETECTION OF THE PATHOGENS IN SEED SAMPLES AND INFECTED PLANT MATERIALS

Collect *Xanthomonas campestris* pv. *campestris*–infected cauliflower leaves/seeds. The brown, streaky stems are dipped in 10 ml sterile PBS for bacterial ooze. Suspected and infected seeds are soaked in sterile saline overnight at 4°C and crushed in sterile PBS. The supernatant fluid obtained after centrifugation at 1500 rpm for 15 minutes at 4°C is aliquoted. Aliquots are examined for the total number of colony-forming units (cfu) by plate assay and also for ELISA reactivity.

Source of Availability of Antibodies for Diagnostics of Phytopathogenic Bacteria

Source Companies	Product Range and Services
Agadia Inc. 30380 County Road 6 Elkhart, Indiana 46514 Telephone: 219-264-2014 Toll free (USA): 800-622-4342 Website: http://www.agadia.com Email: info@agdia.com	Test kits for identification, reagents, testing services, and contract research in agricultural diagnostics.
ADGEN Ltd. Nellies Gate, Auchincruive Ayr, UK KA6 5HW Telephone within UK: 01292 525 275 International: ++44 1292 525 275 Facsimile (UK): 01292 525 275 International: ++44 1292 525 275 Website: http://www.adgen.co.uk Email: info@adgen.co.uk	Test kits for identification, reagents, laboratory, analyses, and contract services.
Dynal Inc. 5 Delaware Drive New Hyde Park, New York 11042-9808 Telephone: 800-638-9416 Facsimile: 516-326-3298	Specializes in serological and molecular-based technical services, product using, biomagnetic separation technology.

30.4   IMMUNODIAGNOSTIC TECHNIQUES (ANTIGEN–ANTIBODIES REACTION)

30.4.1   Precipitation Test

Antigen and antibody precipitation occurs when the reaction between these substances forms a grid-like structure preventing the passage of water molecules (hydrophobic reaction). The formation of the grid structure requires a proportional concentration of reacting substances. For every antigen molecule, a given number of antibodies are needed.

Antibodies usually act as bivalent molecules and the antigens as multivalent molecules. The presence of too many antibodies will make the antibodies act as monovalent molecules only to one particle of the antigen, thus preventing the formation of the grid structure. Too many antigens also prevent the formation of the grid, because the antigens will then act only as monovalent molecules. The antigen–antibody reaction cannot be observed in either case.

1. *Interface ring tests*: In this test, a solution containing both antigens and antibodies is used. A proportional amount of the sample containing the antigen is placed on a given volume of antibodies. Precipitation occurs in the interface and the precipitation plane creates the illusion of a ring.

2. *Tube liquid precipitation test*: This quantitative test is mainly used to determine the titer of antiserum and the antigen concentration. In this test, several (double) dilutions of the antiserum and the antigen are mixed and incubated in two small test tubes for the formation of precipitates. Varying degrees of precipitation allow determination of the concentration of the reacting substances.

3. *Micro precipitation test*: In this test, individual drops of each of the reacting substances are placed on a Petri dish. The plaque is stirred to mix the antibody and the antigen. After incubation, the precipitate forms. A stereoscopic microscope is used to observe the reaction.

30.4.2   GEL DIFFUSION TEST

These precipitation tests are usually done using agar or agarose gels. Their advantage is that the mix of the antigen and its corresponding antibody can be physically separated by the difference in the diffusion coefficients of the components of their gels. Thus, these tests can provide information on the homogeneity and purity of the reacting substances, as well as on their size and relations.

1. *Simple immunodiffusion* (Oudin, 1946): This is based on the use of antibodies immobilized in an agar or agarose matrix melted in a water bath at 50°C. It is precooled for use and put into 10 × 75 ml or smaller tubes. The antigen can diffuse through the agar, thus provoking the appearance of a zone or band migrating along the tube until the concentration of both reacting substances reaches an optimal stage. Precipitation occurs at this point. Some factors affecting band migration are antigen concentration and diffusion coefficient. The latter depends on molecular weight and on the size of the antigen molecule. Lower molecular weight and size mean higher diffusion coefficients. The distance migrated by the diffusion band also depends on time, as well as on the kind, quality, and concentration of the antibody.
2. *Double immunodiffusion* (Ouchterlony, 1948):
   Simple Dimension System: A neutral agar layer is placed between the antigen and antibody solutions. The same principles apply as in the Oudin simple diffusion test. However, both reacting substances are diffusible and a precipitation line appears when they meet. The position and width of the band allow determination of the concentration of the reacting substances.
   Double Diffusion System: Plaques or slab holders are covered with neutral agar at a concentration of 0.7–1.5 percent. A well pattern is designed that suits the purpose of the test, and the antigen and the antibody are placed in separate wells.
   Types of Reactions:
   Identity reaction: The fusion of precipitation lines occurs when both antigens are identical. A compact barrier, the immunospecific barrier, is thus formed.
   Non-identity reaction: In this case, diffusion varies and the precipitation arches do not act as barriers to the antigen, which are not related and therefore intercrops.
   Partial double identity reaction: A partial double identity occurs between antigens. (Almost identical antigens differ in only one antigenic determinant.)
   Partial identity reaction: The formation of a "spur" is observed, indicating partial identity between antigens (e.g., as in the case of different strains of the same virus/bacteria).
3. *Immunoelectrophoresis* (Grabar and Williams, 1953): This is one of the most important analytical tools used in the solution of complex antigen mixes. It is based on electrophoretic mobility and antigenic specificity. The antigen mix is first separated in its components through electrophoresis in an agar gel. Then the antiserum is placed in a parallel channel or electrophoretic migration pathway to allow the formation of precipitation lines. Alkaline buffers between pH 7.5 and 8.6 are generally used because this provides conditions under which the proteins are negatively charged and move toward the anode. This method is used for virus characterization and strain differentiation.

30.4.3   AGGLUTINATION TEST

The main difference between precipitation and agglutination tests is that in the latter the antigen is very often larger than the antibody. For this reason, few antibody molecules are necessary to form a visible particle grouping. The principles governing reactions are similar to those of the precipitation tests.

1. Passive agglutination (indirect or reverse). This test is based on the use of inert substances that carry antigens or antibodies. These substances (latex spheres, bentonite) are several times larger than the reacting substances, thus making it possible to use soluble antigens (bacterial cells) in agglutination tests.
2. Latex test. This test is based on the use of polystyrene spheres (800 nm diameter) covered by immunoglobulin molecules. This technique is 10–100 times more sensitive than the traditional micro precipitation test for detecting plant viruses/bacteria. A reaction may be seen with the naked eye.
3. Neutralization tests. In some cases, the interaction of biologically active antigens with homolog antibodies results in loss of the antigen's biological activity. This reaction is called neutralization. Since the sensitivity of these tests fundamentally depends on the activity of the antiserum, the antigen's activity must be biologically detectable.

30.4.4 IMMUNOLOGICAL TEST WITH MARKERS

These tests use antibodies and antigens marked with independently acting substances called markers that increase test power and sensitivity. The higher the marker's level of activity, the faster the antigen–antibody reaction can be detected.

1. Immunofluorescence (IFA): The technique uses substances transforming light in the ultraviolet range (200 to 400 nm) into longer wavelength radiation. A modified microscope (a fluorescence microscope) allows you to see the light emitted by the fluorescing substance (fluorescein isocyanate: FITC)
2. Radioimmunology: The immunoglobulins are marked with radioactive substances (P32, I128). Their presence is determined by a reaction against photographic material.
3. Immunologic assays with enzymatic conjugates: These tests are based on the property of certain antigens and antibodies to be absorbed into a solid medium allowing them to construct an ordered sequence of biological material (antibody, antigen, antibody conjugated with an enzyme), and which can be seen as the color reaction resulting from the addition of the enzyme-specific substrate conjugated to the antibody, thus allowing adequate quantifying of the antigen. Depending on research needs, the following assays of this kind are frequently used:
   DAS-ELISA (Double Antibody Sandwich)
   NCM-ELISA (Nitro Cellulose Membrane)

30.4.5 IMMUNOLOGICAL ELECTRON MICROSCOPY

This is used for detecting antigens and the positive reaction can be identified with the help of an electron microscope.

Aggregation of bacterial particles. An adequate dilution of specific antiserum is prepared for detecting the suspected bacteria in the tissue suspension. As a result of particle addition, complexes appear that can be seen under the electron microscope.

This technique is particularly useful when low bacterial concentration prevents direct observation. An excess of antibodies will result in reaction inhibition, as in precipitation tests.

30.4.6 ENZYME LINKED IMMUNOSORBENT ASSAY (ELISA)

As the sensitivity of ELISA for detection of plant pathogenic bacteria is usually relatively low (about 10^5–10^6 cfu/ml) because of the lower coating ability of bacteria to the plates as compared to viruses and some other problems (Alvarez, 2004), a preceding enrichment step may be performed before ELISA (Indirect, DAS or DASI formats) to improve the sensitivity of the detection while maintaining the specificity. This is especially necessary when using specific monoclonal antibodies (Lopez et al., 2003). This step is unnecessary for identification of pure cultures. The enrichment should be performed using the most appropriate conditions of incubation (e.g., temperature, oxygen requirements, shaking, and media) specified in the pest-specific diagnostic protocol. Enrichment can fail due to the presence or development of other microorganisms and should be avoided when this is predictable.

1. **Indirect ELISA**
   a. Use 200 μL (sample amount might be reduced down to 100 μL according to the sample volume) aliquots of prepared sample extract, macerate or bacterial suspension (according to the instructions of the protocol) in 1.5–2.0 ml micro vials.
   b. Add an equal volume of double-strength coating buffer and vortex.
   c. Apply an equivalent volume of the sample to two wells of a microtiter plate with good coating characteristics (e.g., Nunc-PolySorp or equivalent) and avoid the borders of the plate for better repeatability. Some techniques are available to prevent border effect such as covering the plate with transparent film to avoid evaporation, in such cases border wells can be used. Incubate at 37°C for 4 hours or overnight at 4°C. Prepare positive and negative controls in the same way. It is also necessary to include negative controls of extraction buffer and of the enrichment medium when appropriate.
   d. Wash the wells gently (e.g., using a wash bottle) three times with PBS-Tween, as washing too strongly with some washing apparatus can affect the coating result and, consequently, the sensitivity, especially in detection.
   e. Prepare the appropriate dilution of specific antibodies in the appropriate blocking buffer. For validated commercially available antibodies use the recommended working dilutions.
   f. Add 200 μL to each well and incubate for 2 hours at 37°C.
   g. Wash as before (d).
   h. Prepare the appropriate dilution of the conjugated, second specific antibodies (usually alkaline phosphatase conjugated, but other enzymes may also be used) in the appropriate blocking buffer. Add 200 μL of this solution to each well. Incubate in the dark at room temperature and read absorbance at 405 nm (or at the recommended wavelength for other enzyme–substrate reactions) at regular intervals within 120 minutes, or as indicated in the protocol.

2. **DAS-ELISA**
   a. Prepare the appropriate dilution of antibodies in carbonate coating buffer pH 9.6. Add 200 μl to each well of a microtiter plates with good coating characteristics, for example, Nunc-PolySorp or equivalent, and avoid the borders of the plate for better repeatability. Some techniques are available to prevent border effect such as covering the plate with transparent film to avoid evaporation, in such cases border wells can be used. Incubate at 37°C for 4 hours or at 4°C overnight.
   b. Wash the wells gently (e.g., using a wash bottle) with PBS Tween, as washing too strongly with some washing apparatus can affect the coating result and, consequently, the sensitivity, especially in detection.
   c. Add 200 μL of each sample (plant extract, macerate, or bacterial suspension) to two wells previously enriched or treated if necessary. Include two wells per plate of positive and negative controls. It is also necessary to include negative controls of extraction buffer and of the enrichment medium used. Incubate 4 hours at 37°C or overnight at 4°C.
   d. Wash as before (b).
   e. Prepare the appropriate dilution of the specific antibodies conjugated with alkaline phosphatase (or linked with other enzyme) in PBS. Add 200 μL to each well. Incubate at 37°C for about 2 hours.
   f. Wash as before (b).

g. Prepare a fresh solution of (1 mg mL/1) p-nitrophenyl phosphate in alkaline phosphatase substrate buffer. For other enzymes, follow the recommendations of the commercial kit. Add 200 μL of this solution to each well. Incubate at room temperature and read absorbance at 405 nm (for other enzyme–substrate combinations follow the recommendation of the supplier of the kit) at regular intervals within 120 minutes, or as indicated in the protocol.

3. **DASI ELISA**

a. Prepare the appropriate dilution of polyclonal antibodies in carbonate coating buffer pH 9.6. Add 200 μL to each well of a microtiter plate with good coating characteristics and avoid the borders of the plate for better repeatability. Some techniques are available to prevent border effect such as covering the plate with transparent film to avoid evaporation, in such cases border wells can be used. Incubate at 37°C for 4 hours or overnight at 4°C.

b. Wash the wells gently (e.g., using a wash bottle) with PBS Tween as washing too strongly with some washing apparatus can affect the coating result and, consequently, the sensitivity, especially in detection.

c. Add 200 μL of each sample (plant extract, macerate, or bacterial suspension) to two wells previously enriched or treated if necessary. Include two wells per plate for both the positive and negative controls. It is also necessary to include negative controls of extraction buffer and of the enriched medium used. Incubate for 4 hours at 37°C or overnight at 4°C.

d. Wash as before (b).

e. Prepare the appropriate dilution of the specific antibodies (preferably monoclonal) in PBS pH 7.2 plus 0.5 percent bovine serum albumin (BSA) and add 200 μL to each well. Incubate at 37°C for about 2 hours and wash as before (b).

f. Prepare the appropriate dilution of the antispecies antibodies (when using monoclonal: Antimouse immunoglobulins) conjugated with alkaline phosphatase (or linked with another enzyme) in PBS. Add 200 μL to each well. Incubate at 37°C for about 2 hours.

g. Wash as before (b).

h. Prepare a solution of 1 mg mL/1 p-nitrophenyl phosphate in alkaline phosphatase substrate buffer. For other enzymes follow the recommendations of the commercial kit. Add 200 μL to each well. Incubate at room temperature and read absorbance at 405 nm (for other enzyme–substrate combinations follow the recommendation of the supplier of the kit) at regular intervals within 120 minutes, or as indicated in the protocol.

4. **Tissue print, squash, or colony dot ELISA**

a. Use a nitrocellulose membrane with tables printed by the manufacturer, or cut a rectangular nitrocellulose sheet and draw (using a ruler and pencil) a table with 12 columns and 8 rows on each. Mark the cell rows A–H and the columns 1–12. An ELISA microplate provides a good template.

   Prepare smooth, freshly cut pieces of symptomatic plant tissues and press them onto a nitrocellulose membrane to obtain a tissue-print.

   Directly squash symptomatic plant material or bacterial exudates onto the nitrocellulose membrane.

   Directly dot the nitrocellulose membrane with a colony or a suspension of the target bacterium. Include as controls positive and negative tissue prints, squashes, or pure cultures (as bacterial growth or in suspension), respectively.

   Let the tissue-prints, squashes, colony dots, or blots dry for at least 10 minutes at room temperature. The imprinted or blotted membranes can be stored in the dark at room temperature for long periods (more than one year) and are durable enough to be mailed, for example, to another laboratory.

b. Block the nitrocellulose membrane using a solution of 1 percent bovine serum albumin (BSA) in PBS pH 7.2. Add enough solution to cover the membrane in an appropriate container. The blocking step can be done overnight at 4°C for 1 hour at room temperature. Remove the BSA solution without washing.

c. Prepare the appropriate dilution of the alkaline phosphatase conjugated specific antibodies in PBS 0.01 M, pH 7.2. Add enough solution to cover the membrane in an appropriate container. Incubate for 2 hours at room temperature under slight agitation (100 rpm).

d. Remove the conjugate and subsequently wash the membrane three times with PBS-Tween, or with another appropriate buffer, with 5 minutes washing steps, while shaking (100 rpm).

e. Prepare a solution of precipitating substrate for alkaline phosphatase (e.g., NBT + BCIP in substrate buffer or distilled water, following the supplier's recommendations. Cover the membrane with this solution in an appropriate container and incubate at room temperature for 10–15 minutes. Stop the reaction by washing under tap water, dry on filter paper, and observe final purple-violet–colored precipitates using a low power (×5) magnification stereomicroscope or a magnifying glass.

30.4.7 Detection of *Xanthomonas campestris* pv. *campestris* through ELISA

Material Required

Antigen: Antigen should be prepared form Xcc culture.
Carriers: Sheep erythrocytes (SRBC).
Preservative: Alsever's solution (citrate saline solution).
Antiserum.
Reagent for ELISA. Coating buffer: 0.05 M carbonate–bicarbonate buffer
(pH 9.6); PBS (pH 7.4); wash buffer; blocking reagent (PBS containing 5 percent
(w/v) skim milk); reagents for alkaline phosphatase conjugates; substrate
buffer; reagent for horseradish peroxidase conjugates.
Substrates: AP substrate, HRPO substrate.

Antigen Preparation

Isolate the bacterium from severely infected cauliflower leaf by incubating the surface-sterilized leaf samples on nutrient agar (NA) plates for 48–72 hours at 27°C. Individual colonies appearing yellow and mucoid are picked up and repeatedly subculture for pure colonies and subjected to confirmatory tests such as pathogenicity, hypersensitivity, and biochemical tests. Culture the Xcc on nutrient broth under static conditions for 48 hours at 27°C and harvest by centrifugation at 3000 rpm for 10 minutes; cells are suspended in 20 mmol l^{-1} phosphate-buffered saline (PBS; pH 7.4) and these intact whole cells are washed three times in the same buffer and use as antigen for immunization and for the ELISA.

Sheep erythrocytes (SRBC) as carriers: Both carbohydrates and protein antigen can easily be adsorbed on to ship erythrocytes to render them agglutinable with antiserum specific for the "add-on" antigens. SRBC can easily be used as a carrier for purified carbohydrate antigens, which by themselves are normally not very immunogenic in rabbits (Diano et al., 1987). By treating the SRBC with 0.005 percent tannic acid, low molecular weight protein antigens can also be adsorbed.

Attachment of crude Lipopolysaccharide (LPS) onto SRBC:

1. Boil the overnight culture of gram-negative bacteria for 1 hour; centrifuge it at 2000 rpm for 10 minutes to sediment cell debris. The supernatant contains crude LPS antigen extract.

2. Mix 4.5 ml of bacterial supernatants extract to 6 ml of 2.5 percent SRBC at a bacterial extract-to-SRBC ration of 3:4 (v/v). Isotonic solutions such as phosphate buffered saline (PBS) or Alserver's solution should be used for all manipulations of the SRBC suspension. Incubate for 30 minutes with occasional shaking.

3. Sediment the SRBC by centrifugation at 200 × g for 10 minutes and wash the cell twice with the 10 ml saline or isotonic buffers.
4. Resuspend the pellet with 6 ml of PBS. The suspension is now ready for injection.

Alsever's solution (citrate saline solution): Alsever's solution is an isotonic, anticoagulant blood preservative that permits the storage of whole blood at refrigerated temperatures for 10 weeks or more.

Dextrose	20.50 g
Sodium citrate (dehydrate)	8.00 g
Citric acid (monohydrate)	0.55 g
Sodium chloride	4.20 g
Distilled water	to 1000 ml

Tanning of SRBC for coating with protein antigens:

1. Add 3 ml of 0.005 percent tannic acid into a centrifuge tube containing 3 ml of 2.5 percent SRBC. Incubate at 37°C for 10 minutes.
2. Centrifuge the cells at 2000 rpm for 10 minutes and wash once in 5 ml of PBS. Centrifuge as before, and resuspend the pellet in 3 ml of PBS.
3. To each centrifuge tube containing 3 ml of tanned SRBC and 3 ml of 0.3 mg/ml soluble protein, mix gently and incubate at 37°C for 15 minutes.
4. Centrifuge and wash twice as above. Resuspend each pellet in 3 ml of PBS diluents. These cells are now ready for injection.

Bacterial cells as carriers: Bacterial smooth LPS containing O-antigen sugars is extremely immunogenic, while it is often difficult to raise antibodies to the LPS-core oligosaccharide epitope of gram negative organism. To elicit a response to epitope of the core region antigens such as pure oligosaccharide, lipid A, or rough LPS can be attached to bacterial cells of a rough strain.

1. Prepare 5×10^9 cfu/ml heat-killed cells of the rough mutant per ml of 1 percent (v/v) acetic acid.
2. Heat the cell suspension to 100°C for 1 hour.
3. Wash three times in distilled water.
4. Lyophilize.
5. Dissolve lipid A or rough LPS in 0.5 percent (v/v) triethylamide at a concentration of 1 mg/ml; then add the lyophilized acid-treated bacteria to a final concentration of 1 mg/ml.
6. Stir slowly for 30 minutes at room temperature.
7. Dehydrate the mixture in vacuo with a speed-vac centrifuge.

Antiserum production: Circulating antibodies against a specific antigen do not appear in significant amounts until at least 7 days after an immunization. *Most of the antibodies of an early (or primary response are of the immunoglobulin M (IgM) class, whereas antibodies from latter (secondary) response are mostly IgG. However, if the antigen used is pure carbohydrate, the secondary response will be mainly IgM.* The amount of antibody formed after a second or booster injection of any given antigen is usually much greater than formed after the first injection. Therefore, for the production of high titer antibodies the following schedule can be used for rabbit.

1. Raise the antibody in a 3-month-old female albino rabbit.
2. Collect pre immune serum (before the first injection) for use as control to detect cross reactive antibodies.

3. Mixed bacterial cells (10^5 ml^{-1}) in a proportion of 1:1 with Freund's complete adjuvant (Robinson et al., 1996).
4. Inject 0.1 ml of cells (1×10^4 ml^{-1}) constituting approximately 100 µg protein at four sites subcutaneously.
5. On the 4th day, repeat the 1 day injection steps but with Freund's incomplete adjuvant. After this step, allow animal to rest for 14 days so that the primary response subsides to a base line level, otherwise a secondary response may not be achieved.
6. On the 18th day, inject 1.0 ml (1:1 mixture of protein antigens with Freund's incomplete adjuvant) intramuscularly into the thigh muscle of one leg of the animal.
7. On the 22nd day, bleed and collect serum.
8. On the 25th day, repeat the above step of intramuscularly injection on another leg of the animal.
9. On the 29th day, bleed and collect serum, and pool the serum with the previous sample.

The intramuscularly injections can be kept up once a month or once every 2 weeks, and serum can be collected 3 to 4 days later. To avoid batch-to-batch variability, the immune sera collected at different times should be pooled. Variations can be made at some of the steps; for example, intravenous injection, without adjuvant, may be given as a booster (on days 18 and 25) to elicit a more rapid response before bleeding to collect the antibodies.

Serum collection: Blood should be collected in a dry, sterile container without any coagulant. Routinely, blood is withdrawn from the marginal vein of the ears of the animal. A small cut (superficial venesection) to a dilated vein with the tip of a fresh scalpel blade will open it up. The droplets of blood that emerge are collected in a suitable container, usually a view cap glass bottle. With sufficient practice 20–30 ml of blood can easily be obtained. Blood withdrawn from animals should be allowed to clot at room temperature for 30 minutes. It can then be kept at 37°C for another 30 minutes. Serum can then be separated and centrifuged to get rid of all blood cells. Alternatively, the blood can be kept to clot at 4°C overnight of complete shrinkage of the clot before removing the serum. The latter procedure will yield more serum per ml of blood and will preserve antibodies from hydrolysis by the action of naturally occurring IgG proteases.

Protocol for ELISA (Joseph and Mutharia, 1994)

Antigen coating or adsorption: Make several dilutions of the antigen in the coating buffer at concentrations between 10 and 100 µg/ml for complex antigens. Mix very well. Distribute in triplicate in the vertical rows, with 50 to 100 µl of each antigen dilution per well. Cover the plates with plastic food wrap (such as Saran wrap or stretch-and-seal) and incubate at 4°C for 16 hours, adsorption of antigens onto the matrix increases with exposure time. Add the appropriate proteases inhibitors to the coating buffer when proteases are suspected in the antigen preparations. Alternatively, if the antibody reacts with the denatured antigens, boil the samples prior to adsorption.

Washing: Shake out the solution from the wells with a strong flick of the wrist. Wash the plates by directing a jet of wash buffer from a wash bottle in to each well. Wash three times with PBS-Tween, incubating for 5 minutes each time. Rinse with PBS. Shake out the entire buffer between each wash.

Blocking: Add 200 µl of 5 percent skim milk to each well. Incubate for 2 hours at 37°C or 4 hours at room temperature. Rinse three times with PBS without incubations.

Test antibody: Prepare dilutions of the test antibody in 2 percent (w/v) skim milk and add to the antigen-coated wells (in a volume equal to that of the antigen). Generally, use primary immune sera at a dilution of 1:10 and hyper immune sera at dilutions of 1:100 to 1:1500 or higher. Aliquot the serum in horizontal rows in such that each serum dilution is tested against all antigen dilutions, that is, a checkerboard titration. Incubate as in Step 2 and repeat the washes.

Second antibody: Make the dilution of the labeled (conjugated) second antibody in the 2 percent skim milk. Repeat the incubation and washing.

Substrate: Prepare the appropriate substrates for the conjugate used and add to the wells in volumes equal to that of the coating antigen. Incubate the plates at 37°C and take the readings (at the appropriate wavelength for the substrate) at 30, 45, and 60 minutes.

Data analysis: Plot the percent absorbance (calculated as a percentage of the maximum reading) against the antigen concentrations. From the graph, calculate the lowest antigen concentration that will give 50 percent of the maximum reading for the time point. This is antigen concentration to use for the subsequent ELISA. The antibody titre at that antigen concentration can then be used in positive control.

Notes: It is important that each plate contain the following control wells:

An antigen blank
A first antibody blank
Homologous and non-related antigen control
The antigen deficient control is used to zero the plate reader. A checkerboard ELISA can also be used to establish the optimum concentrations of conjugates and antibody. In each assay only one parameter is varied at a time. Incubation can also be carried out at 37°C for 2 hours or at room temperature for 4 hours.
Reagent for ELISA
1. Coating buffer: 0.05 M a carbonate–bicarbonate buffer (pH 9.6)

Solution A: Sodium carbonate	21.2 g/L
Solution B: Sodium bicarbonate	16.8 g/L

To prepare the working buffer, mix 20 ml of Solution A and 42.5 ml of Solution B and adjust to 250 ml with distilled water. The pH should be 9.6. Add 0.05 g of NaN_3. Store at 4°C in a dark bottle for up to 4 weeks.
2. PBS (pH 7.4):

Solution A: $NaH_2PO_4 . 2H_2O$:	31.2 g/L
Solution B: Na_2HPO_4:	28.39 g/L
$Na_2HPO_4 . 2H_2O$:	71.7 g/L

The working buffer is prepared as follows: Mix 47.50 ml of Solution A and 202.50 ml of Solution B; adjust to 800 ml with distilled water. Add 8.75 g of NaCl and make up to 1000 ml. The pH should be 7.4.
3. Wash buffer: PBS containing 0.05% Tween 20.
4. Blocking reagent: PBS containing 5 percent (w/v) skim milk. This should be made fresh every day.
5. Reagents for alkaline phosphatase conjugates:
Substrate buffer: Sodium carbonate (0.05 M; pH 9.8) containing 10mM $MgCl_2$. To prepare the substrate buffer, use the buffer stocks made for the coating buffer (Step 1). Mix 27.5 ml of Solution A with 35 ml of Solution B and make up to 250 ml with distilled water. Add 0.05 g of $MgCl_2.2H_2O$. Check the pH before use. Store in a dark bottle at 4°C.
Alternatively use 1 M diethanolamine buffer pH 9.8 containing 0.5 mM $MgCl_2$.
AP substrate: Dissolve in the buffer 1 mg of p-nitrophenyl phosphate per ml. The substrate can be obtained from Sigma in the form of 5 mg tablets. Store substrates at 20°C.
Read the yellow color at 414 nm. Color development can be stopped by addition of 10 μl of 3 M NaOH to the wells.
6. Reagent for horseradish peroxidase conjugates:
Substrate buffer: Make citric acid buffer (pH 4.0) by dissolving 0.2 g of citric acid in 90 ml distilled water. Adjust the pH 4.0 with 1 M NaOH. Store at room temperature.

HRPO substrate: Immediately before use, add to 10 ml of citric acid buffer, 5 μl of 30 percent H_2O_2 and 75 μl of 10 mg/ml [2,2-azino-bis (ethylbenzthiazoline)-6- sulfonic acid. Read the green color at 414 nm. Color development can be stopped by addition of 0.08 M NaF to the wells.

30.4.8 DETERMINATION OF THE ANTIGENIC DETERMINANT IN BACTERIAL PATHOGEN BY IMMUNOBLOT ASSAY

As Xcc positive and specific in ELISA reactivity, an immunoblot assay is performed to identify the antigenic determinant molecule in the pathogen.

Material Required

Nitrocellulose paper, PAb, Xcc, alkaline phosphatase-conjugated swine anti-rabbit IgG chromogenic substrate (5-Bromo-4-chloro-3-indolyl phosphate), alkaline phosphatase.

Protocol

An equal number of bacterial cells should be loaded in duplicate onto 12 percent SDS-PAGE in a mini-gel apparatus (Genei Pvt, Bangalore, India). After electrophoresis, the first half with the marker proteins is silver stained for visualization of the protein bands and the second half of the gel transfer to a nitrocellulose membrane. The nitrocellulose paper should be probed with 1:500 PAb-Xcc followed by 1:1000 alkaline phosphatase-conjugated swine anti-rabbit 1gG. 5-Bromo-4-chloro3-indolyl phosphate is used as the chromogenic substrate, which, on alkaline phosphatase activity, yielded a reduced, stable purple colored complex with a band, which react specifically with PAb-Xcc.

Purification of PAb-Xcc by Sepharcryl-200-HR Column Chromatography

Precipitate the antiserum with 50 percent (w/v) ammonium sulfate and load onto a degassed column packed with Sepharcryl-200-HR (65 × 1.0 cm) 1-ml fractions are collected at a flow rate of 18 ml h^{-1} and fractions monitored at A280 nm spectrophotometrically (U-2000; Shimadzu, Japan). Subsequent peak fractions are pooled as peak 1, PlAb-Xcc, and Peak 2, P2AB-Xcc, and examined for antibody reactivity by ELISA. After lyophilization, they are evaluated for antibody titer, sensitivity, and detection of cells as well as culture filtrates. Crude antibody is also included in the experiment. Appropriate controls, such as detection with pre-immune serum and irrelevant xanthomnads are prepared. The two peaks shows differential reactivity with virulent and avirulent isolates in ELISA.

30.5 USE OF SPECIFIC IMMUNOGLOBULIN UNDER IMMUNOFLUORESCENCE TO DETECT THE LATENT INFECTION

The procedure described by Borkar et al. (1989c) is for *Xanthomonas corylina* strain K-100 and is as follows:

1. Preparation of antigen glycoprotein: Use the test bacterium for the preparation of antigen glycoprotein. Prepare the antigen glycoprotein of test bacterium as described in earlier exercise.
2. Production and extraction of immunoglobulin: Carry out the production and extraction of immunoglobulin as described earlier. Follow the immunization schedule of nine weeks for production and extraction of immunoglobulin. Estimate the presence of effective immunoglobulin content in the antiserum by indirect method of immunofluorescence for the given/test bacterium as described earlier.
3. Extraction of bacterium specific immunoglobulin by affinity. Chromatography: Three ml antigen glycoprotein solution (4.5 mg protein/ml) of test bacterium is adjusted to pH 8 with

1 M Na_2HPO_4 buffer (1M Na_2HPO_4, 483 ml + 1M NaH_2PO_4, 17 ml = pH 8 buffer) and mix with 0.8 ml p-benzoquinone solution (30 mg/ml in absolute cold alcohol). This mixture is agitated for activation in dark chamber at 4°C for 1 hour. The activated glycoprotein is precipitated thrice with 6 ml of cold ethanol (by subsequent centrifugation at 4000 rpm for 8 minutes). The precipitated glycoprotein is dissolved in 8 ml $NaHCo_3$ (0.1 M, pH 8.2) and mixed with 4 g gel saphrose 6 B overnight at 4°C on an agitator. Put this mixture in a column (B. P. 16 mm × 130). The filled column is washed with 0.1 M $Na_2 HPO_4$ buffer pH 7 (161 ml Na_2HPO_4 (0.5 M) + 89 ml NaH_2PO_4 (0.5 M) and ultrapure water is added to make vol 11 to make pH 7.0) to adjust the pH (7.0) of the glycoprotein in the column. The column is further washed with HCl–glycerin buffer pH 2.7 (HCl 0.1 N, 700 ml + glycerin, 300 ml (15.014 g/1)) to eliminate the non-fixed glycoprotein (3 vol of HCl–glycerin buffer is required at a speed of 25 ml/hour of flow regulator to eliminate the non-fixed glycoprotein). The non-fixed glycoprotein is collected during washing of column with two buffers. After the collection of non-fixed glycoprotein, the pH of the column is adjusted to 7 (with 0.01 M Na_2HPO_4 buffer) and 3 ml antiserum of test bacterium is pass into the gel for fixation of specific immunoglobulin to the fixed glycoprotein in the gel. The gel is washed with 0.01 M Na_2HPO_4 buffer (pH 7) to elute nonspecific non-fixed immunoglobulin. When the peak of non-fixed immunoglobulin is over, the coulomb is washed with HCl–glycerin buffer (pH 2.5) to elute specific fixed immunoglobulins. Specific immunoglobulins is collected in 1M K_2HPO_4 solution (pH 7) to avoid denaturation (generally 3 ml K_2HPO_4 solution is required for 2 ml of immunoglobulin solution collected from column). pH of immunoglobulin is adjusted to pH 7 with 0.01 M p.Na/Na_2 buffer. This immunoglobulin solution is concentrated by immersing a dialysis tube containing specific immunoglobulin into 20 percent polyethylene glycol (generally 2 hours are sufficient to absorb 10 ml of water from dialysis tube) and the protein is estimated by UV absorption at 280 nm from concentrated specific immunoglobulin solution.

4. Determination of specific immunoglobulins and their effective dilution to be used in the detection of latent infection: Make serial dilution of log phase suspension (10^7 C.F.U/ml) of test bacterium. Add 10 µl suspension of each dilution (10^{-2} to 10^{-7}) in the well of immunofluorescence slide and dry at 26°C. This is fixed with alcohol for 10 minutes. Add 10 µl specific immunoglobulin of various dilution (1/100 to 1/6400) separately in each well and allow to react for 30 minutes. Excess of immunoglobulin is flooded twice with phosphate buffer (NaCl, 8 g; Na_2HPO_4, 2.7 g; NaH_2PO_4, 2.4 g; distilled water, 1.1; pH 7.2) for 5 minutes each. Add 10 µl of fluorescein conjugated goat anti-rabbit immunoglobulin at 1/300 dilution to each well and allowed to react for 30 minutes. Remove the excess of fluorescein conjugated by washing the slide with phosphate buffer thrice for 3 minutes and dry with blotter paper. Add A drop of glycerin buffer on the slide and observe under oil immersion in immunofluorescence light.

5. Detection of latent infection of test bacterium from healthy buds/leaves of host plant by using specific immunoglobulin: Do the sampling from healthy buds, young leaves and twigs from natural host of test bacterium. Macerate 25 g sample of each in 250 ml of distilled water and centrifuged at 4000 rpm for 20 minutes. Collect the supernatant and filtered through Whatman filter paper no. 42. Collect the filtrate and make serial dilution. Put 10 µl solution of serial dilution (10^{-2} to 10^{-7}) of each sample separately in the well of immunofluorescence slide and dry at 26°C. This is fixed with alcohol for 10 min. Add 10 µl specific immunoglobulin of 1/400 dilution in each well and allow to react for 30 min. Excess of immunoglobulin is flooded twice with phosphate buffer for 5 minutes each. Add 10 µl of fluorescein conjugated goat anti-rabbit immunoglobulin at 1/300 dilution and allowed to react for 30 minutes. Excess of fluorescein conjugated is removed by washing the slide with phosphate buffer thrice for 3 minutes and dry with blotter paper. Add a drop of glycerin buffer on the slide and observed under oil immersion in immunofluorescence light.

31 Identification of Bacteria by Using Molecular Techniques

Identification of various strains of plant pathogens has traditionally been based upon microscopic examination of morphological and biochemical characters and growth characteristics of the pathogen on specific media. Variations in the morphological characters exist and lead to difficulties in accurate identification by traditional methods. In addition, identification based on pathogenicity assays or growth characteristics are time-consuming. Accurate and rapid characterization is important for precise diagnosis of related plant pathogens and understanding of the population structure and genetics of the pathogen for several reasons.

Molecular markers, on the other hand, are highly reliable for cataloging variability in plant pathogens since they are not related to biologically important characteristics. Molecular techniques have also been used to study genetic diversity and evolutionary origins in populations of many different genera. Molecular detection can generate rapid accurate results enough to be useful for disease management decisions.

In plant pathogenic bacteria the variation exits at molecular or genetic level in a given pathovar itself and therefore the races exit in the given pathovar of the bacterium, which cannot be identified by cultural, physiological, or biochemical characteristics. To know this variability the molecular techniques are most important and appropriate tools. The knowledge of the pathogenic variability is important for the development of crop varieties resistant to that pathogen. To study the molecular variability the molecular techniques like hybridization-based techniques, PCR-based techniques, and guanine-plus-cytosine-ratio–based techniques are generally used. This chapter deals with all these techniques along with other molecular based techniques which can be employed in the study of plant pathogenic bacteria.

31.1 ISOLATION OF BACTERIAL DNA

31.1.1 ISOLATION OF BACTERIAL GENOMIC DNA

Material Required

Test bacterium, sodium dodecyl sulfate, proteinase-k, CTAB–NaCl solution, chloroform, isoamyl alcohol, phenol, isopropanol, ethanol, TE butter, centrifuge, and so on.

Protocol

Centrifuge the bacterial culture at 5000 rpm for 5 minutes at room temperature. Collect the pellet by discarding culture medium completely. Add 10 percent sodium dodecyl sulfate and 20 mg/ml proteinase-K to a final concentration of 10 μg/ml proteinase-K in 0.5 percent SDS. Mix thoroughly and incubate for 1 hour at 37°C. Add 2.5 volume of CTAB–NaCl solution. Mix thoroughly and incubate for 10 minutes at 65°C. Add an approximately equal volume of chloroform: isoamyl alcohol (mixed in equal ratio), mix thoroughly, and spin for 5 minutes in a microcentrifuge at room temperature. Collect aqueous phase to a fresh microcentrifuge tube. Add an equal volume of phenol: chloroform: isoamyl alcohol (in 25:24:1 ratio), and spin for 5 minutes.

Transfer the supernatant to a fresh tube and add 0.7 volume of isopropanol to precipitate the genomic DNA. Spin it for 10 minutes at room temperature. Wash the DNA with chilled 70 percent ethanol to remove residual salts and spin again for 2 minutes at room temperature without disturbing the pellet. Redissolve the pellet in suitable volume of TE buffer (10mMTris + 1mM EDTA) or sterile distilled water and quantify it using spectrophotometer at 260 and 280 nm wavelengths. The extracted genomic DNA is quantified accordingly, diluted to get an equimolar working concentration of 40–100 ng/μl, which is used as template for PCR amplification.

31.1.2 ISOLATION OF BACTERIAL PLASMID DNA

The method includes lysis of bacterial cell by suitable detergent (SDS). Exposure of bacterial suspension to the strongly ionic detergent at high pH causes rupturing of cell wall, the genomic DNA breaks into pieces (being attached to cell wall) and denaturation of proteins. The alkali (NaOH) used causes the denaturation of both plasmid and chromosomal DNA. Plasmid being covalently closed strands remains interwoven even after denaturation, and makes renaturation easy when the conditions revert back to normal after adding neutralization solution. Bacterial proteins, broken cell wall, and denatured chromosomal DNA become enmeshed in large complex that are coated with dodecyl sulfate. These complexes of protein debris along with genomic DNA are efficiently pelleted from solution by centrifugation. Further, the native plasmid DNA can be recovered from the supernatant by precipitating with two volume of ethanol in the presence of 0.3 M potassium acetate, followed by incubation in deep freezer and centrifugation.

Materials Required

Buffers and Solutions: Stock solutions. 50 percent glucose (200 ml). Dissolve 100 g of glucose (dextrose) in 150 ml of distilled water. Make up to 200 ml and autoclave.

1 M Tris–HCl pH 8.0 (500 ml): Dissolve 60.55 g of Tris in 400 ml of distilled water. Adjust the pH to 7.5 with concentrated HCl. Make up the volume to 500 ml.

0.5 M EDTA pH 8.0 (100 ml): Dissolve 16.8 g of anhydrous disodium salt in distilled water. Adjust the pH to 8.0 using NaOH pellets. More EDTA will dissolve as the pH approaches 8.0. Make up the volume to 100 ml and autoclave.

10 percent SDS (100 ml) (w/v): Dissolve 10 g in 100 ml distilled water. Heat gently to get SDS into solution.

1 N sodium hydroxide (100 ml): Dissolve 4 g in 100 ml distilled water.

5 M potassium acetate (500 ml): Dissolve 245.5 g in 400 ml distilled water. Adjust the pH to 5.5 with glacial acetic acid. Make up the volume to 500 ml.

Solution I (100 ml)

50 mM glucose:	1.8 ml from 50 percent glucose
25mM Tris –HCl pH 8.0:	2.5 ml from 1N Tris–HCl, pH 8.0
10mM EDTA pH 8.0:	2.0 ml from 0.5M EDTA pH 8.0
Water:	93.7 ml

Solution II (100 ml): Solution II should be freshly prepared and used at room temperature.

1 percent SDS:	10 ml from 10 percent SDS
0.2 N NaOH:	20 ml from 1 N NaOH
Water:	70 ml

Solution III (100 ml)

5 M Potassium Acetate:	60 ml from 5M potassium acetate
Glacial acetic acid:	11.5 ml
Water:	28.5 ml

(Continued)

Tris–acetic acid–EDTA buffer (TE) (100 ml)

10 mM Tris–HCL, pH 8.0:	1 ml from 1M Tris HCl, pH 8.0
1mM EDTA:	0.5 ml from 0.5 M EDTA pH 8.0
Water:	98.5 ml

RNase can be added to TE at final concentration of 20 µg/ml.

Ethanol.

Phenol:chloroform (1:1, v/v).

Media.

10 g/l Tryptone.

5 g/l Yeast extract.

10 g/l NaCl.

Protocol

Preparation of cells: Inoculate 2 ml of LB medium containing the appropriate antibiotic with a single colony of transformed bacteria. Incubate the culture overnight at 27°C with vigorous shaking to ensure that the culture is adequately aerated: The volume of the culture tube should be at least four times greater than the volume of the bacterial culture. The tube should be loosely capped. The culture should be incubated with vigorous agitation. Pour 1.5 ml of the culture into a microfuge tube. Centrifuge at maximum speed for 30 seconds at 4°C in microfuge. Store the unused portion of the original culture at 4°C. When centrifugation is complete, remove the medium, leaving the bacterial pellet as dry as possible.

Lysis of cells: Resuspend the bacterial pellet in 100 µl of ice-cold alkaline lysis solution I by vigorous vortexing. Make sure that the bacterial pellet is completely dispersed in alkaline lysis solution I. Add 200 µl of freshly prepared alkaline lysis solution II to each bacterial suspension. Close the tube tightly, and mix the contents by inverting the tube rapidly five times. Do not vortex. Store the tube on ice. Make sure that the entire surface of the tube comes in contact with alkaline lysis solution II.

Add 150 µl of ice-cold alkaline lysis solution III. Close the tube and disperse alkaline lysis solution III through the viscous bacterial lysate by inverting the tube several times. Store the tube on ice for 3–5 minutes.

Centrifuge the bacterial lysate at maximum speed for 5 minutes at 4°C in a microfuge. Transfer the supernatant to a fresh tube.

Optional: Add an equal volume of phenol:chloroform. Mix the organic and aqueous phases by vortexing and then centrifuge the emulsion at maximum speed for 2 minutes at 4°C in a microfuge. Transfer the aqueous upper layer to a fresh tube.

Recovery of plasmid DNA: Precipitate nucleic acids from the supernatant by adding 2 volumes of ethanol at room temperature. Mix the solution by vortexing and then allow the mixture to stand for 2 minutes at room temperature.

Collect the precipitated nucleic acids by centrifugation at maximum speed for 5 minutes at 4°C in a microfuge.

Remove the supernatant. Stand the tube in an inverted position on a paper towel to allow all of the fluid to drain away.

Add 1 ml of 70 percent ethanol to the pellet and invert the closed tube several times. Recover the DNA by centrifugation at maximum speed for 2 minutes at 4°C in a microfuge. Again remove all of the supernatant. Take care with this step, as the pellet sometimes does not adhere tightly to the tube.

Remove any beads of ethanol that form on the sides of the tube. Store the open tube at room temperature until the ethanol has evaporated and no fluid is visible in the tube (5–10 minutes).

Dissolve the nucleic acids in 50 μl of TE (pH 8.0). Vortex the solution gently for a few seconds. Store the DNA solution at –20°C.

Troubleshooting Plasmid DNA Isolation

Symptoms	Possible Causes	Comments
Low plasmid yield	Bacterial strain is endA	When working with bacterial strains containing the wild-type endA gene, low yields can result due to nuclease digestion of the plasmid, especially in the presence of co factors such as Mg^{2+}. Consult the genotype of your bacterial strain and, if possible, switch to a strain which has an inactive end A gene.
	Plasmid copy number is low	Plasmid yield is directly related to plasmid copy number.
	Cells did not lyse completely	Do not exceed the recommended culture volumes for the procedure. If the cells are not properly lysed, the plasmid DNA is not released and is pelleted along with the cell debris. After pelleting the cells from the bacterial culture, be sure to disperse the cell clumps completely during the resuspension step. At the lysis step, the solution should clear (some particulates may be present) such that no cell mass is visible.
	Old antibiotic was used	It is critical to use fresh antibiotic during overnight growth of cultures and when preparing plates. Most antibiotics are sensitive to light and also degrade during long term storage at 4°C.
Disappearing DNA	Bacterial strain is end A positive	See comments above under "low plasmid yield."
	DNA quantification was done by spectrophotometric analysis, not gel analysis	To accurately determine yields, the DNA should be serially diluted and analyzed on an agarose gel with appropriate mass standards immediately after isolation of the DNA. Spectrophotometric determinations tend to overestimate yields due to the presence of substances in the DNA preparation which will absorb at 260 nm.
DNA cannot be digested or sequenced	Volume of DNA added to the reaction was too high	When setting up an enzymatic reaction, the volume of DNA added should be minimal compared to the total reaction volume. Even if present in minimal amounts, contaminants in the DNA preparation can be carried over to the reaction in sufficient quantities to inhibit enzymes. Minimizing the DNA volume minimizes the inhibitory effect. DNA can be concentrated by alcohol precipitation, followed by a 70 percent ethanol wash.
Chromosomal DNA contamination	Sample was over vortexed during alkaline lysis preparation	Overvortexing the preparation after addition of the lysis buffer should be avoided, since the bacterial chromosomal DNA is susceptible to shearing. This sheared DNA renatures during the neutralization step along with plasmid DNA, then binds to the resin and elutes with the plasmid DNA.
DNA contamination	Isopropanol volume was too high or cold isopropanol was used to precipitate the plasmid DNA	Increasing the proportion of isopropanol can result in increased RNA co precipitation. Using room temperature isopropanol and centrifugation will decrease the amount of co-precipitated RNA.

31.2 PLASMID PROFILING BASED TECHNIQUES

Plasmids are nonessential extra-chromosomal circular DNA molecules present in prokaryotes which provides them additional advantage such as antimicrobial resistance or virulence genes. They replicate autonomously and have size range varying from a few to several hundred kilo base pairs (kb).

31.2.1 PLASMID DNA RESTRICTION ANALYSIS

Principle

Most plant pathogenic bacteria are known to harbor plasmids carrying genes responsible for virulence, antibiotics resistance, and exopolysaccharide production (Sundin, 2007). These plasmids are often large in size, making it difficult to characterize the plasmids by simple agarose gel electrophoresis. By plasmid DNA restriction analysis, a large plasmid DNA is reduced to series of smaller fragments of defined size on digestion with restriction end nuclease. Restriction end nucleases are bacterial enzymes that bind and cleave DNA at specific target sequences. They bind DNA at a specific recognition site, consisting of a short palindromic sequence, and cleave within this site yielding either sticky or blunt-ended cleavage products. The number of fragments thus produced and size of each fragment reflect the distribution of restriction sites and the total size of the plasmid DNA. The fragment thus produced will be specific for each target plasmid DNA/restriction enzyme combination and can be used as fingerprint-specific for a given target plasmid DNA. The plasmid DNA fingerprinting reveals higher polymorphism and is quicker than RFLP analyses of genomic DNA.

Plasmid profiling procedure can simultaneously targets pathogenicity associated loci therefore this method is sensitive in detecting pathogen. Plasmid profiling also has proved their role in the following studies such as diversity, epidemiology, differentiation, host–pathogen interaction, virulence, and antibiotic resistance/susceptibility. Algeria et al. (2005) observed that *Xanthomonas axonopodis* pv. *citri* contains two *virB* gene clusters, one on the chromosome and another one on a 64-kb plasmid, each of which codes for a previously uncharacterized type IV secretion system.

Indigenous mega plasmids have been reported from various species of *Xanthomonas* with sizes ranging from 7 to 100 kb (Pruvost et al., 1992). Kale et al. (2012) observed that on plasmid profiling of *Xanthomonas axonopodis* pv. *punicae* revealed that the approximate composite size of these plasmids is in between 94 to 96 kb, which coincides with that of reports of two plasmids, that is, ~100 kb (60–65 kb and 34–40 kb) from other *Xanthomonas axonopodis* pathovars (Satyanarayana and Verma, 1993; da Silva et al., 2002). Satyanarayana and Verma (1993) found that highly virulent races of *X. axonopodis* pv. *malvacearum* harbor at least five plasmids, among which two large plasmids of 60 kb and 40 kb size were common. There was greater variability in smaller plasmids that were likely involved in extra aggressiveness. *X. axonopodis* pv. *citri* has two plasmids (64.9 kb and 33.7 kb) (da Silva et al., 2002); *X. axonopodis* pv. *vesicatoria* 85–10 that causes bacterial spot of pepper and tomato have 4 plasmids (182 kb; 38 kb; 19 kb, and 1.9 kb); while in *X. oryzae* pv. *oryzae* no plasmids were observed (Lee et al., 2005).

Material and Reagents Required

Agarose; bovine serum albumin (BSA; 10 mg/ml); dithiothreitol, 20 × (DTT; 20 mM); ethidium bromide (10 mg/ml); ice; HindIII size markers, concentration provided by the manufacturer (~0.5 µg/ml); microcentrifuge tubes (0.5 ml); PCR reaction samples; pRY 121 plasmid DNA from maxi-prep; pRY 121 plasmid DNA from mini-prep; restriction enzyme reaction buffers, 10 × (manufacturers will supply a recommended buffer with the enzyme).

Restriction enzymes: HindIII, *Pst*I, *Bam*HI, *Eco*RI (in glycerol).

Stop buffer: 200 mM EDTA (pH 8.0). TAE buffer (25×: 1M Tris, 15 mM EDTA, 125 mM sodium acetate (pH 7.8). To 750 ml of distilled H_2O add 121 g of Tris base, 10.2 g of sodium acetate, and

18.6 g of EDTA. Adjust the pH to 7.8 with glacial acetic acid, and bring the volume to 1 L. Store at 4°C. Tracking dye, 6 × (for 100 ml of a 6 × stock solution, add 40 g of sucrose to 60 ml of distilled H$_2$O. Add 250 mg of bromophenol blue. Mix by stirring for 15–20 minutes. Store at 4°C. This 6 × dye solution is added at one-sixth of the final volume of the DNA solution to be electrophoresed).

Equipment and plastic ware: Erlenmeyer flasks (250 ml), gloves, laboratory tape, micropipette and tips, Pasteur pipette, Polaroid film (type 667), plastic wrap, Tupperware containers (or equivalent) for staining, heat block at 65°C, horizontal gel electrophoresis apparatus, microcentrifuge, microwave oven or Bunsen burner, ultraviolet (UV) transilluminator, water bath.

Protocol

Guidelines for preparation of plasmid DNA for restriction digestion: Determine the quantity of plasmid DNA to be digested (i.e., one μg). On the basis of the stock concentration (units per microliter) of restriction endonuclease, calculate the microliters of enzyme required to achieve digestion of the desired quantity of DNA. For this experiment, 2 units of enzyme is sufficient to digest 1 μg of plasmid DNA. (*Note*: 6 units/μg is the rule of thumb for digestion of chromosomal DNA.) As use of excess enzyme ensures complete digestion. Determine the total volume of the digestion reaction to attain an appropriate volume in which the glycerol content in the reaction will be less than or equal to 5 percent (v/v). Usually restriction nucleases are supplied in 50 percent (v/v) glycerol so that a 1:10 dilution of enzyme in the final reaction mixture would suffice. For some enzymes bovine serum albumin (BSA) is required as a stabilizer for optimal digestion. It is prudent to check whether an enzyme requires BSA or not prior to digestion.

Calculate the amounts of the ingredients to be used in each of the following restriction digests: (1) pRY121 (mini-prep) cut with *Bam*HI, (2) pRY121 (maxi-prep) cut with *Hind*III, (3) pRY121 (maxi-prep) cut with *Eco*RI, (4) pRY121 (maxi-prep) cut with *Pst*I, and (5) pRY121 (maxi-prep) cut with *Bam*HI. (See Table 31.1.)

Use a micropipette to dispense ingredients into a 0.5-ml microcentrifuge tube. Use a fresh sterile tip for each ingredient to avoid cross-contamination. The enzyme should be at –20°C until use and *should be added last*. All other ingredients should be kept on ice. The enzyme storage buffer is very viscous and will stick to the outside of the micropipette tip. To ensure accuracy, the very tip of the pipette should just touch the surface when drawing up the enzyme solution. Mix the digest by gently flicking the microcentrifuge tube. Do not vortex. Pellet any droplets by spinning the tube for 3 seconds in a microcentrifuge at top speed. Incubate the digests at 37°C in a water bath for 1 hour. Remove the reaction tubes from the water bath and pellet the condensation in the tubes by centrifuging for 3 seconds at top speed in a microcentrifuge.

This step is optional. Add 1 μl of 200 mM EDTA (pH 8.0) stop buffer. The inactivation of restriction enzymes by chelation of metal cofactors prevents anomalous effects due to overexposure of DNA to active enzyme. For some restriction end nucleases other inactivating steps (i.e., heating at 65°C for 10 minutes) are recommended. Prepare a molecular weight standard by adding 1 μl of (0.5 μg/ml) stock to 19 μl of distilled water. Add 4 μl of 6 × tracking dye solution to all samples. Spin the tubes for 3 seconds at top speed in a microcentrifuge. Heat the samples at 65°C in a heat block for

TABLE 31.1

Preparation of Reaction Mixture for DNA Digest by a Restriction Endonuclease

Tube	Concentration of DNA	DNA Solution (1 μg of DNA)	Concentration of Enzyme	Enzyme Solution (2 U)	10× Buffer	10× BSA	Distilled Water	Total Volume
1	0.5 μg/μl	2 μl	*Eco*RI, (10U/μl)	1 μl of a 1:5 dilution	2 μl (medium)	2 μl	13 μl	20 μl

Note: Restriction enzymes may be diluted in 1× reaction buffer (not water).

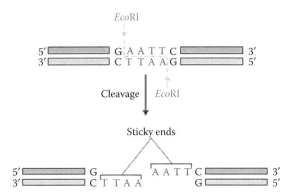

FIGURE 31.1 Cleavage of DNA by the restriction enzyme *Eco*RI.

two minutes to dissociate DNA aggregates. Run the restriction digested samples on 0.8 percent Agarose gel. (See Figure 31.1.)

31.2.2 REPLICON TYPING

Principle of PCR-based replicon typing (Carattoli et al., 2005): Plasmids are autonomously replicating circular extra-chromosomal DNA in prokaryotes. The plasmid replication system, which dictates the plasmid's behavior (host range, copy number) is used for plasmid classification and identification. Plasmids are originally classified in incompatibility (Inc) groups based on commonality of replication controls (Novick, 1987).

The standard procedure for determining Inc groups is laborious requiring multiple conjugation or transformation assays (Datta and Hedges, 1971). This methodology cannot be easily applied to a large number of strains as their application has been limited by the laborious and time-consuming work required. If the resident plasmid is eliminated in the progeny, the incoming plasmid is assigned to its same Inc group. Plasmids with the same replication control are "incompatible," whereas plasmids with different replication controls are "compatible." On this basis two plasmids belonging to the same Inc group cannot be propagated in the same cell line (Datta and Hughes, 1983; Couturier et al., 1988).

Inc group identification has been frequently used to classify plasmids. In 1988, Couturier et al. developed a new method recognizing the FIA, FIB, FIC, HI1, HI2, I1-Ig, L/M, N, P, W, T, A/C, K, B/O, X, Y, F, and FIIA replicons, representative of the major plasmid incompatibility groups circulating among the *Enterobacteriaceae*. They tested the specificity of the method on 61 reference plasmids by PCR-based inc/rep typing. A PCR-based detection of plasmids has also been devised on the basis of published sequences but it is limited to the IncP, IncN, IncW, and IncQ plasmids (Gotz et al., 1996).

PCR-based Inc/rep typing method: Template plasmid DNA is prepared by the method described previously. It requires approximately 200 ng total DNA per reaction.

Specific primer pairs are designed for the different replicons using software Primer3 based on published sequences of various Gen Bank accessions (see http://www.ncbi.nlm.nih.gov). These primer pairs recognize either the coding sequence of the replicase *repA* genes, or the cis-repeats of the *ori* site (origin of replication) or the counter transcript RNAI; or the *parA–parB* genes, which control plasmid partitioning; or the replicons of virulence plasmids.

The PCR-based Inc/rep typing method consists of combination of different multiplex-PCRs recognizing different replicon types as well as simplex-PCRs of each replicon type. All PCR amplifications are performed with the following amplification scheme: 1 cycle of denaturation at 94°C for 5 minutes, followed by 30 cycles of denaturation at 94°C for 1 minute, annealing at 60°C for 30 seconds and elongation at 72°C for 1 minute. The amplification is concluded with an extension program of 1 cycle at 72°C for 5 minute. Results of the Inc/rep PCR-based method on plasmids of known incompatibility groups are to be noted.

31.3   HYBRIDIZATION-BASED TECHNIQUES

The basic concept of a target complementary probe that signals target presence in the case of hybridization is a widely used target visualization technique. The classic techniques Southern and Northern blot transferred gel-electrophoresis fragments to a membrane filter with target sequence-specific labeled probes for DNA and RNA detection, respectively. Similarly, Western blot technique has been developed for detection of proteins blotted using labeled probes showing affinity for targeted proteins such as complementary antibody.

31.3.1   SOUTHERN BLOTTING TECHNIQUE

Southern blotting is a technique used to identify specific DNA sequences. Usually, larger fragments of DNA (genomic DNA) are made into conveniently small-sized fragments by restriction analysis. Then they are run through an agarose gel. The separated fragments in the gel are transferred to a nylon membrane by capillary transfer or vacuum blotting. The sequences of interest can be identified by hybridization using radioactive or non-radioactive probes and visualized by autoradiography or staining.

Materials

Reagents: Depurination solution (500 ml)	
0.25 N HCl:	10.775 ml
Make up to 500 ml with distilled water.	
Denaturation buffer (500 ml)	
1.5 M NaCl:	43.85 g
0.5 M NaOH:	10.0 g
Dissolve in 500 ml of distilled water.	
Neutralization buffer (500 ml)	
1 M Tris-HCl:	60.55 g
1.5 M NaCl:	43.85 g
Dissolve in 400 ml of distilled water.	

Adjust the pH to 7.4 with HCl. Make up the volume to 500 ml.

20× SSC (Sodium saline citrate)	
3 M NaCl:	175.3 g
0.3 M sodium citrate:	88.2 g
Dissolve in 800 ml of distilled water.	

Adjust the pH to 7.0 with NaOH. Make up the volume to 1 L.

Protocol

Preparation of gel: Cleave the genomic DNA with restriction endonuclease into thousands of fragments. The fragments are separated according to size by gel electrophoresis. Distains the agarose gel in sufficient amount of water with shaking under room temperature. Carefully transfer the gel to 0.2 M HCl and shake for 10 minutes at room temperature.

DNA fragments larger than 10 kb do not transfer to blotting membrane efficiently. In order to facilitate their transfer, these fragments are reduced in size, either by depurination or UV irradiation. During this period the color of the marker dye (bromophenol blue) will change from blue to yellow, indicating that the gel has been completely saturated with the acid. This depurination step should not be too long, since small fragments attach less firmly with the membrane. Depurination is recommended only for fragments larger than 10 kb or otherwise these yield fuzzy bands or smears in the final autoradiograph, presumably because of increased diffusion of DNA during transfer. (See Figure 31.2.)

Denature the double stranded DNA in order to create suitable hybridization targets by incubating the gel with denaturation buffer containing 0.5 M NaOH and 1.5 M NaCl for 15–30 minutes with gentle shaking. (During denaturation the bromophenol blue will regain its color.) Incubate the gel with neutralization solution containing 1.5 M NaCl and 0.5 M Tris-HCl pH 7.5 for 30 minutes. Set up a Southern transfer as in Figure 31.3.

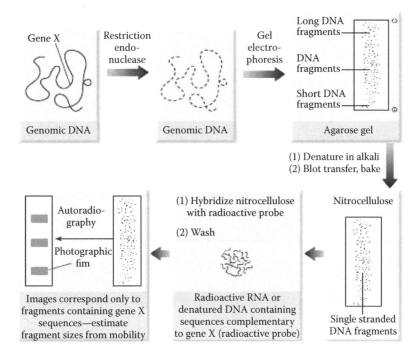

FIGURE 31.2 Southern blotting technique.

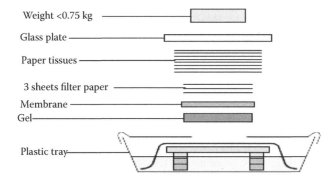

FIGURE 31.3 Southern blotting transfer.

Place the gel on the platform. Remove any air bubbles trapped between the gel and the plat-form by rolling a pipette several times back and forth over the gel. Surround the gel with plastic wrap. This ensures that the transfer buffer (10× SSC) moves only through the gel and not around it. Place the blotting membrane (nylon membrane) on the top of the gel so that it covers the entire surface. Remove any air bubbles between the membrane and gel as described earlier. Place three precut sheets of Whatman 3 mm paper in 10× SSC, and place them on the top of the nylon membrane. Again remove any trapped air bubbles as described.

Place a 15–20 cm stack of dry paper towels or paper tissues on top of the filter paper. (Transfer efficiency is improved by removing the wet paper towels and replacing them with dry papers at least once during transfer.) Place a glass plate on the top of the paper towels. Place a weight (about 500 g) on top of the plate. Allow the transfer for 12–18 hours.

Fixing the DNA: Immediately after transfer, remove the weight, paper towels, and filter papers. Turn over the gel and blotting membrane together (gel side on the top) on a filter paper towel. Mark the position of gel lanes on the membrane using a ball point pen. Remove carefully the membrane from the gel. Fix the DNA to the blot, either by baking or UV cross-linking. (UV cross-linking generally gives better results. To bake, first let the blot dry on a sheet of filter paper, then place between sheets of filter papers, and bake it at 80°C. To UV cross-link, protect the surface of the DNA side of the membrane with plastic wrap and then expose this side to the UV source for 3 minutes.) Preserve the blot if not used immediately at room temperature covered in plastic wrap. Hybridized the nylon membrane with radioactive probe visualized by autoradiography. (See Table 31.2.)

31.3.2 Western Blotting Technique

Western blotting or immunoblotting or protein blotting involves transferring protein bands from an acrylamide gel to a more stable and immobilizing medium such as nitrocellulose membrane. Once transferred, detection of specific proteins with antibodies can be possible. The approach is similar to the method used to transfer DNA from agarose gels to nylon membrane (Southern blotting).

Material Required

Transfer buffer (100 ml): Tris, 3 g; glycine, 14.4 g; distilled water, 800 ml. Adjust the volume to 1000 ml with methanol.

Blocking solution (TBS) (100 ml). Dissolve 5 g milk powder in 10 mM Tris-HCl, pH 7.5. Add 900 mg (0.9 percent) NaCl and 50 µl Tween 20.

Substrate solution (50 ml). Dissolve 30 mg 4-chloro-1-naphthol in 10 ml of methanol. Add 30 µl H_2O_2 in 40 ml of TBS. Mix these solutions thoroughly and add to the blot (nitrocellulose membrane).

Primary antibody 1:1000 dilutions. Use 100 µl for 100 ml. Do ELISA for determining the dilution factor.

Secondary antibody. 1:500 dilutions (use 200 µl for 100 ml). Follow the company-recommended dilution factor.

Protocol

Wash the unstained protein gel following electrophoresis (SDS-PAGE) in distilled water. Transfer the gel to transfer buffer for 15–30 minutes. Cut pieces of Whatman filter paper no. 3 and nitro-cellulose sheet to the size of the acryl amide gel. Immerse them in distilled water and soak in transfer buffer for 15–30 minutes. (Whatman filter papers are used as supporting material. Care should be taken while handling nitrocellulose membrane, and avoid touching the paper as finger prints lead to unbinding of peptides during transfer.)

Assemble a sandwich as follows: Place the wet filter paper on the negative side of the blotter. Pour transfer buffer over the filter paper little by little. Remove any air bubbles by rolling pipette over the paper. Place the gel on the paper. Put transfer buffer over the gel and remove air bubbles. Place the

TABLE 31.2
Troubleshooting for DNA Blotting and Hybridization Analysis

Problem	Possible Cause	Solution
Poor signal	Probe specific activity too low	Check labeling protocol if specific activity is $<10^8$ dpm/μg.
	Inadequate depurination	Check depurination if transfer of DNA >5 kb is poor.
	Inadequate transfer buffer	1. Check that 20× SSC has been used as the transfer solution if small DNA fragments are retained inefficiently when transferring to nitrocellulose. 2. With some brands of nylon membrane, add 2mM Sarkosyl to the transfer buffer. 3. Try alkaline blotting to a positively charged nylon membrane.
	Not enough target DNA	Refer to text for recommendations regarding amount of target DNA to load per blot.
	Inefficient transfer system	Consider vacuum blotting as an alternative to capillary transfer.
	Probe concentration too low	Check that the correct amount of DNA has been used in the labeling reaction. Check recovery of the probe after removal of unincorporated nucleotides. Use 10 percent dextrin sulphate in the hybridization solution. Change to a single-stranded probe, as reannealing of a double-stranded probe reduces its effective concentration to zero after hybridization for 8 hours.
	Incomplete denaturation of probe	Denature as described in the protocols.
	Incomplete denaturation of target DNA	When dot or slot blotting, use the double denaturation methods, or blot onto positively charged nylon.
Poor signal	Blocking agents interfering with the target-probe interaction	If using a nylon membrane, leave the blocking agents out of the hybridization solution.
	Final wash was too stringent	Use a lower temperature or higher salt concentration. If necessary, estimate Tm.
	Hybridization temperature too low with an RNA probe	Increase hybridization temperature to 65°C in the presence of formamide (see Alternate Protocol).
	Hybridization time too short	If using formamide with a DNA probe, increase the hybridization time to 23 hours.
	Inappropriate membrane	Check the target molecules are not too short to be retained efficiently by the membrane type.
	Problems with electro blotting	Make sure no bubbles are trapped in the filter-paper stack. Soak the filter papers thoroughly in TBE before assembling the blot. Used uncharged rather than charged nylon.
Spotty background	Particles in the hybridization buffer	Filter the relevant solution(s).
	Agarose dried on the membrane	Rinse membrane in 2× SSC after blotting.
	Baking or UV cross linking when membrane contains high salt	Rinse membrane in 2× SSC after blotting.
Patchy or generally high background	Part of the membrane was allowed to dry out during hybridization or washing	Avoid by increasing the volume of solutions if necessary.

(Continued)

TABLE 31.2 (CONTINUED)
Troubleshooting for DNA Blotting and Hybridization Analysis

Problem	Possible Cause	Solution
	Membranes adhered during hybridization or washing	Do not hybridize too many membranes at once (ten mini-gel blot for a hybridization tube, two for a bag is maximum).
	Bubbles in a hybridization bag	If using a bag, fill completely so there are no bubbles.
Patchy or generally high background	Walls of hybridization bag collapsed on to membrane	Use a stiff plastic bag; increase volume of hybridization solution.
	Not enough wash solution	Increase volume of wash solution to 2 ml/10 cm^2 of membrane.
	Hybridization temperature too low with an RNA probe	Increase hybridization temperature to 65°C in the presence of formamide (see Alternate Protocol).
	Formamide needs to be deionized	Although commercial formamide is usually satisfactory, background may be reduced by deionizing immediately before use.
	Labeled probe molecules are too short	Use a 32 P-labeled probe as soon as possible after labeling, as radiolysis can result in fragmentation. Reduce amount of DNase I used in nick translation.
	Probe concentration too high	Check that the correct amount of DNA has been used in the labeling reaction.
	Inadequate prehybridization	Prehybridize for at least 3 hours with nitrocellulose or 15 minutes for nylon.
	Probe not denatured	Denature as described in the protocols.
	Inappropriate membrane type	If using a nonradioactive label, check that the membrane is compatible with the detection system.
	Hybridization with dextrin sulphate	Dextrin sulphate sometimes causes background hybridization. Place the membrane between Schleicher and Schuell no. 589 WH paper during hybridization, and increase volume of hybridization solution (including dextrin sulphate) by 2.5 percent.
	Not enough SDS in wash solutions	Check the solutions are made up correctly.
Extra bands	Final wash was not stringent enough	Use a higher temperature or lower salt concentration. If necessary, estimate Tm.
	Probe contains nonspecific sequences (e.g., vector DNA)	Purify shortest fragment that contains the desired sequence.
	Target DNA is not completely restriction digested	Check the restriction digestion.
	Formamide not used with an RNA probe	RNA-DNA hybrids are relatively strong but are destabilized if formamide is used in the hybridization solution.
Nonspecific background in one or more tracks	Probe is contaminated with genomic DNA	Check purification of probe DNA. The problem is more severe when probes are labeled by random printing. Change to nick translation.
	Insufficient blocking agents	Use of alternative blocking agents.
	Final wash did not approach the desired stringency	Use a higher temperature or lower salt concentration. If necessary, estimate Tm.
Cannot remove probe after hybridization	Membrane dried out after hybridization	Make sure the membrane is stored moist between hybridization and stripping.

nitrocellulose membrane on the gel and remove air bubbles. Wet the 3 MM filter paper and place it over the nitrocellulose sheet on the positive side of the blotter. Pour transfer buffer over the filter paper and remove air bubbles. Place the positive side of the blotter over the filter paper. Put rubber bands around the sandwich.

If it is a semi-dry blotting apparatus, run at 60 volts constantly for 2–6 hours (for mini gel) at 4°C (follow manufacturer's instructions). If it is a wet blotting apparatus, keep the sandwich in the blotting tank containing transfer buffer and run at 60 volts constantly for 2–4 hours (for mini gel) at room temperature (follow manufacturer's instruction). After transfer remove the nitrocellulose membrane and air dry it. Soak the membrane (blot) in a blocking solution (Tris buffered saline, TBS) containing milk powder, preferably 5 percent milk powder for 3 hours with shaking. Wash the blot 2–3 times with TBS-Tween 20 for 10 minutes with shaking. Incubate the blot with primary antibody (primary antibody dilution factor can be estimated by ELISA) in 1:1000 dilutions with TBS in 3 percent milk powder for 2–3 hours under shaking. Wash 2–3 times in TBS-Tween 20 and each 10 minutes with shaking. Transfer the blot to secondary antibody (anti-rabbit IgG) in 1:500 dilutions (follow the company's recommendation) with TBS-Tween 20 and 3 percent mild powder for 1 hour under shaking. Wash two times in TBS-Tween 20 and two times in TBS alone. Develop the color by adding substrate solution. Stop the reaction by transferring the blot to distilled water.

31.3.3 Fluorescent *In Situ* Hybridization (FISH)

FISH assay technique involves combination of cytology and hybridization tools. Earlier *in situ* hybridization (ISH) was developed as a method that involved hybridization of radiolabel DNA (Pardue and Gall, 1969; John et al., 1969).

This method allowed for whole-cell detection via hybridization with nucleic acids within the target cell without altering the morphological integrity of the cells (Moter and Göbel, 2000). FISH was introduced to bacteriology by Giovannoni et al. (1988). DeLong et al. (1989) introduced the use of fluorescent labeled probes which replaced the radioactive labeled probes for *in situ* detection of bacteria, hence the name *fluorescence in situ hybridization* (FISH). Fluorescent probes supplied a number of advantages compared to radioactive labeled probes, among which enhanced safety, improved resolution, reduction in detection steps, and due to the possibility of using dyes of different excitation and emission spectra, possibility of detection of several target sequences in one sample.

FISH utilizes fluorescent-labeled probes to locate complementary target *in situ* to get an image. Probes used are synthetic polynucleotide strands or short DNA fragments that bear sequences known to be complementary to specific target sequences at specific chromosomal locations. FISH probes often target sequences of ribosomal RNA or mitochondrial genes because they are abundant in sequence databases and in multiple copies in each cell. The probes can be designed by computer-assisted search from target organisms for "signatures." These signature regions can be specific at different phylogenetic levels depending on the variability of the target molecule sequences. The probe comprises a short sequence, ranging from 15 to 20 nucleotides, that is specific for one or several taxa at species, genus, or higher taxonomic ranks. The probe or oligonucleotide is then labeled with a fluorochrome, for example a carboindocyanine dye (CY3), that can be detected by a fluorescence microscope.

FISH enables the position of a marker on a chromosome or extended DNA molecule to be directly visualized. FISH has thus far been used for a multitude of applications ranging from DNA, RNA, cytogenetic chromosome analysis, and bacterial 16S rRNA detections. FISH is a powerful method for the *in situ* detection of active growing organisms in environmental samples.

Protocol

FISH-based assays are performed by following these common steps: (1) fixation, (2) preparation, (3) hybridization, (4) washing, and (5) visualization and documentation. (See Figure 31.4.)

Fixation: Is performed in order to permeabilize the cells allowing for the entry of the probes as well as to prevent nucleic acids from degradation. This procedure is carried on using cross-linking

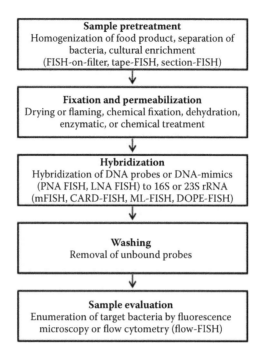

FIGURE 31.4   FISH-based assays.

fixatives like aldehydes (e.g., formaldehyde or paraformaldehyde) or precipitating fixatives such as methanol or ethanol. Precipitating agents are the fixative of choice when fixation of cells possessing a thick cell wall such as Gram-positive bacteria is desired; while aldehyde based fixatives perform very well with Gram-negative microorganisms. (Fixation causes loss of cell viability, and even though FISH of live cells has been shown by Silverman and Kool [2005], the possibility of uptake of oligonucleotide probes by live cells was contradicted by later work of Amann and Fuchs [2008].)

Incubate target tissues/cell suspension (5 μl) to be fixed on 10-well glass slides in 4 percent paraformaldehyde overnight at 4°C, further wash 3–4 times with 100 percent methanol. PFA fixed cell suspension can be immobilized by drying at 60°C for 10 minutes.

Sample preparation: The second step may involve treatment of Gram-positive bacteria with compounds which improve the permeability of cells to probes (e.g., lysozyme or lysostaphin; Schönhuber et al., 1997; Wagner et al., 1998; Moter and Göbel, 2000) or for example simply coating of slides with gelatin (Amann et al., 1990) or other coating agents, ethanol dehydration of samples air-dried onto slides or de-waxing of the paraffin-embedded preparations (Boye et al., 1998). Cells can be dehydrated in 50 percent, 80 percent, and 96 percent ethanol (2 minutes each). The preparation step might not be necessary if no special treatments are needed to conduct a complete FISH procedure.

All tissues/cells should be permeabilized, either by heat–shock treatment or proteinase K treatment. Heat–shock treatment involves incubating tissues/cells at 95°C degrees, 2 minutes, and quick-chill on very wet ice. Proteinase K treatment (10 μg/ml) for 15 minutes followed by incubation in 4 percent PFA for 20 minutes.

Hybridization: The third step basically involves annealing of the probe to its target sequence in the cell rRNA. Hybridizations are conducted under stringent (high degree of homology between probe and target sequence) conditions. Basically hybridization is conducted using preheated hybridization buffer containing the probe. The stringency of hybridization can be varied by varying the formamide concentration in the hybridization buffer, the temperature of hybridization, or the salt concentration. In order to assure a successful hybridization, the temperature of the hybridization must be maintained below the melting temperature of the probe in order for annealing of the probes to target sequences to occur. An empirical formula can be used for calculating the Tm of an oligouncleotide in relation to its GC content: Tm (in °C) = 2(A + T) + 4(G + C) (Suggs et al., 1981). Formamide in the hybridization buffer weakens the hydrogen bonds of the target sequence, practically lowering the Tm, while decreasing its concentration will increase Tm. Formamide can play another role in hybridizations, by destabilizing the secondary structures of rRNA in the ribosomes and improving accessibility of the probes to the target sequences.

Give brief Pre-hybridization treatment with hybridization buffer (containing 75 mM NaCl, 0.01% SDS, 20 mM Tris, 10–15% formamide) by incubating for 3 hrs (or overnight) at 65°C. Replace the pre-hybridization mix with the 12μl hybridization mix containing 3 μl (1 μg/μl) polynucleotide probes and continue incubating at 65–80°C overnight. To avoid evaporation of the buffer during the following incubation steps, the slides are stored in isotonically equilibrated chambers. Terminate hybridization by rinsing the slide with distilled water.

Washing: Is performed in order to mediate removal of the unbound probe. It is performed in wash buffer containing elements which were previously described for hybridization buffers while keeping stringency under control.

Brief washing with hybridization buffer is followed by three initial highly stringent washing treatments with prewarmed 2 × SSC buffer (with 75 percent/50 percent/25 percent hybridization mix, respectively) for 10 minutes each at 65°C. This is followed by later less stringent washing treatments with 0.2 × SSC buffer (with 75 percent /50 percent /25 percent hybridization mix, respectively) for 5 minutes each at room temperature.

Visualization: Can be performed by either fluorescence microscopy or following mounting of the samples in a mounting medium with or without added agents which prevent fading of the fluorochromes under intense illumination or by cytometry.

If fluorescently labeled probes are used, the cells could be analyzed directly with an epifluorescence microscope. Keep the plates in the dark as much as possible (color reaction is light-sensitive). Stop color development by washing with 4 percent paraformaldehyde. Slides are mounted in antibleaching mounting and analyzed with an epifluorescence microscope (Axioplan, Zeiss, Germany) and appropriate filtersets for Cy3 and fluorescein. Images are taken using a CCD camera (CCD 1035 × 1317) in combination with the WinView image software (version 2.1.7.6., Princeton Instruments, United States) and pseudo-colored with Adobe Photo Deluxe, BE 1.0.

31.3.4 MICROARRAY TECHNOLOGY

Microarrays with key features of speed, reliability, robustness, ease of use, and cost-effectiveness, have already made a marked impact on many fields of plant biology and are likely to become a standard tool of the microbiology/pathology laboratory. They may be based on nucleic acid detection (either cDNA or oligo based) or protein detection array with potential of recognizing thousands of spots on a single array chip/slide. DNA microarray is a solid support on which lots of DNA capture probe are immobilized in high density. Hybridization patterns in between DNA probes and samples can be analyzed for detecting the differences at nucleotide or transcriptional level.

Development of microarray technology concept is loosely based on a reversed northern blot strategy, where an array of target sequence-specific probes attached to a solid surface by hybridizing with labeled target molecules that would detect signal. What began back in 1995 with 45 complementary DNA probes spotted on a glass slide to create a microarray, has since evolved rapidly to allow for fabrication of microarrays with over million features probes. DNA Microarrays can either be constructed based on the standardized open microscope slide format or alternative integrated strategies, such as those employed by Affymetrix in a closed "kit" format.

Microarrays have been utilized for a multitude of applications including genome-wide genotyping, expression profiling, RNA detection, protein arrays, and pathogen nucleic acid detections. They have potential to increase diagnostics throughput, automate the process while simultaneously reducing unit cost. At present, DNA microarray technology is the most suitable technique for high-throughput detection and identification, as well as quantification, of multiple pathogens in a single assay. Lievens et al. (2006) demonstrated the utility of DNA array technology to detect single nucleotide polymorphisms (SNPs) that may be targeted for pathogen identification and should be pursued in any diagnostic assay. Pathogen diagnosis by microarray technique has been reviewed by Yoo et al. (2009).

Continued commercial interest in microarray technology promises increasing array element density, better detection sensitivity, and cheaper, faster methods. The exponential growth of pathogen nucleic acid sequences available in public domain databases has invited their direct use in pathogen detection, identification, and surveillance strategies. DNA microarray technology has offered the potential for the direct analysis of a broad spectrum of pathogens of interest. However, to achieve the practical attainment of this potential, numerous technical problems need to be addressed.

The key unifying principle of all microarray techniques is that labeled nucleic acid/antibodies molecules in solution hybridize, with high sensitivity and specificity, to complementary sequences immobilized on a solid substrate, thus facilitating parallel quantitative measurement of many different sequences in a complex mixture. They are composed of densely packed probe arrays based either on nucleic acid or antibodies coated on optically flat glass plates containing 96 wells. Main steps involve generation of ssDNA fragment to be spotted on glass slides or membranes and subsequently hybridized with labeled probes. Resulting hybridization image is scanned and data analyzed. Although several methods for building microarrays have been developed, two have prevailed.

In one method, DNA microarrays comprise of physically attached DNA fragments such as library clones or polymerase chain reaction (PCR) products to a solid substrate. The main advantage of this method is relatively low cost and substantial flexibility (which explain its wide implementation in the academic setting) in addition, to no prior requirement of primary sequence information to print a DNA element. By using a robotic array and capillary printing tips, we can print above 40000 elements on a microscope slide. In this system scanning is done by dual lasers capable of recognizing Cy3 and Cy5 dyes.

In the other method, arrays are constructed by synthesizing single stranded 25-mer oligonucleotides *in situ* by use of photolithographic techniques. Advantages of the latter method include higher density (>280,000 features on a 1.28×1.28-cm array) and elimination of the need to collect and store cloned DNA or PCR products. This microarray scanning system is based on single laser baser detection of pyroerthyrin dye.

31.4 PCR BASED TECHNIQUE

31.4.1 PCR Technology

Polymerase chain reaction (PCR) is a primer mediated enzymatic amplification of DNA sequences. Amplification of a selected region from a complex DNA mixture is carried out *in vitro* by thermo stable DNA polymerase 1 enzyme isolated from a bacterium *Thermus aquaticus*.

This amplification is achieved by a repetitive series of cycles involving three steps:

Denaturation: Denaturation of a template DNA duplex by heating at 94°C.
Annealing: Annealing of oligonucleotide primers to the target sequences of separated DNA strands at 35°C–65°C.
Synthesis or extension: DNA synthesis from the 3′–OH ends of each primer by DNA polymerase at 72°C.

By using repetitive cycles, where the primer extension products of the previous cycle serve as new templates for the following cycle, the number of target DNA amplified.

TABLE 31.3
Composition of PCR Mixture

Sr. No.	Constituents	Stock Concentration	Volume of PCR Reaction Mixture per Tube (25 µl)
1	*Taq* DNA polymerase buffer 100 mM Tris (pH 9.0) 500 mM KCl and 0.1 percent gelatin	10×	2.5 µl
2	MgCl$_2$	25 mM	2.5 µl (2.5 mM)
3	dNTPs (2.5 mM each)	10 mM	2 µl (200 µM each)
4	Primer	1 µM	5 µl (0.200 µm)
5	*Taq* DNA polymerase	3 U/µl	0.33 µl (1 U)
6	Template DNA	7 ng/µl	3 µl (21 ng)
7	Sterile distilled water	–	9.67 µl
8	Total volume	–	25 µl

Material Required

PCR thermal cycler, agarose gel electrophoresis apparatus and accessories, gel documentation system, microfuge tubes, micropipettes, mineral oil, PCR tubes, pipette tips and boxes, vortex, reagents, dNTPs, primers, Taq DNA polymerase, agarose gel electrophoresis chemicals and buffers. (See Table 31.3.)

Protocol

Add the reagents in 0.2 ml micro centrifuge tube placed on ice. The 25 µl reaction mixture is gently vortexed and spinned down. About 5 µl mineral oil is overlaid on the top of the reaction mixture. The DNA amplification was carried out on a thermal cycler (Eppendorf, Master Cycler Gradient, Germany). The PCR conditions set for amplification are tabulated in Table 31.4.

Agarose Gel Electrophoresis of Amplified PCR Products

Requirements

Electrophoresis unit (gel casting through gel combs, power pack), UV transilluminator.

Solutions Required

Ethidium bromide, 10 mg/ml; Tracking dye, 1 percent (w/v) BPB + 20 percent (w/v) Ficoll + 10mM EDTA; agarose; stock solution of 10 × TBE, 121 g Tris (1 M), 51.3 g boric acid (830 mM), and 7.6 g

TABLE 31.4
PCR Program Set in Thermal Cycler

Step No.	Name of the Steps Followed	Temp.	Time	Cycles
First step	Denaturation	95°C	5 min	1 cycle
Second step	Denaturation	94°C	1 min	
	Annealing	39°C	1 min	45 cycles
	Extension/Elongation	72°C	1 min	
Third step	Final extension	72°C	5 min	1 cycle
Fourth step	Final hold	4°C	Till retrieval	–

EDTA (10 mM) to make final volume 1000 ml and adjust pH 8.0; working solution of 1 × TBE, 10 ml of 10 × TBE is diluted to 100 ml using Milli-Q water.

Procedure

Agarose (1.5 g) is added to 100 ml of 1 × TBE buffer and agarose is melted by heating the solution in microwave oven. Solution is cooled to about 55–60°C and 5 μl of ethidium bromide (0.5 μl/ml) is added in it. The agar solution is poured into the gel casting unit after keeping the gel comb in the proper place. The gel is allowed to solidify at room temperature. Gel is placed in the electrophoresis apparatus in such a way that the end with wells is in line with the cathode. The apparatus is filled with 1 × TBE buffer in order to submerge the gel in the buffer to prevent the entry of air bubbles while removing the gel combs. The 10 μl PCR product to be analyzed are mixed with 2 μl tracking dye and loaded carefully in the wells of the gel. The unit is connected to a power pack and electrophoresis is carried out at 60 volts. The amplified PCR products are observed under UV transilluminater in gel documentation system (Flour Chem.™ Alpha Innotech, United States) and image is captured. (See Table 31.5.)

31.4.2   PRIMER IDENTIFICATION/DESIGNING

Primer designing requires that certain parameters are considered into account as given below. Free designing primers are available on websites such as the NCBI gene bank (http://www.ncbi.nlm .nih.gov>guide>all), which is based on Primer3 software. We need to consider the targeted region that has to be amplified while designing primers. (See Table 31.6.)

31.4.3   ARBITRARY/RANDOM PRIMERS BASED PCR METHODS

31.4.3.1   Randomly Amplified Polymorphic DNA Markers (RAPD)

RAPD technology is based on use of a single random decamer primer for PCR amplification at low annealing temperatures (32–38°C). Since this technology is not based on any previous sequence knowledge, its universal applications make it popular for initial molecular marker work in any new laboratory.

Protocol

Primers: Any random decamer sequence without prior sequence knowledge can be used for amplification, provided that it fulfills essential criteria for any primer. The working concentrations of random primers (50 picoMoles per μl) are prepared using the autoclaved sterile distilled water.

PCR reaction mixture: Amplification reaction mixture is prepared in 0.2 ml thin walled flat cap PCR tubes, containing the following components with the total volume of each reaction mixture being 20 μl. (See Table 31.7.)

PCR amplification: The PCR amplification for RAPD analysis is performed according to Williams et al. (1990) with certain modifications. The amplification conditions used are as given in Table 31.8.

Separation of amplified products by agarose gel electrophoresis: Agarose gel of 1.2 percent is prepared using the electrophoresis grade agarose in a volume of electrophoresis buffer (1 × TAE) sufficient for setting a gel (300 ml for 18 × 30 cm gel). Then ethidium bromide is added at concentration of 0.5 μl per 10 ml (10 mg/ml) of gel. The gel is allowed to set fully before removing the comb and loading the sample. Five μl of loading dye is added to 20 μl of PCR products and mixed well before loading into the wells. Care is taken to prevent mixing of samples between the wells. A voltage of 1.5 V/cm is given for a period of three hours for separation of PCR fragments. After the run, the gel is viewed under UV light and the DNA banding pattern is observed directly in gel documentation unit (gel doc). The banding pattern itself is noted from the digital image of the gels and analysed further for identifying polymorphic bands.

TABLE 31.5
Troubleshooting PCR

Symptoms	Possible Causes	Comments
Low yield or No amplification Product (PCR)	Insufficient number of cycles	Return reactions to thermal cycler for 5 more cycles.
	Template degraded	Verify the integrity of the template by electrophoresis after incubation.
	Thermal cycler programmed incorrectly	Verify that times and temperatures are correct.
	Temperature too low in some positions of thermal cycler	Perform a set of control reactions to determine if certain positions in the thermal cycler give low yields.
	Top of thermal cycler open	The top must be closed for correct heating and cooling.
	Inhibitor present in sample	Reduce the volume of sample in the reaction. Ethanol precipitate to remove inhibitors.
	Improper reaction conditions	Reduce the annealing temperature or allow longer extension times for longer amplicons.
	Missing reaction component	Check the reaction components and repeat the reaction.
	Mineral oil problem	The reaction must be overlaid with high quality, nuclease-free light mineral oil. Do not use autoclaved mineral oil.
	Reaction tubes not autoclaved	Autoclaving of tubes eliminate contaminants that inhibit amplification.
	Poor primer design	Make sure primers are not self-complementary or complementary to each other. Try a longer primer.
	Incorrect primer specificity	Verify that the primers are complementary to the appropriate strands.
	Primer concentration too low	Verify primer concentration in the reaction. increase primer concentration in the reaction.
	Suboptimal reaction conditions	Optimize Mg2+ concentration, annealing temperature and extension time. Verify that primers are present in equal concentration.
	Nucleotides degraded	Keep nucleotides frozen aliquots, thaw quickly and keep on ice once thawed. Avoid multiple freeze/thaw cycles.
	Target sequence genuinely not present in target DNA	Redesign experiment or try other sources of target DNA.
Multiple nonspecific amplification production (PCR)	Suboptimal reaction conditions	Optimize $MgCl_2$ concentration, annealing temperature, size, extension time and cycle number to minimize nonspecific priming.
	Poor primer design	Make sure primers are not self-complementary or complementary to each other, especially near the 3′ends. Try longer primer. Avoid using three G or C nucleotides in a row at the 3′end of a primer.
	Primer concentration too high	Verify primer concentration in the reaction. Try a lower concentration in the reaction.
	Contamination by another target RNA/DNA	Use positive displacement pipettes or aerosol resistant tips to reduce cross- contamination during pipetting. Use a separate work area and pipette for pre- and post-amplification. Wear gloves and change them often.
	Multiple target sequences exist in target DNA	Design new primers with higher specificity to target sequence.

TABLE 31.6
Parameters Taken into Account While Designing Primers

Sr No.	Primer Description	Suggestions
1	Primer length:	18–30 bases
2	Product position:	Primer sequence close to 3′end most preferred
3	Primer melting temperature Tm:	55–60°C
4	Annealing temperature:	Lower than Tm of primer
5	Primer pair Tm:	Difference should be less than 5°C
6	GC content:	Total GC content of primer should be 40–60 percent
7	GC clamp:	At 3′end priming efficiency increases however more than 3Gs or Cs should be avoided in the last 5 bases at 3′end of the primer
8	Poly (A) or (T):	Avoid 4 or more
9	T:	At 5′ or 3′end avoided
10	Amplification region length:	100–150 bp
11	Amplification cycles:	30 ~ 95°C 40 ~ 88°C
12	Amplification time:	≤ 1kb – 30 second – 1 minute
13	Annealing time:	~ 30 sec
14	Product Tm:	Should be less than 92°C
15	Degeneracy:	512 folds avoided
16	Hairpin:	3′end with $\Delta G = -2$ Kcal/mol internal hairpin with $\Delta G = -3$ Kcal/mol
17	Self dimer:	3′end self dimer $\Delta G = -5$ Kcal/mol internal self dimer $\Delta G = -6$ Kcal/mol tolerated
18	Cross dimer:	3′end self dimer $\Delta G = -5$ Kcal/mol internal self dimer $\Delta G = -6$ Kcal/mol tolerated
19	Runs:	Maximum number of runs accepted is 4 bases
20	NaCl or KCl:	More than 50 mM inhibit *Taq* polymerase activity
21	Reaction buffer:	
	Tris HCl:	10–50 mM
	KCl:	50 mM
	$MgCl_2$:	1.5 mM or higher
	Primer:	0.2 to 1 mM
	Gelatin or BSA:	100 mg/ml
	dNTPs:	50–200 mM
	Detergent:	0.05–0.10 percent

TABLE 31.7
Various Chemical Components Required for the PCR Reaction

Sr No.	Components	In μl
1	Template DNA (15 ηg/μl)	2.0
2	Primer (50 picoMoles/μl)	2.0
3	10 × buffer with 15mM $MgCl_2$	2.0
4	"dNTPs" mix (2.5 mM each)	2.0
5	$MgCl_2$ (25mM)	0.5
6	Sterile distilled water	11.17
7	Taq polymerase (3U/μl)	0.33

TABLE 31.8
Stepwise Cycles of Temperature and Duration Required for the PCR Amplification

Sr No.	Steps	Temperature (°C)	Duration (min)	Cycles
1	Denaturation	94	5	1
2	Denaturation	94	1	
3	Annealing	36 (32–38)	1	40
4	Extension	72	2	
5	Final Extension	72	20	1
6	Dump	4	Until removed	

31.4.3.2 Inter Simple Sequence Repeats Markers (ISSR)

ISSR primers are designed to amplify region in between two SSR/microsatellite repeat regions. They amplify such regions which are clustered within amplifiable distance in opposite orientation. ISSR primers are 16–18 base long consisting of random repeats (di-tetramer) sequence with/without extra one or two bases randomly added. Such primers are universal since their designing needs no prior sequence information and can be used in wide range of organisms. They are dominant markers which identify significant level of variation at several loci in unique regions of genomes.

PCR amplification mix: Amplification reaction mixture is prepared in 0.2 ml thin-walled flat-capped PCR tubes, containing the following components. The total volume of each reaction mixture is 20 μl. The 20 μl reaction mixture is gently vortexed and spinned down. (See Table 31.9.)

PCR Program: The DNA amplification is carried out in a Thermal Cycler (Eppendorf, Master Cycler Gradient, Germany). The PCR conditions set for amplification are tabulated in the Table 31.10. The sample is placed in thermal cycler until it reached to 50°C.

TABLE 31.9
Composition of PCR Reaction Mixture

PCR Reaction Component	Initial Conc.	Volume Used	Final Conc.
Genei *Taq* DNA polymerase Buffer A	10×	2 μl	1×
dNTP mix	2.5 mM each	1.6 μl	0.25 mM
Primer	10 pmole	1 μl	10 pmole
Genei *Taq* DNA polymerase	3 units/μl	0.3 μl	1 unit
Sterilized distilled water	–	14.07 μl	–
Template DNA	20 ng	1 μl	20 ng
Final volume of PCR mix	–	20 μl	–

TABLE 31.10
PCR Program Set in Thermal Cycler

Name of the Steps Followed	Temperature	Time	Cycles
Initial denaturation	94°C	5 min	1 cycle
Denaturation	94°C	60 sec	
Annealing	As per primer	60 sec	40 cycles
Extension/elongation	72°C	90 sec	
Final extension	72°C	10 min	1 cycle
Final hold	04°C	Till retrieval	

Agarose Gel Electrophoresis of Amplified PCR Products

Material Required

Electrophoresis unit (gel-casting trough, gel combs, power pack); gel imaging system or UV transillumination (Gel Logic 1500 Imaging System, United States); ethidium bromide, 10 mg/ml; 6 × gel loading dye; agarose, 1 × TAE buffer.

Procedure

Agarose (2 g) is added to 200 ml of 1 × TAE and agarose is melted by heating the solution in microwave oven. Allow the gel solution to cool to about 55–60°C and add 4 μl of Ethidium bromide (0.5 μl/ml) in it. The agarose solution is poured into the gel casting unit after keeping the gel comb in the proper place. The gel is allowed to solidify at room temperature. Gel is placed in the electrophoresis apparatus in such a way that the end with wells is in the line with cathode. The apparatus is filled with 1 × TAE buffer in order to submerge gel in the buffer to prevent the entry of air bubbles while removing gel comb.

The 20 μl PCR product to be analyzed are mixed with 1/6 volume of loading dye (M/s. Bangalore Genei Ltd.) and loaded carefully in the wells of gel. The unit is connected to power pack and electrophoresis is carried out at 80 volts. The power supply is switched off when the dye front is about 2 cm away from positive end (anode). A 100 bp low-range DNA ruler is used as a molecular size reference for band yielded from PCR. The amplified PCR product is observed under UV transilluminator in gel documentation system (Gel Logic 1500 Imaging System, United States) and image is captured.

Data Analysis

The clearly resolved PCR amplified bands are scored manually as binary matrix for their presence (1) and absence (0) in the data sheet. The analyses are carried out using the computer packages NTSYSpc 2.02i (Rohlf, 1994).

31.4.4 SEQUENCE SPECIFIC PRIMERS BASED PCR METHODS

31.4.4.1 Simple Sequence Repeats (SSR) or Microsatellite Markers

Microsatellites or simple sequence repeats (SSRs) markers are tandemly arranged repeats of different lengths of repeat motifs. Simple microsatellites contain only one kind of Repeat sequences $(GT)_n$ or $(AC)_n$ or $(AG)_n$; while complex microsatellites or composite microsatellites contain more than one type repeats, for example $[(GT)_2(AC)_1(AAG)_3]_n$. The DNA sequences flanking the SSRs are unique sequences and are conserved. These conserved flanking sequences have been exploited to designing loci specific primers suitable for amplification of highly polymorphic SSR loci using PCR. The SSRs are widely distributed throughout the plant and animal genomes that display high levels of genetic variation based on differences in the number of tandemly- repeating units at a locus. Microsatellite markers being single locus, codominant markers have advantages over dominant multilocus (RFLP and AFLP) markers because each allele of the markers can be detected and data on frequency of various homozygotes and heterozygotes can be used to characterize the breeding system.

This technique utilizes primer to amplify genomic DNA taken as template. The amplified fragment profile is generated by agarose gel electrophoresis or polyacrylamide gel electrophoresis. The differences in the fragment profile of different genotypes can be used as genetic marker for variety or genotypic identification, genome mapping and gene tagging. The uniqueness and value of microsatellite arise from their multi allelic nature, codominant transmission, ease of detection by PCR, relative abundance, extensive genome coverage, and requirement of only a small amount of starting DNA.

TABLE 31.11
Composition of PCR Mixture

Sr. No.	Constituents	Stock Concentration	Volume of PCR Reaction Mixture per Tube (20 μl)
1	*Taq* DNA polymerase buffer A 100 mM Tris (pH 9.0) 500 mM KCl 15 mM MgCl$_2$ and 0.1 per cent gelatin	10×	2 μl
2	MgCl$_2$	25 mM	0.66 μl (2.5 mM)
3	dNTPs	10 mM (2.5 mM each)	2 μl (200 μM each)
4	Primer forward	1 μM	1 μl (0.200 μm)
5	Primer reverse	1 μM	1 μl
6	*Taq* DNA polymerase	3 U/μl	0.33 μl (1 U)
7	Template DNA	40 ng/μl	0.5 μl (20 ng)
8	Sterile distilled water	–	12.51 μl
9	Total volume	–	20 μl

Material Required

Equipment and other material: PCR thermal cycler; agarose gel electrophoresis apparatus and accessories; gel documentation system; microfuge tubes; micropipettes; mineral oil; PCR tubes; pipette tips and boxes; vortex.

Reagents: dNTPs =; microsatellite primers (forward and reverse); Taq DNA polymerase; agarose gel electrophoresis chemicals and buffers. (See Table 31.11.)

Protocol

Add the reagents in 0.2 ml microcentrifuge tube placed on ice. The 20 μl reaction mixture is gently vortexed and spinned down. About 5 μl mineral oil is overlaid on the top of the reaction mixture. The DNA amplification is carried out on a thermal cycler (Eppendorf, Master Cycler Gradient, Germany). The PCR conditions set for amplification are tabulated in Table 31.12.

Agarose Gel Electrophoresis of Amplified PCR Products

Procedure

Agarose (2.5 g) is added to 100 ml of 1 × TBE buffer and agarose is melted by heating the solution in microwave oven. (Since most of the fragments generated after PCR are below 5 kb in size, a high percentage gel is used to achieve a better resolution of fragments). Solution is cooled to about 55–60°C and ethidium bromide (1 μl/10 ml) is added in it. The agar solution is poured into the gel casting unit

TABLE 31.12
PCR Program Set in Thermal Cycler

Step No.	Name of the Steps Followed	Temp.	Time	Cycles
First step	Denaturation	94°C	5 min	1 cycle
Second step	Denaturation	94°C	30 sec	
	Annealing	Depends upon Tm of primer	30 sec	40 cycles
	Extension/Elongation	72°C	30 sec	
Third step	Final extension	72°C	10 min	1 cycle
Fourth step	Final hold	04°C	Till retrieval	–

after keeping the gel comb in the proper place. The gel is allowed to solidify at room temperature. Gel is placed in the electrophoresis apparatus in such a way that the end with wells is in line with the cathode. The apparatus is filled with 1 × TBE buffer in order to submerge the gel in the buffer to prevent the entry of air bubbles while removing the gel combs. The 20 µl PCR product to be analyzed are mixed with 2 µl tracking dye and loaded carefully in the wells of the gel. The unit is connected to a power pack and electrophoresis is carried out at 60 volts. The power supply was switched off when the dye front is about 2 cm away from positive end (anode). The amplified PCR products are observed under UV transilluminator in gel documentation system and capture the image.

31.4.4.2 Single Nucleotide Polymorphism (SNP)

SNPs are single base–pair differences in DNA due to point mutations (substitutions or insertions/ deletions). With recent developments in genome sequencing technologies and available sequences in public domain they will most likely be the ideal marker for population analysis for the immediate future. They represent most common variation and provide genome-wide view of population. However, SNP development is time consuming and requires technically skilled manpower.

31.4.5 MODIFIED PCR METHODS

31.4.5.1 Multiplex PCR

Multiplexed-tandem PCR (MT-PCR) is a technology developed for multiplexed gene expression profiling and the rapid diagnosis of important pathogens (Stanley and Szewczuk, 2005). It involves parallel amplification of different targeted PCR fragments in single-tube reactions for the multiple pathogen detection. It consists of two rounds of amplification. In the first step, a multiplex PCR is performed for 10 to 15 cycles to allow enrichment of target DNA without creating competition between amplicons (Lau et al., 2008). This product is diluted and used as template for the second amplification that consists of multiple individual quantitative PCR reactions with primers nested within those used in the multiplex PCR. Up to 72 different PCR reactions can be multiplexed and performed simultaneously.

Fluorescence is measured by SYBR green technology at the end of each extension cycle, and melt-curve analysis provides species-specific or gene-specific identification. The incorporation of two sets of species-specific primers for each target ensures correct amplification and detection, thus avoiding the expense of DNA probes. SYBR green detection also increases the multiplexing and quantitative capacity of real-time PCR systems, which are usually limited by the availability of fluorescent channels and the need to optimize each individual multiplex PCR.

Though multiplex PCR are advantageous in terms of time and cost, but difficulties arise when more than one primer pairs are used in a single tube for PCR reaction. Engineering an efficient multiplex PCR requires laborious strategic and mathematical modelling on primer design, primer-template ratios, fragment lengths, nucleotide concentrations, optimal salt and buffer conditions, and chemical adjuvants. One has to modify the reaction conditions like annealing conditions, amount of $MgCl_2$ and polymerase extension time.

31.4.5.2 Nested PCR

It involves two rounds of PCR amplification, the second of which is performed using primers that hybridize within the fragment amplified during the first round of PCR. This approach offers the advantage of increasing both the sensitivity and specificity of PCR. A large number of pathogen specific primers are available on the web. Specificity of PCR depends upon the uniqueness of the sequences selected for primers and probes. Improvements in sequencing technique are making selection of reliable primers a routine.

31.4.5.3 Real-Time PCR

Real-time PCR can detect plant pathogens accurately and rapidly (Schaad and Fredreick, 2002). Real-time PCR is a kinetics-based quantitative PCR technique, where the amount of newly synthesized DNA is measured after each cycle throughout the PCR amplification process. In practice, a video camera detects the light emitted by a fluorchrome incorporated into the newly synthesized PCR product. It may be based on fluorescent light emission that is proportional to amplicon amounts generated. These fluorescent signals maybe generated by the use of either nonspecific double-strand DNA (dsDNA) intercalating SYBR Green dye or fluorescently labeled target sequence-specific probes. In terms of signal intensity curves, ten times the standard deviation from the average baseline signal is defined as the threshold cycle (CT value). Real-time PCR can use Taqman probes, fluorescent resonance energy transfer (FRET) probes, or molecular beacons to detect the production rate of the amplicons. These methods are based upon the hybridization of fluorescently labeled oligonucleotide probes sequences to a specific region within the target amplicon that is amplified using traditional forward and reverse primer.

Taqman probes: This method utilizes 5′-3′ exonuclease activity of *Taq* DNA polymerase to clean a dual-labeled fluorogenic hybridization probe during extension that is quantitatively measured by using a combined thermocycler fluorescence detector. Three primers are used together in a single real-time PCR reaction, a primer pair anneal to sites flanking the sequence of interest and the third fluorescently labeled primer anneals in between them. As PCR elongation step takes place the flanking primers extend, the labeled primer is released and fluorescence detection occurs. This system eliminates the time consuming process after PCR, and is also relatively resistant to carry over contamination of PCR amplification.

FRET probe: Principle relies on the non-radioactive transfer of energy from an excited donar (fluorophore) to an acceptor (fluorophore) by means of intermolecular long range dipole-dipole coupling.

31.4.5.4 Reverse Transcription–PCR (RT–PCR)

Reverse transcription–PCR (RT–PCR) involves a reverse transcription (RT) step before PCR amplification process (RT–PCR). It is mostly use for diagnosis of the plant viruses and viroids having RNA genome. Reverse transcription requires use of the reverse transcriptase enzymes isolated either from Avian Myoblastosis Virus (AMV) or from Maloney Marine Leukemia virus (M-MuLV). Single tube RT–PCR assays developed involve use of hot start DNA polymerase, which get activated only after subjecting them to denaturing conditions for around 5–10 minutes following the reverse transcription. All components of RT and of PCR reaction can thus be added together. This form of RT–PCR is also referred as one-step RT–PCR (Sellner et al., 1992). Single tube RT–PCR assays can also involve use of a single thermo-tolerant enzyme from *Thermus thermiphilus* with both reverse transcriptase (RT) and DNA polymerase activity. This format has advantage of requiring fewer hands on time to set up an assay and also to reduce possibility of false positive results through contamination, since it involves only a single series of pipetting. Besides this, many variants of original PCR protocols have been described which are described below.

31.4.5.5 Quantitative Real-Time Reverse Transcription–PCR (RT–PCR)

Quantitative real-time reverse transcription–PCR (RT–PCR) is a combination of real-time PCR technology with reverse transcription–PCR (RT–PCR). Reverse transcribed single stranded cDNA is further PCR amplified and its amplification is monitored simultaneously in real time after each cycle throughout the PCR amplification process. This method is suitable for monitoring the differential transcript levels/level of transcription during particular physiological process either in the host/or the pathogen.

31.4.5.6 Loop Mediated Isothermal Amplification (LAMP)

Isothermal amplification strategies are carried out at a uniform temperature without any need for repeated thermocycling. They involve use of strand displacing polymerases that are capable of displacing downstream duplex DNA segments during primer extension, which constitute a key strategy in numerous isothermal amplification schematics.

Isothermal amplification could also be achieved through a system of sophisticated primer designs. Loop-mediated isothermal Amplification (LAMP) utilizes a set of four specially designed primers, recognizing a total of six distinct sequences on the target DNA. External primers strands displace the internal primer extension templates. The internal primers are designed with regions complementary to both the sense and antisense target sequence, and thus form feedback loop formations that continue to extend independently of the original target. The final product is a stem-loop DNA structure with multiple inverted repeats of the target, resembling a cauliflower-like structure.

Loop-mediated isothermal amplification (LAMP) has been recently described as a rapid, cost-effective and easy-to-use method for specific DNA amplification. LAMP products are monitored in real-time by adding SYBR Green I after amplification. The LAMP method does not require a thermal cycler, making it suitable for on-site applications. Ghosh et al. (2015) developed a loop-mediated isothermal amplification assay to detect *Fusarium oxysporum f.sp ciceri*.

31.4.5.7 Cooperational PCR

Cooperational PCR is a high-sensitivity PCR technique for the amplification of viral RNA or bacterial targets from plant material (Olmos et al. 2002). The method has been patented as Co-PCR (Spanish patent P20002613; October 31, 2000). The Co-PCR (cooperational amplification) technique can be performed easily in a simple reaction based on the simultaneous action of three to four primers. The reaction process consists of the simultaneous reverse transcription of two different fragments from the same target, one internal to the other, the production of four amplicons by the combination of the two pairs of primers, one pair external to the other, and the cooperational action of amplicons for the production of the largest fragment. Cooperative primers are the first technology to prevent primer–dimer formation and propagation during the actual amplification rounds. The polyethylene glycol linker connecting the primer and the capture sequence prevents the polymerase from extending through the capture sequence, retaining the primer specificity in each round of amplification. Nonspecific amplicons that do not have a complementary region to the capture sequence, such as primer–dimers, are not propagated. A co-operational PCR technique was developed to detect *Ralstonia solanacearum* in water (Caruso et al., 2003).

31.4.5.8 Eric and Box PCR Amplification

Genomic fingerprinting is carried out using primer sets corresponding to Eric and Box elements (Versalovic et al., 1994). The Eric-1R (5′ATGTAAGCTCCTGGGGAT-3′) and Eric-(5′AAGTAA GTGACTGGGGGTGAGC-3′) as well as Box 1A (5′CTACGGCAAGGCGACGCTGACG-3′) are used to amplify putative Eric and Box-elements, respectively. These primers are synthesized by Inqaba Biotech (Pretoria, South Africa).

The PCR protocols are perform as previously described (Versalovic et al., 1994). PCR amplification reactions are performed using the following conditions: An initial denaturation step at 95°C for 5 minutes followed by 30 cycles consisting of 94°C for 1 minutes and annealing at 40 or 50°C for 1 minute with either Eric- or Box-primers, respectively. Extension is at 65°C for 8 minutes, this is followed by a single final polymerization step at 65°C for 15 minutes before cooling to 4°C.

An Icyler thermal cycler (Bio-Rad, United Kingdom) is used to amplify the DNA. In both PCRs the final reaction volume of 25 µl consist of 12.5 µl double strength PCR master mix (0.05U/µl Taq DNA Polymerase in reaction buffer, 0.4mM of each dNTP (dATp, dCTP, dGTP, dTTP), 4mM $MgCl_2$

(Fementas Life Science, United States), additional 1 mM $MgCl_2$, PCR-grade water (Fermentas Life Science, United States), 50 ng sample DNA and 25 pmole of the primer for Eric-PCR and 20 pmol for Box-PCR.

Gel Electrophoresis of PCR Products

Agarose gel electrophoresis, Tris-boric acid EDTA (TBE-PAGE) and sodium dodecyl sulfate, polyacryamide gel electrophoresis (SDS-PAGE). For all the gel systems, eight microliters (µl) of each PCR product of the amplified Eric- and Box-sequences are mixed with 2 µl of 6× orange loading dye (10 mM Tris-HCL (pH7.6, 25°C), 0.15 percent orange G, 0.03 percent xylene cyanol FF, 60 percent glycerol and 60 mM EDTA) and load onto gel wells. A molecular weight marker (Fermentas Technologies, Carlsbad, California) is included in each of the gels.

Agarose gel (1.5 percent w/v) is prepared in 1× TBE (100mM Tris, 100mM boric acid, 2mM EDTA, pH8, 25°C) gel buffer. The mixtures are heated in a microwave until complete melting. Ethidium bromide (1 µg/ml) is added to the cooled mixture before casting. After complete poly-merization, gels are electrophoresed in 1 × TBE buffer at 45 V for 4 hours 30 minutes

Polyacrylamide gels consisted of the following; 1 to 1.5 cm, 4 percent (w/v) staking gel and a 15 cm, 7.5 percent (w/v) resolving gel. The TBE gels are buffered with 1 × Tris-Borate-EDTA (TBE) (100mM Tris, 100mM boric acid, 2mM EDTA, pH8, 25°C). While SDS-PAGE (Laemmli, 1970) staking gel is comprised of (4 percent (w/v) acrylamide in 0.5M Tri-HCl, (pH 6.8, 25°C) 0.4 percent w/v SDS) and the 7.5 percent (w/v) resolving gel are prepared in gel buffer containing 1.5M Tri-base (pH 8.8, 25°C) and 0.4 percent SDS.

TBE- and SDS-PAGE gels are stacked for 40 minutes at 60 V and resolved at 120 V for 4 hours. This is done in a vertical electrophoresis unit (Protean II XL Bio-RAd, United Kingdom) with 1 mm gel spacers. The gels are stained for 15 minutes with EtBr (1µg/ml) and visualize using a Gene Genius Bio-Imaging System (Syngene, Synoptics, United Kingdom) and GeneSnap version 6.00.22 software.

Analysis of Eric and Box PCR Profiles

DNA profiles are analyzed using a trial version of Phoretix 1D and Phoretix 1D Pro (version 10) software packages from Total Lab Limited (United Kingdom). Pearson's coefficient with Ward's algorithm is used to create the dendograms. Pearson's coefficient is a more stable measurement of similarity than band-based methods because whole densitometirc curves are compared, omitting subjective band scoring steps (Rademaker and Bruijn, 1997). Additionally, by applying Pearson's coefficient to a densitometric graph, artefacts differences between gels can be normalized and removed so that they do not alter the results (van Ooyen, 2001). Ward's clustering method is a hierarchical agglomerative method whose objective is to create clusters that gives the minimum increase in the total within group error sum of squares (Ward, 1963).

31.5 HYBRIDIZATION–PCR COMBINATION-BASED TECHNIQUES

31.5.1 AMPLIFIED FRAGMENT LENGTH POLYMORPHISM MARKERS (AFLP)

AFLP combines restriction digestion and PCR amplification. To begin with, the genomic DNA is cleaved with two restriction enzymes—a rare cutter, for example, *EcoR1*, and a frequent cutter, such as *Mse1*. Adapters are added to the ends of the DNA fragments to provide known sequences for PCR amplifications. These adapters are necessary because the known base sequences at the end of the restricted fragments are insufficient for primer design. Short stretches of known sequences are added to the fragment ends using DNA ligase. If PCR amplification of such fragments is then carried out, all the fragments would be amplified which would be difficult to resolve on a single gel. Primers are, therefore, designed in such a way that they incorporate the above adapter sequences plus 1 to 3 additional base pairs at their 3′ end. In such cases PCR amplification will only occur where the primers are able to anneal to the fragments which have the adapter sequence plus complementary base pairs to the additional nucleotides. The additional base pairs are referred to as selective nucleotides. If one selective nucleotide is used, more fragments will be amplified than if two or more are used. Normally two selective rounds of PCR are carried out. In the first round only one selective nucleotide is used, whereas in the second round the same selective nucleotide plus one or two additional nucleotides are used. In order to detect the amplified fragments, one of the primers has a radioactive or fluorescent label attached. The amplified products are then resolved in a sequencing gel and visualized by autoradiography or by scanning through a specialized fluorescent detector.

Since stringent conditions for primer annealing and long (18–20 bp) primers are used, the technique is very consistent. The AFLP technique, like RAPDs, could be easily automated for high sample throughput. The drawbacks of AFLPs are complexity of the protocol, preparation of sequencing gel, and its expensive nature. Therefore, full potential of this technique can be achieved only after proper automation. AFLPs are quantitative in the sense that heterozygous and homozygous genotypes can be differentiated by the intensity of the amplified bands. Although AFLPs are anonymous markers they are useful for the saturation of regions around loci of interest and should facilitate map-based cloning efforts.

Protocol

The AFLP reactions can be carried out using custom designed/synthesized Primers-adapters combinations, involving use of P32 radio labeled adapters for selective amplification and resolving selectively amplification products in a sequencing gel. Other option is use of readymade company supplied kits like ABI AFLP kit system (PE Applied Bio systems) following the instructions supplied with the fluorescently labeled fragments resolved by semiautomated analysis.

    I. Restriction Ligation: Simultaneous restriction-ligation reactions are performed overnight at 25°C by double digesting genomic DNA (250 ng) with *Mse*I and *Eco*RI restriction enzymes (1.0 unit each); *Eco*RI and *Mse*I adapters are ligated to the digested DNA samples by T4 DNA ligase (0.25 unit) to generate template DNA for amplification. A fraction of the digested DNA is checked on a 1.5 percent (w/v) agarose gel to ensure check for complete DNA digestion. The restricted-ligated samples are incubated at 75°C for 15 minutes to inactivate the restriction endonucleases.

    II. Pre-Selective Amplification: Pre-selective amplification is carried out with adapters +1– primers each carrying one selective nucleotide (*Eco*RI adapter + A and *Mse*RI adapter + C) in a thermocycler for 20 cycles set at 94°C denaturation (30 seconds), 56°C annealing (30 seconds), and 72°C extension (30 seconds). The initial denaturation is done at 94°C for 30 seconds and the final extension at 72°C for 8 minutes. The preamplification products are diluted 20-fold in TE buffer and stored at –20°C; and a fraction of each samples was checked on a 1.5 percent (w/v) agarose gel to check for preselective amplification.

    III. Selective Amplification: Selective AFLP amplification is carried out with 5′ fluorescently labeled *Eco*RI adapters + 3 primers (6-FAM- *Eco*RI ACT; NED-*Eco*RI AAC and JOE-*Eco*RI ACG) and unlabeled *Mse*I adapters +3 primer (*Mse*I CAG) and 5 µL of the diluted PCR products from the preamplification. Three AFLP primer pairs (*Mse*I CAG × *Eco*RI ACT; *Mse*I CAG × *Eco*RI AAC and *Mse*I CAG × *Eco*RI ACG) are selected from ABI AFLP Selective amplification kits (Applied Biosystems) on the basis of their informativeness, reported previously. The PCR amplifications are carried out as follows: 13 cycles of touchdown PCR in which the annealing temperature is decreased by 0.7°C every cycle until a touchdown annealing temperature of 56°C was reached; that is, 94°C for 30 seconds, 65°C for 30 seconds with a decrease of -0.7°C per cycle, and 72°C for 1 minute. Once annealing temperature of 56°C is reached, another 20 cycles are conducted as described above for preamplification.

IV. Polymorphism Analysis: After selective amplification, JOE-labeled reactions are diluted 100-fold, FAM-labeled reactions by 5-fold, and NED-labeled reactions are not diluted. For multiplexing and semi-automated analysis, 12 µl of deionised formamide, 0.4 µl of Genescan 500 ROX size standard (Applied Biosystem) and 1 µl of each diluted sample are mixed together in a PCR tube. Capillary Electrophoresis is performed on the ABI Prism 310 Genetic analyzer (Applied Biosystems) to resolve the fragment polymorphism.

31.6   rRNA/DNA BASED TECHNIQUES

Typically, each ribosomal operon is polycistronic consisting of three cotranscribed genes encoding the structural rRNA molecules, 16S, 23S, and 5S. Among bacterial species, the average lengths of the 16S, 23S, and 5S structural rRNA genes are 1,522 bp, 2,971 bp, and 120 bp, respectively. The copy numbers, overall ribosomal operon sizes, gene sequences, and secondary structures of the three rRNA genes are highly conserved within a bacterial species due to their fundamental role in polypeptide synthesis. Because the 16S rRNA gene is the most conserved of the three rRNA genes, 16S rRNA gene sequencing has been established as the "gold standard" for identification and taxonomic classification of bacterial species. Knowledge of interspecies conservation of the 16S rRNA gene sequence and basic 16S-23S-5S ribosomal operon structure led Grimont and Grimont to the first insights into its usefulness in developing rib typing for bacterial classification.

The diversity of polymorphic ribotype fragments relies on the assumption that evolution of strains involves acquiring random mutations throughout their genome and, as such, is dependent on the rate of point mutations occurring in the genes flanking the ribosomal operons. A single base pair change in a typical 6-bp restriction endonuclease recognition site results in the loss of the cutting site and consequently in a change in the RFLP fingerprint profile. Given that ribosomal operons are flanked by ~50,000 bp of DNA carrying neutrally evolving genes, ribotype RFLP variation is a reflection of neutral gene evolution.

31.6.1   RIBOTYPING OR RIBOSOMAL DNA RESTRICTION ANALYSIS

Ribotyping involves the fingerprinting of restriction fragments that contain all or part of the genes coding for the 16S and 23S rRNA. Ribotype polymorphisms result from sequence variability in neutral housekeeping genes flanking rRNA operons, with rRNA gene sequences serving solely as conserved, flank-linked tags. As depicted in Figure 31.5, conventional ribotyping is based on restriction digestion of total genomic DNA followed by electrophoretic separation, Southern blot transfer, and hybridization of transferred DNA fragments with a labeled ribosomal operon probe. Following autoradiography, only those bands containing a portion of the ribosomal operon are visualized. It forms a unique pattern for each species and can be used, almost like a barcode. Rationales for the application of ribotype-based differentiation of independent isolates within a species have included taxonomic classification, epidemiological tracking, geographical distribution.

Primers: 5′-GTGCGGCTGGATCACCTCCT-3′ (16S primer) and 5′-CCCTGCACCCTTAATA ACTTGACC-3′ (23S primer) and corresponded to bases 1482–1501 of the 16S ribosomal RNA gene and bases 1–24 of the 23S ribosomal RNA gene.

Protocol

Amplification reactions are performed in 100 µl volume containing 10 mM Tris-HCl (pH 8.8), 50 mM KCl, 1.5 mM $MgCl_2$, 200 µM of each dNTP, 50 pmol of each primer, 2.5 units of *Taq* polymerase, and 10 µl of DNA extract (or distilled water as negative control). A comparison is performed with and without the addition of 5 percent DMSO (dimethyl sulfoxide) to the reaction mix. Amplifications is carried out in the thermal cycler (Perkin Elmer Cetus 480) for 1 cycle of 6 minutes at 94°C for denaturation, followed by 35 cycles (1 minutes at 94°C, 1 minute at 57°C, and

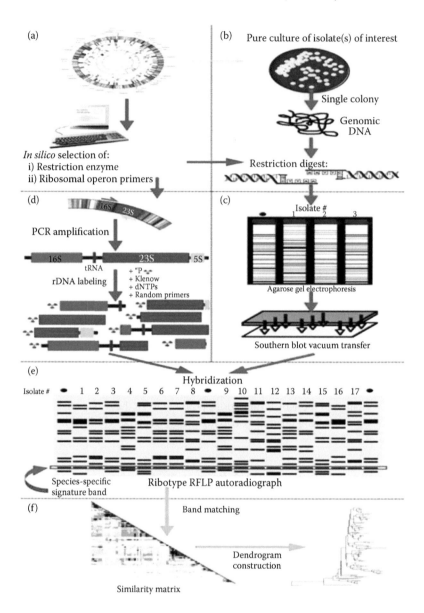

FIGURE 31.5 Ribotyping or ribosomal DNA restriction analysis. (a) Selection of restriction enzyme and ribosomal operon primers, (b) restriction digest of genomic DNA of bacterial colony, (c) agarose gel electrophoresis, (d) PCR amplification, (e) hybridization and RFLP autoradiograph, and (f) similarity matrix and dendrogram.

1 minute at 72°C) and a final extension of 7 minutes at 72°C. Fractionate amplification products through 3 percent Metaphor™ agarose electrophoresis for 6 hours at 80 V in TBE after ethidium bromide staining. Gel images are acquired by CCD camera and analyzed. Reproducibility is studied by repeated PCR assays of the same DNA extract.

31.6.2 16S rRNA Sequencing

16s ribosomal RNA is a component of the smaller subunit of prokaryotic ribosomes. A key marker in the diagnostics of bacteria is the 16S rRNA gene, as it is present in all bacteria and encodes for the same product. Due to the conservedness in its gene sequence it is used extensively at taxonomic levels for species identification for isolates. Phylogeny can be defined based on its nucleic acid differences, which can be sufficient for species identification. 16s rRNA gene sequencing provides genus and species level identification for isolates that do not fit any recognized biochemical profiles for strains generating only a low likelihood or acceptable identification according to commercial systems.

Protocol

Universal 16s rRNA gene specific primers are used for PCR amplification. An equimolar working concentration of 40–100 ng/µl is used as template for PCR. PCR amplification mix (20–50 µl) is prepared on ice in 0.2 ml by adding 1 unit Taq DNA polymerase, its 1× buffer, dNTP mix, $MgCl_2$ (if absent in buffer) 10 picomoles of each primer, and ~40–100 ng template DNA, and water to make up final volume (20–50 µl). The PCR tubes are placed in a thermal cycler, PCR amplification is carried out by programming the thermal regime as Initial denaturation: 94°C for 5 min followed by 40 cycles of each of denaturation (94°C); annealing (58°C) and extension (72°C). This is followed by a final hold of extension at 72°C for 5 min given to ensure complete extension of PCR product. Amplified DNA is purified to remove free primer and nucleotides from the PCR reaction. It can be custom sequenced on paid basis using the universal primers that are used for PCR amplification from both the sides. Sequence is analyzed to check for error if any and is subjected for homology search with previously reported sequence using the Basic Local Alignment Search Tool (BLAST) program available at ncbi.nlm.nih.gov. BLAST is a web-based program that is able to align search sequence to thousands of different sequences in a database (that you choose) and show a list of the top matches. This program can search through a database of thousands of entries in a minute.

31.6.3 INTERNAL TRANSCRIBED SPACER (ITS) rRNA SEQUENCING

The nuclear small subunit rRNA genes evolves relatively slowly and are useful for studying distinctly related organisms, whereas ITS regions and intergenic repeat evolves fastest and may vary among species and populations (White et al., 1990). New data is being added to international DNA databases at an exponential rate, but already around thousands of sequences for the well-studied ITS DNA region exist. ITS-rRNA amplification is combined with restriction end nuclease digestion for species identification and to further develop species–specific primers (Ristaino et al., 1998).

Protocol

Universal ITS rRNA specific primers are used for PCR amplification. Rest of steps are similar as in case of 16S rRNA Sequencing.

31.7 TERMINAL RESTRICTION FRAGMENT PATTERN (TRFP) METHOD

Terminal restriction fragment length polymorphism (TRFLP or sometimes T-RFLP) is a fingerprinting technique for profiling of microbial communities. It is based on finding the position of a restriction enzyme site closest to a fluorescently labeled end of an amplified gene. It involves restriction digestion of PCR amplified variants of a single gene and detecting the polymorphism in the size of the different terminal fragments using laser detection based DNA genotype. The method was first described by Liu and colleagues in 1997, which employed the amplification of the 16S rRNA gene from the DNA of several isolated bacteria as well as environmental sample. Community 16S rRNA genes are amplified and fluorescently labeled by PCR. The labeled products are digested by a restriction enzyme and the labeled, terminal restriction fragments (TRFs) are separated by electrophoresis and detected by laser-induced fluorescence on an automated genotype/sequencer.

When the target of PCR is 16S ribosomal DNA (rDNA), then the TRF pattern reflects the taxonomic diversity of the bacteria in the sample. TRF pattern analysis allows rapid monitoring of environments over time and space. This method has been used to compare bacterial communities from different environments. Comparing TRF patterns taken at different times can also monitor temporal changes in bacterial community structure.

Protocol of Creating TRF Patterns for Analysis

Each 100 mg sample is homogenized by pulverization under liquid nitrogen. DNA is extracted from the sample by using an Ultrapure Soil DNA Kit by the manufacturer's protocol. Amplify the template DNA by using a 5′-FAM-labeled primer, 46f (5′-FAM-GCYTAACACATGCAAGTCGA), and unlabeled primer 536r (5′-GTATTACCGCGGCTGCTGG). Carry out reactions in triplicate with the following reagents in 50-μl reaction mixtures: Template DNA, 10 ng; 1× PCR buffer; deoxynucleoside triphosphates, 0.6 mM; bovine serum albumin, 0.8 μg/L; MgCl$_2$, 3.5 mM; 46f, 0.2 μM; 536r, 0.2μM; and *Taq* DNA polymerase, 2 U. Reaction temperatures and cycling for samples should be as follows: 94°C for 2 minutes; 35 cycles of 94°C for 2 minutes; 48.5°C for 1 minute; 72°C for 1 minute; and one cycle of 72°C for 10 minutes. Purify the PCR product with any suitable PCR purification kit to remove the primers and concentrate the sample. Fluorescently labeled DNA (200 ng) is cut with one restriction end nuclease enzyme—*Msp*I, *Dpn*II, or *Hae*III (2.0 to 4.0 U)—in the manufacturer's recommended reaction buffers by incubating reaction mixtures for 5 hours at 37°C and then stop digestion by immersing in a 65°C water bath for 20 minutes. Ethanol precipitate the restriction digested DNA and dissolve it in 18 μl of formamide, with 1 μl each of Gene scan Rox 500 and Rox 550-size standards. The DNA is denatured at 95°C for 5 minutes and snap-cooled on ice. Samples are run on an automated genotyping analyzer and quantify the electropherogram output. Sample data consist of the size (base pairs) and peak area for each TRF peak in a pattern. The peak detection threshold is normalized by creating artificial detection thresholds for each sample.

31.8 PULSE FIELD GEL ELECTROPHORESIS TECHNIQUE (PFGE)

Pulse field gel electrophoresis is used for the separation of large sized DNA molecules in a gel by applying an electric field that periodically changes direction. Standard gel electrophoresis technique is unable to separate DNA molecules larger than 15–20 kb which essentially move together in a size-independent manner. Schwartz and Cantor (1984) developed the PFGE technique which expanded the range of resolution for DNA fragments by as much as two orders of magnitude.

PFGE uses restriction enzymes to digest bacterial genomic DNA to generate a small number of DNA pieces that can be separated based on size. Usually these restriction fragments are large and need to be specially treated and separated to generate a DNA fingerprint. First the bacteria are loaded into an agarose suspension, then the bacterial cell is lysed to release the genomic DNA. Once the DNA is released then the agarose and DNA suspension, also known as a plug, is treated with restriction enzymes. The treated plugs are then loaded onto an agarose gel and the restriction fragments are separated based on size using an electric field. What makes PFGE different from how scientists usually separate DNA is because PFGE can separate several large restriction fragments by constantly changes the direction of an electric field applied to the gel. It is considered a gold standard, a universal generic method of bacterial sub typing in pathogenic organisms. DNA restriction patterns generated by PFGE are stable and reproducible.

Protocol

Pulsed-field gel electrophoresis (PFGE) typing uses restriction digested chromosomal DNA (5 to 20 fragments) of various lengths from approximately 10 kb to 900 kb. All isolates are prepared for PFGE by individually suspending the bacterial cells, grown on culture medium at suitable incubation temperature, into a cell suspension buffer (100 mM Tris-HCl, 100 mM EDTA, pH 8.0) to a spectrometer absorbance of 0.7 ± 0.05 at 612 nm. Proteinase K (20 µl) is added to 400 µl of the bacterial suspension along with 400 µl of molten (54°C) 1 percent SeaKem Gold Agar. Mix quickly and dispense 300 µl into prepared plug molds. Once solidified, the plugs are placed into 1.5 ml cell lysis buffer (50 mM Tris-HCl, 50 mM EDTA, pH 8.0, 1 percent Sarcosyl) and 40 µl of Proteinase K and incubated for 1.5 hour at 54°C in a shaking water bath.

Wash the plugs twice in ultrapure water for 15 minutes in a 50°C water bath followed by four washings in Tris-EDTA (TE) buffer (10 mM Tris-HCl, 1 mM EDTA, pH 8.0). Store the washed plugs in TE buffer at 4°C. For PFGE, the plugs are cut into 3-mm by 9-mm pieces and then digested in 173 µl of sterile water, 2 µl of bovine serum albumin, 20 µl of 10× *Xba*I buffer, and 5 µl of *Xba*I (10 U/µl) at 37°C in a shaking water bath for 1.5 hour. The plugs are run in a 1 percent agarose gel using a CHEF III Pulsed-Field System (Bio-Rad) in 0.5 percent Tris-borate-EDTA buffer (Sigma) at 10°C. The parameters are set with the initial switch time at 2.2 seconds, the final switch time at 64 seconds, a voltage of 6V/cm, and a duration of 21 hours. The gel is stained with ethidium bromide and recorded the image on a Gel Doc system (Bio-Rad Laboratories, Inc., Hercules, California). These file images are processed by BioNumerics software (Applied Maths BVBA, Kortrijk, Belgium). Check the isolates within a PFGE subtype that have identical bands, therefore one isolate from each PFGE subtype is randomly selected as a representative and cluster analysis is done on the subtypes by Dice similarity coefficient and 0.8 percent band position tolerances as recommended by BioNumerics.

31.9 GUANINE-PLUS-CYTOSINE-BASED ANALYSIS

Guanine-Plus-Cytosine-Based Analysis is a DNA-based, direct method for initial characterization of the total bacterial community. DNA-based community-level molecular analyses have been used to obtain information on microbial community diversity, structure, and function. The highly efficient bacterial extraction and lysis helps in recovering DNA that accurately reflects the bacterial communities. Total bacterial DNA samples are fractionated according to their percent G+C content. The profiles reflecting the composition of the bacterial community were reproducible. Fractionation of total bacterial community DNA based on percent G+C content has previously been used to analyze bacterial communities and how they respond to changing conditions in soils and, bioreactors. Guanine-plus-cytosine based analysis is also used to identify the group or species in the genus of the bacterial plant pathogen.

Protocol

1. Bacterial extraction: Sample (1 g) is suspended in 30 ml of wash buffer and then shaken for 10 minutes on a reciprocating horizontal platform shaker at moderate speed at room temperature. The resulting suspension is subjected to centrifugation at $30,000 \times g$ for 15 minutes to collect the bacterial fraction. The pellet is resuspended and washed three more times in 30 ml of fresh wash buffer, and samples of the suspended bacteria were taken at each step for direct microscopic enumeration. Microscopic enumeration of bacteria is accomplished by fluorescence microscopy of DAPI (4′,6-diamidino-2-phenylindole)-stained cells.
2. Bacterial Lysis and DNA Recovery: Effective lysis of complex mixtures of bacteria often requires a combination of various treatments known to be effective for lysing individual bacterial population. Samples of the bacterial fractions (0.3 g of sample bacterial pellet and 0.7 g of bacterial pellet) is resuspended in 3 ml of TE buffer (10 mM Tris [pH 8], 1 mM EDTA). These suspensions are subjected to five freeze-thaw cycles of incubation at −70°C for 60 minutes followed by 40°C for 15 minutes. Lysozyme is then added to each sample as 0.7 ml of a 200-mg/ml stock solution (in TE buffer) followed by incubation at 37°C for 3 hours. After adding 0.2 ml sodium dodecyl sulfate (10 percent [wt/vol]) and 20 µl of proteinase K solution

(20 mg/ml in TE buffer), the mixture is incubated at 37°C for an additional hour. Following this incubation, 0.72 ml of 5 M NaCl, 0.6 ml of 10 percent CTAB (hexadecyltrimethyl ammonium bromide) in 0.7 M NaCl, and 1 g of 1,000-μm-diameter glass beads (Sigma Chemical Company, St. Louis, Missouri) are added. The mixtures are then incubated at 65°C for 20 minutes with vortexing for 30 seconds after every 5 minutes. Samples of the suspended bacterial fraction are taken before and after lysis to assess lysis efficiency by direct microscopic enumeration of bacteria. The cell lysate is extracted with an equal volume of chloroform-isoamyl alcohol (24:1) and then is subjected to centrifugation at 6,000 × g for 10 min at room temperature to separate the aqueous and organic phases. After transfer of the aqueous phase to a clean centrifuge tube, the DNA is precipitated by adding of 0.6 volume of isopropanol and incubating it for 1 hour at room temperature. The precipitated DNA is collected by centrifugation at 10,000 × g for 15 minutes at room temperature. Wash the DNA pellet briefly with 70 percent ethanol, vacuum dry it, and dissolve it in 2 ml of TE buffer. Purify the extracted DNA by two rounds of caesium chloride–ethidium bromide equilibrium gradient centrifugation. The DNA bands from these gradients are subjected to ethidium bromide extraction, desalting, and ethanol precipitation for concentrating DNA. DNA concentration and purity are estimated based on $A_{260/280}$.

3. Percent G+C profiles of the digestal communities. To obtain a profile of bacterial communities based on percent G+C content, 100 μg of each DNA sample is subjected to cesium chloride-bisbenzimidazole gradient analysis. Since there is no residual protein in these highly purified DNA preparations, DNA quantification is based on A280, which minimizes background absorbance resulting from the cesium chloride gradient itself and unbound bisbenzimidazole. Determination of the percent G + C content represented by each gradient fraction is accomplished by regression analysis (r2 > 0.99) of data obtained from gradients containing standard DNA samples of known percent G + C composition.

31.10   FLOW CYTOMETRY FOR BACTERIAL DIAGNOSIS

Flow cytometry is based on simultaneous multiparametric evaluation of physio/chemical properties of single cells flowing through a stream of fluid. The technique enables rapid identification and quantification of thousands of cells/particles per second. We can measure numerous cell properties (cytometry) as the cells move in a single file (flow) in a fluid column and interrupt a beam of laser light. Flow cytometry has applications in detection of plant pathogenic bacteria and viruses; and in viability assessment of spores and bacteria (Bergervoet et al., 2007). Fluorescent microspheres are conjugated to different antigens/antibodies, thereby constituting the solid phase for detecting antibodies or antigens in biological samples. These particles pass individual through a sensor in liquid stream, get sorted individually and therefore they are detected. Multiple parameters (relative size, granularity, and up to three/nine colors of emitted fluorescence) can be measured simultaneously and correlated particle by particle. The signals can be instantly collected as histograms for immediate results or stored as raw data for experimental analysis later. Various uses of flow cytometry include measurements of auto-fluorescent proteins, DNA and RNA content, antigen or ligand density, enzyme activity, membrane potential, viability obtained from cells, isolated nuclei, organelles or microorganisms. These assays seem to be more sensitive than traditional immunoassays, have a high throughput capacity and provide a wide analytical dynamic range. Additionally, they have multiplexing ability, that is, are capable of measuring multiple antibodies or antigens simultaneously. Flowcytometer consists of three parts:

Sample Handling – Flowcell and associated fluidics
Light Sensing – Light source, optics and detectors of light scatter and fluorescence
Signal Processing – Data collection and analysis

Instrumentation

Fluidics

Sheath fluid: This consists of fluid column with laminar flow. A flow rate of is 10 m/sec generated when an isotonic fluid under pressure is passed through a tubing into a flow cell also called the "Heart of the Flowcytometer."

Sample core: Sample is introduced into the flow cell by a computer driven syringe in the center of the sheath fluid creating a coaxial stream within a stream. The pressure of the sheath stream hydro-dynamically aligns the cells, so that they are presented to the light beam one at a time.

Light source and light beam: Laser light of argon type acts a sensing system. The sensitivity and resolution depend on shape and intensity of laser beam determined by two cylindrical lenses. The first lens (vertical) results in an elliptical beam with a longer vertical dimension (perpendicular to the axis of the flow) and shorter horizontal dimension (along the axis of the flow). The second lens (horizontal) changes the beam so that its horizontal dimension is longer than the vertical. Length-ening the horizontal axis increases the resolution of the instrument. Shortening the axis will enhance sensitivity. The vertical axis controls the discrimination between two closely spaced cells such as doublets (Wood et al., 1993).

Light scattering: The light scattered in the forward direction is called *forward scatter* (FSC) or *forward angle light scatter* and depends on the cell size or volume. The light scattered to the side is *side scatter* (SSC) or right angle light scatter or 90° light scatter. FSC gives the measure of the cell size and SSC gives the surface topography and granularity of the cell. FSC and SSC subsequently gives the cell population of interest which can be gated on the computer using a mouse.

Fluorescent light and signal processing: Certain dyes or fluorochromes absorb laser light at a particular wavelength and emit light at longer wavelengths, a phenomenon known as *fluorescence*. The fluorescent wavelength must be longer than the excitation wavelength for separation. The difference is called stroke shift. When the fluorescent dye tagged to the cell, passes through the laser beam, it emits light that is collected, amplified and transformed into voltage pulses. The height of the pulse is related to the intensity of fluorescence and the width of the pulse is related to the distribution of the dye on the surface. A small gate is selected by defining a region on a prepared to select the population of interest. A gate is selected by defining a region on a univariate histogram or a cytogram (bivariate hostogram). Only cells falling within the gate can pass through to the next stage of analysis. Gates are also used to select desired populations for cell sorting.

31.11 DEOXYRIBONUCLEIC ACID CONTENT OF BACTERIUM, ITS RELATION TO VIRULENCE AND DRUG RESISTANCE

Materials Required

Races of *Xanthomonas campestris* pv. *malvacearum*, exopolysaccharide production medium, DNA isolated from these races.

Protocol

Use most virulent race-32 (capable to attack five bacterial blight resistant genes, viz., polygene + B_7 + B_2 + B_{In} + B_N), less virulent race-8 (capable to attack polygene + B_2) and race-1 (not capable to attack any bacterial blight resistant gene), and avirulent streptomycin resistant mutant of race-32 and race-1 (race-1 also lost its virulence due to repeated subculturing and as used as avirulent strain) for the extraction and estimation of exopolysaccharide and DNA content.

Harvest the exopolysaccharide (EPS) from above *Xanthomonas campestris* pv. *malvacearum* races as described in earlier experiments and estimate its protein content by Folin–Lowry method.

Extraction, Detection, and Quantitative Estimation of DNA

Xanthomonas campestris pv. *malvacearum* races were suspended in 5 ml tris buffer (0.05 M, pH 8.0) and adjusted to the extinction of 1.0 at 620 nm. One ml lysozyme (5 mg/ml of water) and 1 drop of $CHCl_3$ were added to the suspension and allowed to stand for an hour. Ten percent sodiumlauryl sulphate (0.2 ml) was added to this and the tubes were kept in water bath at 40°C for 10 minutes. A solution of NaCl (5 M) was added in the bacterial lysate to make a final concentration of NaCl to 1M, and was stored overnight at 4°C. This was centrifuged at 15000 rpm for 15 minutes. The supernatant was decanted and the residues was dissolved in 5 ml of sterilized water and were screened under UV range (235 to 290 nm) of spectrophotometer to detect the presence of DNA.

Perform the quantitative estimation of DNA by method of Burton.

Correlate the amount of DNA with virulence of Xam isolate. Corelate the amount of EPS with virulence of Xam isolate. Corelate the amount of DNA with EPS production. Correlate the amount of DNA of streptomycin resistant (avirulent) mutant and streptomycin susceptible (virulent) xam race-32.

32 Bacteriophages of Plant Pathogenic Bacteria

Bacteriophages are the viruses which infect and lysed or kill the bacteria. They are widely prevalent in nature wherever the host bacteria exist and may be of two types.

1. *Specific phages*: These attack only one or a group of closely related bacteria and are often found in high concentrations, where their host bacteria are present, viz., infected tissue, rhizosphere of the infected plant, field water, and soil. In some cases they are also carried in the pure culture of the host bacteria.
2. *Nonspecific phages*: These are of universal occurrence, abounding in soil, river water, and often in healthy plant materials. They usually occur in very low concentrations. The infection and process of bacterial lysis occurs in two ways and based on this the bacteriophages are known either as lytic phages or lysogenic phages.
   a. *Lytic phages*: The bacteriophage infect its host bacterium, and takes control over the DNA synthesizing machinery of the bacterial cell to synthesise its own virus particles in the host bacterial cell. The synthesis and thereby presence of large amount of virus particles in the bacterial cell causes the bacterial cell to rupture and thus causes the lysis of the bacterial cell.
   b. *Lysogenic phages*: The bacteriophage infects its host bacterium and release its DNA material into the bacterial cell which integrates with the bacterial chromosomal DNA and the bacteria multiplies in the same state. Under specific conditions like radiations or heat/cold shocks, the synthesizing of the virus particles takes place in the bacterial cell and lysis occur.

Bacterial genus, species, or pathovars specific bacteriophages can be used in the identification of the bacterial species or pathovars. The lytic phages can be used for the detection of bacterial presence in the quarantine plant material. Similarly these can be used in the management of bacterial population responsible for disease initiation and spread.

32.1 ISOLATION OF LYTIC PHAGES

32.1.1 BY DIRECT METHOD

From infected plant material: One gram of material is chopped finely and suspended in 50 ml of sterile water in 100 ml conical flasks and incubated at 25–28°C for 48 hours.

From soil: 5–10 g of infested soil is suspended in 25–50 ml of water, shaken well, and allowed to stand for a few hours at room temperature.

From field water: Water from field is collected and allowed to settle for a few minutes.

From cultures: Stock cultures as such or after irradiation with ultraviolet light are grown in liquid medium for 48 hours.

The supernatants in the above treatments are decanted. If turbid, they are centrifuged at 6000 rpm and filtered through bacteria proof filters. The filtrate is tested for presence of phage.

Filtration through bacteria proof filters can be substituted with chloroform treatment. Ten ml of the supernatant is mixed with 0.5 ml of chloroform in a tube and vigorously shaken. The chloroform is then allowed to settle down and the supernatant is tested for presence of phage.

32.1.2 INDIRECT METHOD OR ENRICHMENT METHOD

This method is used for isolating phages present in low concentrations in soil or infected planting material. To build up phage population at detectable levels, the material or soil sample is to be suspended in water and inoculated with a suspension of actively growing host bacterium of the phage and is incubated for 48–72 hours for multiplication of phage titre. This enriched material is used for isolating phages as described in direct method.

32.1.3 DETECTION OF PHAGE

The presence of phage is determined by spot test. Mix 1 ml of bacterial suspension with 25 ml of melted cooled nutrient agar medium and pour in a sterilized Petri dish previously marked into four equal sectors. When the agar solidifies, place 0.5 ml of the test liquid on each sector. Incubate the plates at an optimum growth temperature of the host bacterium for 48 hours. If the phage is present in the test liquid, the spotted area will be clean and devoid of the bacterial growth because of the lysis of the bacterium in the test area due to lytic action of the phage.

Cut agar bit (2 mm) from the lytic area and transfer into 5 ml sterile distilled water in a tube. The water is agitated so as to get the phage into suspension from the agar bit. The phage in this suspension is purified by plaque test.

Purification of Phage

For this mix 1 ml of suitable dilution of the stock-phage with 1 ml suspension of the homologous bacterium in a sterile tube. Add this phage-bacterium mixture to 25 ml melted nutrient agar medium in a 100 ml flask. Mix the contents thoroughly and pour into a sterile plate. After the agar solidifies, incubate the plate in inverted position at optimum growth temperature of host bacterium for 48 hours. Clear lytic areas, referred to as plaques, develop within 24 hours. After 48 hours cut the plaques different in sizes and other morphological characters along with the medium and transfer separately into 5 ml sterile distilled water in tubes. The phage from these single plaques is again purified twice by plaque test. Transfer the plaque containing purified phage in 5 ml sterile water column and stored at 5–7°C as stock phage solution.

32.1.4 CHARACTERIZATION OF PHAGE

The phage should be studied for the following properties.

1. Host range: Test the phage for its ability to attack different bacterial species and strains by the spot test as described earlier to determine its host range and host specificity. (See Figure 32.1.)
2. Plaque morphology: The plaque's morphology is studied on the basis of plaque size, type of margin, and whether they are clear or turbid.
3. Electron microscopy: Suspend a plaque in 2 ml of sterile distilled water. Put a drop of this suspension on formvar-coated copper grid. Shadow-cast these mounted grids under high vacuum with gold-palladium alloy (60:40) at a suitable (250) angle and study the shape of the phage, for instance, T shape or otherwise under the electron microscope.
4. Plaque formation time: Dilutions of the stock phage are placed against the indicator host bacterium for a single plaque development. The time taken for the appearance of the plaque is noted. For a given system, under a given set of conditions, plaque-formation time is constant.
5. Number of plaque formation particles per ml: Dilutions of phages derived from single fully formed plaques are plated for single plaque development. The number of plaque forming particles per ml is determined by multiplying the average number of plaques per plate and the dilutions factor.

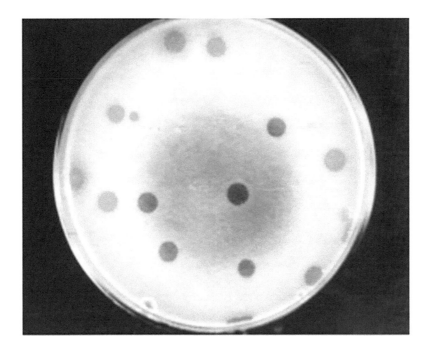

FIGURE 32.1 Plaques form by bacteriophage on host bacterium. (Courtesy of Dr. S. G. Borkar and Dr. J. P. Verma, Division of Plant Pathology, IARI, New Delhi.)

6. Thermal death point: Take the phage suspension (0.5 ml) in a sterile capillary tubes. The tubes are subjected to different temperatures in thermostatistically controlled water baths for 10 minutes each. After the treatment, transfer the tubes to cool water to cool them. The suspension is spotted on indicator host bacteria to determine the viability of the phage by formation of plaque and correlate with the temperature treatment to determine thermal death point.

7. Dilution end point: Plate tenfold serial dilutions of the stock-phage suspension with the indicator bacteria. Incubate the plates and observed for plaque development. Dilution end point is that dilution after which no plaques are observed.

32.2 BACTERIAL PHAGE TYPING

Material Required

Nutrient agar plates, phage solution, test bacterium, water blank, and so on.

Protocol

Prepare the suspension of test bacterium (0.1 OD) and add to lukewarm sterilized nutrient agar medium (1 ml of bacterial suspension in 25 ml of sterile medium) and pour in sterilized plate. Allow the medium to solidify under laminar flow. Add 0.1 ml of phage solution on the test bacterial medium and incubate the plates at $27 \pm 2°C$ in incubator for 48 hours. Observe the plates for formation of clear plaques.

Observation

Formation of clear plaques on the bacterial lawn indicates the bacteriophage is specific to the bacteria and thus the bacterial phage typing is studied. (See Figure 32.2.)

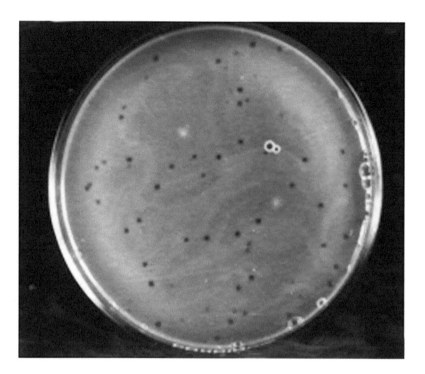

FIGURE 32.2   Bacterial phage typing to determine host specificity of bacteriophage. (Courtesy of Dr. S. G. Borkar and Dr. J. P. Verma, Division of Plant Pathology, IARI, New Delhi.)

32.3   DETERMINATION OF PHAGE TITER

Material Required

Stock phage solution, bacterial culture specific to phage, nutrient agar medium, water blank, and so on.

Protocol

Prepare serial dilution of stock phage solution up to 10^{-7}. Prepare the suspension of phage specific bacterial culture (0.1 OD at 620 nm) and add 1 ml of bacterial suspension to 25 ml of lukewarm sterilized nutrient agar medium. Now add 1 ml of known serial dilution of stock phage solution. Mix the content by rotating on a working table and pour in sterile Petri plate. Repeat the process for each serial dilution of the phage culture. Incubate the plates for 48 hours at $27 \pm 2°C$ and count the number of plaques.

Observation

Count the number of plaques in the last dilutions and calculate the phage titer by multiplying the dilution factor. The phage titer is expressed as plaque forming unit/ml of stock solution.

32.4   ISOLATION OF LYSOGENIC PHAGES

Material Required

Bacterial isolates of *Xanthomonas campestris* pv. *vesicatoria*, *Xanthomonas campestris* pv. *glycineae*, *Xanthomonas campestris* pv. *erythrinae*, *Xanthomonas campestris* pv. *azadirachtae*, *Xanthomonas campestris* pv. *clerodendri*, *Xanthomonas campestris* pv. *rhynchosia*, and *Xanthomonas campestris* pv. *malvacearum* known to harbor lysogenic phages (Borkar, 1997).

Protocol

To isolate the temperate phages from lysogenic bacterial cultures, suspend the exponential growth (10^7 cfu/ml) of these bacterial isolates separately in 50 ml nutrient broth and incubate at ordinary cultural condition (at 28°C) for 48 hours. After the full growth is observed keep one set at ordinary growth condition while another set at 4°C to give shock treatment to the bacteria for release of phages. Incubate these cultures for a week under these conditions. Read the growth of the bacteria in the broth solution of the two sets in small aliquot at 620 nm in spectronic-20 and also observe visually for the turbidity of bacterial growth. The bacterial growth where turbidity is lost indicate the lysis of bacteria due to release of lysogenic or temperate phages. For collection of temperate phages, filter the lysed growth of respective bacterium in nutrient broth through bacteriological filter (0.45 µ) and collect the filtrate containing temperate phages and stored under aseptic condition in freezer.

Test the reaction of these temperature phages on different pathovars of *Xanthomonas campestris*. To test the reaction of temperate phages of *Xanthomonas campestris* pv. *erythrinae*, *Xanthomonas campestris* pv. *azadirachtae*, and *Xanthomonas campestris* pv. *malvacearum* on different pathovars of *Xanthomonas campestris*, mix 0.1 ml phage suspension separately and individually with cultures of *Xanthomonas campestris* pv. *vesticatoria*, *Xanthomonas campestris* pv. *rhynchosia*, *Xanthomonas campestris* pv. *glycineae*, and *Xanthomonas campestris* pv. *clerodendri* in the nutrient broth in flask. Incubate the flask at 28 ± 2°C temp for the growth of the bacteria. After the full growth is achieved, within 48 hours give the shock treatment to the grown culture either by UV radiation or under cold treatment at 0 to 4°C for release of phages to cause lysis of infected bacteria.

Observation

Release of temperate phages from the bacterial growth cause the lysis of bacteria, loss of turbidity, and clear growth solution of the otherwise bacterial growth in the broth.

32.5   PRESERVATION OF BACTERIOPHAGES

Procedure

Inoculate the 48-hour-old growing culture of phage-specific bacteria into nutrient broth (Peptone, 5 g; beef extract, 3 g; water 1 L) and incubate the flasks at 28°C for 48 hours on a to-and-fro shaker. Add 1 ml phage suspension prepared from single phage to the nutrient broth containing bacterial growth and incubate for 48 hours. Filter the broth through bacterial proof filter (ultrafine, G-5) to separate phages. Label the filtrate containing phage suspension as "stock phage." Keep the bacteriophage containing vials in deep freeze for preservation of phages.

33 Determination of Bacterial Sensitivity to Antibiotics and Pesticide

Bacteria, being microorganisms, are sensitive to various environmental conditions, including chemical pesticides. Chemical pesticides inhibit the bacterial metabolism and also affect the various cellular component of bacteria. The sensitivity of bacteria to various pesticides are determined by various techniques.

33.1 DETERMINATION OF BACTERIAL SENSITIVITY BY POISON FOOD TECHNIQUE

Material Required

Test chemical (antibiotic/bactericide, pesticide), test bacterial culture, nutrient agar medium, sterilized Petri plates, sterile pipettes, sterile water blank, and so on.

Protocol

Add the measured quantity of test chemicals into measured quantity of sterilized lukewarm nutrient agar medium to obtain a required concentration of test chemical. Pour this poisoned medium into the sterile Petri plate and allow to solidify. Prepare the bacterial suspension of test bacterium (0.1 OD) and add 0.2 ml in the central of the plate and spread with the help of glass rod. Incubate the plate at $27 \pm 2°C$ in incubator for 24 hours to record the growth of the bacterium. Maintained the unpoisoned simple nutrient agar plate inoculated with test bacterium as control.

Observation

Record the growth of the test bacterium on poisoned medium and compare with control plates. No growth in the poisoned medium indicate the sensitivity of the test bacterium to given chemical at that concentration. (See Figure 33.1.)

33.2 DETERMINATION OF BACTERIAL SENSITIVITY BY DISC ASSAY METHOD

Material Required

Test chemical (antibiotic/bactericide/pesticide), test bacterial culture, nutrient agar medium, sterilized Petri plates, sterilized water blank, sterilized disc of filter paper no. 24.

Protocol

Prepare the bacterial suspension of test bacterium (0.1 OD) and add 2 ml of it into 50 ml of sterile lukewarm nutrient agar medium. Pour this medium into sterile Petri plates and allow to solidify. Prepare the suspension of known concentration of test chemical in distilled sterile water. Charge the sterilized disc of filter paper with the test chemical solution and put on the solidified

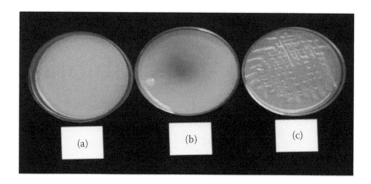

FIGURE 33.1 (a) Sensitivity of bacterial culture to test antibiotic/pesticide. (b) Formation of mutant of bacteria against test antibiotic/pesticide. (c) Nonsensitivity of bacterial culture to test antibiotic/pesticide. (Courtesy of Dr. S. G. Borkar and Swapanali Kadam, Department of Plant Pathology, Mahatma Phule Krishi Vidyapeeth, Rahuri.)

medium. Incubate the plates at $27 \pm 2°C$ in an incubator for 24 hours to record the zone of inhibition formed around the filter paper disc on bacterial lawn. Maintain proper control, that is, plates with filter paper disc without charge with test chemical solution.

Observation

Formation of zone of bacterial inhibition around the paper disc indicate the sensitivity of the test chemical to the bacterium. Measure the zone of inhibition formed around the filter paper disc on bacterial lawn. Measure the zones obtained with different concentration of test chemical solution. (See Figure 33.2.)

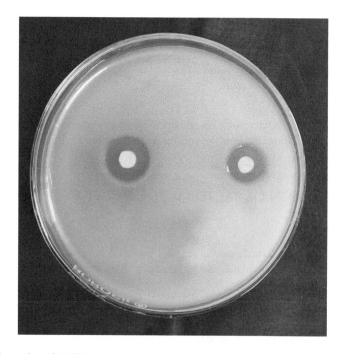

FIGURE 33.2 Formation of inhibition zone by test antibiotic/pesticide against bacterial growth. (Courtesy of Dr. S. G. Borkar and Ashwini Bhosale, Department of Plant Pathology, Mahatma Phule Krishi Vidyapeeth, Rahuri.)

33.3 DETERMINATION OF MINIMAL INHIBITORY CONCENTRATION (MIC) OF ANTIBIOTICS/PESTICIDES

Material Required

Test chemical (antibiotics/bactericide/pesticide), test bacterial culture, nutrient agar medium, sterilized Petri plates, sterile water blank, and so on.

Protocol

Prepare a stock solution of test chemical in distilled sterilized water. Measure the appropriate quantity to add in nutrient agar medium to obtain a required concentration of test chemical and add to the sterilized lukewarm nutrient agar. Thus prepare nutrient agar plates with different concentrations viz., 10, 25, 50, 100, 250, 500, 1000, and 2000 ppm of test chemical and allow the medium to solidify. Prepare the suspension of test bacterial/culture (0.1 OD) and put 0.2 ml on each plate and spread with glass rod. Incubate the plate at $27 \pm 2°C$ for 48 hours to observe the growth/inhibition of growth at a given concentration.

Observation

Observe the plates for bacterial growth of test bacterium. The lowest concentration at which the bacterial growth is inhibited is known as minimal inhibitory concentration of that chemical for the test bacterium.

33.4 ISOLATION OF ANTIBIOTICS/PESTICIDE RESISTANT MUTANT OF BACTERIA

Material Required

Test pesticide (antibiotics/bactericide/pesticide), nutrient agar, Petri plates, water blank, and so on.

Procedure

Prepare a minimal inhibitory concentration and a slightly higher concentration of test pesticide in nutrient agar and pour the plate. Solidify the medium by keeping the plate at $10°$ angle. Prepare the suspension of test bacterium (0.1 OD) and plate 0.2 ml on the solidified medium and spread with glass rod. Incubate the plates at $27 \pm 2°C$ for 5 days for the appearance of pesticide resistant mutant.

Observation

Record the development of bacterial colonies on MIC or next higher concentration, as mutant against the said pesticide. Pick up the bacterial colonies and maintained on the same concentration of the pesticide on nutrient agar medium. (See Figures 33.3 and 33.4.)

33.5 STUDIES ON PESTICIDE CROSS RESISTANCE IN BACTERIA

Material Required

Test pesticide (antibiotic/bactericide/pesticide), nutrient agar medium, sterile Petri plates, sterile water blank, and so on.

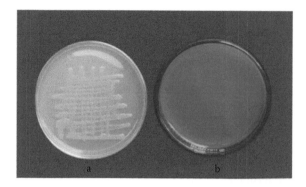

FIGURE 33.3 (a) Pesticide resistant bacterial culture formed growth on pesticide containing media. (b) The growth of pesticide sensitive bacterial culture was inhibited on pesticide containing media. (Courtesy of Dr. S. G. Borkar and Ashwini Bhosale, Department of Plant Pathology, Mahatma Phule Krishi Vidyapeeth, Rahuri.)

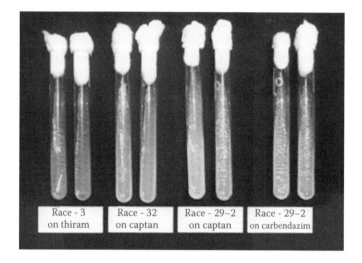

FIGURE 33.4 Growth of *Xanthomonas axonopodis* pv. *viticola* pesticide-resistant races on different fungicides. (Courtesy of Dr. S. G. Borkar and Swapanali Kadam, Department of Plant Pathology, Mahatma Phule Krishi Vidyapeeth, Rahuri.)

Procedure

Prepare the individual plates of nutrient agar medium with minimal inhibitory concentration of test chemicals against the test bacteria.

On these plates, streak the pesticide resistant mutant of the known pesticide and incubate the plates at 27 ± 2°C for 4–5 days.

Observe the growth of pesticide resistance mutant on various pesticide plates to determine the cross-resistance.

Observation

Observe the growth of known pesticide resistant mutant on the plates of the other pesticide. If the known pesticide mutant form the growth on other pesticide plates, it is assumed that the mutant has cross-resistance for that particular pesticide.

33.6   STUDIES ON MULTIDRUG RESISTANCE IN BACTERIA

Material Required

Test drug (antibiotics/bactericide/pesticide), test bacterial cultures, nutrient agar medium, sterilized Petri plates, sterile pipettes, sterile water blank, and so on.

Protocol I

Prepare the nutrient agar plates with minimum inhibitory concentrations of the testing drug. Prepare plate number 1 with one drug, plate number 2 with two drugs, plate number 3 with three drugs, and so on. Prepare the bacterial suspension of test bacterium (0.1 OD) and add 0.2 ml on each plate and spread with the glass rod. Incubate the plates at $27 \pm 2°C$ for 3 days and observe for the development of bacterial colonies in each plate. Maintain a proper control of nutrient agar plate without any drug.

Observation

Record the growth of bacterium in the respective plate and determine the resistance developed against number of drugs.

Protocol II

Prepare nutrient broth and suspend in test tubes at appropriate quantity and sterilized. Add minimum inhibitory concentration of the respective drug in the prescribed test tube. Prepare 1st tube with one drug, 2nd tube with two drugs, 3rd tube with 3 drugs, and so on. Prepare the bacterial suspension of test bacterium (0.1 OD) and add 0.2 ml in each test tube and shake well. Incubate the tubes at $27 \pm 2°C$ temp for 2 days in the incubator. Take a loopful of the suspension from each tube and streak on nutrient agar plates and then incubate them at $27 \pm 2°C$ for 3 days. Observe the plates for bacterial growth.

Observation

Record the growth of the bacterium in the respective plate and determine the resistance developed against number of drugs.

33.7   STUDIES ON PATHOGENICITY OF MULTIDRUG RESISTANCE BACTERIA

Material Required

Test bacterial drug resistant culture, native pathogenic culture, host plant, hypodermic syringe, distilled water, and so on.

Procedure

Prepare the suspension of test bacterial drug resistance culture (0.1 OD) and native pathogenic culture (0.1 OD) separately. Infiltrate these suspensions (0.3 ml) into the host plant leaves separately in a localized spot and tag them properly. Maintain the inoculated plant at $27 \pm 2°C$ with more than 87 percent relative humidity for the development of disease symptoms.

Observation

Observe the inoculated leaves with both the bacterial culture and compare the symptoms developed after the 4th day onward.

33.8   CURING OF BACTERIAL MUTANT FROM ANTIBIOTIC RESISTANCE

Resistance to particular antibiotic poses serious threat in the management of bacterial population with the same antibiotic or with other antibiotics if the bacteria develop cross resistance to these. Therefore, curing of bacteria for antibiotic resistance is very important for their management.

Hiramasty et al. (2012) described a procedure for curing bacteria of antibiotic resistance and called such antibiotic *reverse antibiotic*.

Material Required

Antibiotic resistance bacterial strain, Nabomycin, nutrient agar medium, antibiotic susceptible strain of the same bacterium, and so on.

Protocol

Take a quinolone resistant bacterial strain (staphylococcus aureus or a plant pathogenic if available) with mutated gyr A gene. Take a quinolone susceptible bacterial strain with intact gyr A gene. Grow these culture on Nabomycin antibiotics. The cultures grown on Nabomycin antibiotic are to be tested against quinolone for their inhibition of growth.

Observation

A quinolone resistant mutant when grown on Nabomycin antibiotic reverts back to its natural culture which is then susceptible to quinolone antibiotic, thus curing it of antibiotic resistance.

33.9   CURING OF PLASMID BORNE ANTIBIOTIC RESISTANCE

Material Required

Various antibiotics, nutrient agar plates, ethidium bromide, SDS, test bacterial culture, and so on.

Protocol

Test the bacterial isolates for their resistance/sensitivity for different antibiotics like gentamycin, ampicillin, vancomycin, chloramphenicol, tetracycline, streptomycin, penicillin, erythromycin, and other antibiotics. Select the antibiotic resistance mutant and obtain their cultures. Take 24-hour-old cultures of antibiotic resistant mutant and grow them in sterile nutrient broth having ethidium bromide (100 mcg/ml) and 2 percent SDS. Incubate the broth tubes at $27 \pm 2°C$ for 48 hours. Isolate the bacterial colonies from this broth on nutrient agar and then inoculate into simple nutrient broth and incubate for 48 hours. Isolate the bacterial colonies, multiply them on slant, and test against the respective antibiotic by poison food technique.

Observation

The antibiotic resistant mutant, if inhibited or killed by the same antibiotic, means the plasmid borne antibiotic resistance is cured.

34 Determination of Mutation in Bacterial Plant Pathogen

Mutation in bacterial plant pathogens are known to occur for various characters, such as mutation for pathogenicity (loss of virulence and the bacterial cultures becoming avirulent), mutation for pigmentation (loss of pigment particularly *Xanthomonadia* in *Xanthomonas* to form the albino colonies), mutation for antibiotic resistance (the mutant colonies resist to antibiotic and grow in their presence), mutation for morphological characters, physiological characters or biochemical characters, and so on.

Due to environmental conditions or growth conditions, a few cells in the bacterial population mutate for a particular genetic character. An abrupt change in the genetic character is known as a *mutation* and occurs in one-in-a-million populations. This change in bacteria is either temporary or permanent. Temporary variations are caused due to environmental factors and does not involve the restructuring of DNA. Such variations may be morphological or physiological and disappear as soon as the environmental changes resumes to the original position. On the other hand, variations in bacteria that involve alteration of the DNA are termed as *permanent variations*. Such variations can occur by mutations (a change in the nucleotide sequence of the organism's genome) and recombination (the addition of genes from an outside source, such as other cells). Mutations are permanent heritable changes in the genetic information of the cell and can be either monotonous (if they occur randomly) or induced (if they are due to directed chemical or physical agents called mutagens).

A microorganism that exhibits a natural, nonmutated inheritance is termed as a *wild type* or *wild strain* while a *mutant strain* refers to the altered version. Mutant strains can show variance in morphology, nutritional characteristics, genetic control mechanisms, resistance to chemicals, temperature preference, and nearly any type of enzyme function. Applications of mutant strains are to trace genetic events, unraveling genetic organization and pinpointing genetic markers.

Genetic and biochemical investigations in bacteriology are often initiated by the isolation of mutants. In order to detect mutants of a particular organism, one must know the normal or wild characteristics so as to recognize an altered phenotype. Since mutants are generally rare, about one per 10^7 cells, it is important to have a very sensitive detection system so that rare events will not be missed. Detection systems in bacteria and other haploid organisms are straight forward because any new allele should be seen immediately, even if it is a recessive mutation.

34.1 ISOLATION OF ANTIBIOTIC RESISTANT MUTANT BY GRADIENT PLATE TECHNIQUE

Material Required

Twenty-four-hour nutrient broth culture of test bacterium, two nutrient agar deep tubes (10 ml per tube/culture), streptomycin sulfate solution (1 percent, i.e., 10 mg/100 ml), a beaker with 90 percent ethanol, sterile Petri plates, sterile 1-ml pipette, glass rod spreader, water bath, glass marking pencil.

Protocol

Gradient plate preparation: Melt two nutrient agar deep tubes in a water bath maintained at 96°C and cool to 50°C. Pour the contents of one agar tube into a sterile Petri plate and allow the medium to solidify in a slanted position by placing a pencil under one edge.

After the solidification of the agar medium, remove the pencil from the plate and place it in a horizontal position.

Pipet 0.1 ml streptomycin solution (100 µg/ml) into the second tube of the melted nutrient agar medium, rotate the tube between the palms, and pour contents to cover the gradient agar layer and allow to solidify on a level table.

Label the low and high concentration area on the bottom of the plate.

First Inoculation

Pipet 0.2 ml of the test bacterium suspension onto the gradient-plate after 24 hours of its preparation. With an alcohol-dipped, flamed, and cooled bent glass rod, spread the inoculums evenly over the agar surface by rotating the plate. Incubate the inoculated plate in an inverted position at $28 \pm 2°C$ for 48 to 72 hours.

Observe the plate for the appearance of test bacterium colonies in the area of low streptomycin concentration (LSC) and high streptomycin concentration (HSC) and count these and record the results.

Results

The colonies of test bacterium that appear in the area of high streptomycin concentration will be streptomycin-resistant mutants. Select and mark an isolated colony of test bacterium in the HSC area.

Second Inoculation

Pick the selected colony with a sterile inoculating loop, and streak onto a second gradient-plate toward the high concentration portion. Repeat this step with two or three colonies of streptomycin resistant mutants from the HSC area. Incubate the inoculated plates in an inverted position at $28 \pm 2°C$ for 24 to 72 hours. Examine each plate for a line of growth from the streaked colonies into the area of high streptomycin concentration.

Results

Development of test bacterium colonies in the HSC area indicates the successful isolation of streptomycin-resistant mutants.

Precautions

Always label the bottom of the plate to show the low and high streptomycin concentration areas.

The gradient-plate must be prepared at least 24 hours before inoculation to allow the establishment of antibiotic gradient in the plate. Always grow the organism overnight in a rich broth. Don't disturb the nutrient agar plate when kept in a slanted position until complete solidification has occurred.

34.2 ISOLATION OF AVIRULENT MUTANTS (MUTATION FOR PATHOGENICITY CHARACTER)

Material Required

Virulent culture of bacterial pathogen, nutrient agar medium, nutrient agar slants, streptomycin sulfate antibiotic, host plant of the bacterial pathogen.

Protocol I

Subculture the virulent bacterial culture on nutrient agar slant and incubate at $28 \pm 2°C$ for 48 hours for its growth. After 8–10 days, subculture the bacterium from above the tube and allow it to grow as above. Repeat the same subculturing process for 8 to 9 generations. Take the growth from 9th generation and use it for a virulence test on the host plant by hypodermic syringe.

Observation

Observe the reaction developed by the isolate on the host plant and compare it with the susceptible reaction developed by original isolates. Nondevelopment of disease reaction indicate the conversion of virulent isolate to avirulent isolate due to frequent subculturing.

Protocol II

Plate the virulent bacterial isolate on 50 ppm streptomycin containing plates and incubate for 5 days. Select the colonies developed on this plate, subculture, and test for virulence. Plate the 50 ppm streptomycin resistant isolate on 100 ppm streptomycin containing plate and incubate for 5 days. Select the colonies developed on this plate, subculture, and test for virulence. The same procedure has to be followed for higher concentration of streptomycin.

Observation

Observe the susceptible or lack of reaction developed by the streptomycin resistant isolate on the host plant and compare with the susceptible reaction developed by original isolate. Non development of disease reaction indicate the conversion of virulent isolate to avirulent through antibiotic resistance or antibiotic resistant mutant.

Maintain the avirulent isolate on the streptomycin containing medium of appropriate concentration.

Protocol III

Infiltrate the 48-hour growth of virulent test bacterium in the resistance cv of the host plant. Isolate the bacterium from the infiltrated areas on 2nd day and grow them on nutrient agar media. Infiltrate this culture again on resistance cv of the host plant and isolate from the infiltrated areas on the 2nd day of infiltration and grow the culture on nutrient agar media. Follow the same process for 6–7 generations of infiltration and isolation.

After 7 generations of isolations from resistance cvs, prepare the bacterial culture and inoculate on susceptible host cvs of the bacterium.

Observation

Observe for the development/nondevelopment of disease symptoms on susceptible cv. of the host. Non development of disease symptoms indicate the avirulent mutant development from virulent culture through resistance host cvs.

34.3 ISOLATION OF ALBINO MUTANT (MUTATION FOR PIGMENTATION)

Material Required

Test cultures of *Xanthomonas*, nutrient agar medium, different phenol. Phenols are known to cause albinism in *Xanthomonas* (Borkar and Verma, 1984a).

Protocol

Prepare 0.007, 0.005, 0.003, and 0.001 ppm concentration of phenolic substances particularly pyrogallol, catechol, resorcinol, phenol, and hydroxyquinone in nutrient agar media plates. Inoculate the respective plate with test culture of *Xanthomonas* bacterium and incubate up to 120 hours at $28 \pm 2°C$ in an incubator for the appearance of albino colonies.

Observation

Observe the development of albino colonies for respective phenol and its concentration and period required for development of albino colonies.

34.4   DIFFERENTIATION OF VIRULENT AND AVIRULENT MUTANT

Virulent and avirulent colony types of *Pseudomonas solanacearum* and certain other phytopathogenic bacteria like *Erwinia carotovora*, *Pseudomonas phaseolicola*, *Corynebacterium insidosum*, and so on can be distinguished on tetrazolium chloride agar medium (Kelman, 1954; Friedman, 1964; Smale and Worley, 1956; Carroll and Lukezic, 1971).

Protocol

Prepare a basal medium containing peptone, 10 g; casamino acid, 1.0 g; agar–agar, 20 g; distilled water, 1000 ml. Sterilize the medium in 200 ml quantities in flasks. Cool it to 45°C and in each flask add one ml of sterile 1 percent solution of 2,3,5 triphenol tetrazolium chloride (TZC). This will give a 0.005 percent concentration of TZC in the agar medium. TZC solution should be sterilized by filtration or by autoclaving for 8 minutes, and should be stored in dark.

Pour the medium in plates. After the agar solidifies, streak a dilute suspension of the bacterial culture. Incubate for 24–48 hours at optimum growth temperature and examine the colonies.

Observation

The characters of virulent and avirulent colony types will be as follows:

Bacterium	Virulent	Avirulent
P. solanacearum	Irregularly round, fluidal, white with pink center	Round, butyrous deep red with a narrow bluish border
P. phaseolicola	Intense red	Reddish white
E. carotovora	Larger colonies, reduce dye more intensely	Small colonies, reduce dye slightly

In *P. solanacearum* within the virulent colonies, more virulent can be distinguished from less virulent on the amount of reddening present in the center. The more virulent has less reddening and less virulent has more reddening.

35 Transfer of Genetic Factor in Plant Pathogenic Bacteria

Genetic variability is created in all living beings by transfer and recombination of genetic material of two different genetic makeups, may they be human beings, animals, plants, or microorganisms. Bacteria are unicellular organism which multiply by division through binary fission and transfer their own genetic material from one cell to another during binary fission to possess the same genetic makeup and characteristics. Creation of variability is a phenomenon in nature and to create this variability in unicellular organism like bacteria, at least three methods of transfer of genetic material from unrelated or closely related species are known.

35.1 TRANSFER OF GENETIC MATERIAL THROUGH TRANSFORMATION

Transfer of genetic material through transformation was demonstrated by Borkar and Verma (1991b) in isolates of *Xanthomonas campestris* pv. *malvacearum*.

Transformation of the pathogenicity character among *Xanthomonas campestris malvacearum* strains in plants was studied by using DNA of *Xanthomonas campestris malvacearum* race-32 virulent (streptomycin-sensitive) and a viable exponential–phase culture of *Xanthomonas campestris malvacearum* race-32 avirulent (streptomycin-resistant mutant), and *Xanthomonas campestris malvacearum* race-1 avirulent (obtained by repeated subculturing).

Material Required

DNA of virulent race, bacterial culture of avirulent race, susceptible host cv.

Protocol

Isolation of DNA: DNA of the *Xanthomonas campestris malvacearum* strain is isolated as described earlier (Borkar and Verma, 1988). Suspend exponential phase culture of virulent *Xanthomonas campestris malvacearum* strain in 5 ml of tris buffer (50 mmol/L, pH 8.0) and adjust to absorbance of 1.0 at 620 nm. Add one mL of lysozyme solution (concentration 5 g/L) and one drop of $CHCl_3$ in the bacterial suspension, and allow to stand for 1 hour. Add 0.2 ml of 10 percent sodium dodecyl sulfate and keep tubes in a water bath at 40°C for 10 minutes. Add 5 ml of NaCl solution to this suspension to bring the final concentration of NaCl to 1 mol/L. Store the suspension at 4°C for overnight and centrifuged (250 Hz, 15 minutes). Decant the supernatant (containing plasmid DNA) and dissolve the residues (containing chromosomal DNA) in 5 ml sterilized water. Screen the supernatant and the residue in the UV range to determine the presence of DNA and observe an absolute peak for DNA. Store the DNA at 4°C for transformation experiments.

Transformation of pathogenicity character from virulent to avirulent strain: The experiment is to be conducted with total DNA, plasmid DNA, and chromosomal DNA of the virulent strain. Cotton cultivar Soneville 2BS9 (showing susceptibility to virulent but hypersensitivity to avirulent race-1 and no reaction to avirulent race-32) is used as a medium for transformation.

Infiltrate Exponential phase culture (3×10^6 CFU per infiltration per spot) of avirulent race-1 and avirulent race-32 separately mixed with 1 ml of total, chromosomal, and plasmid DNA of virulent race-32 respectively in the leaf tissues of cotton. Maintain proper controls with total, chromosomal and plasmid DNA alone, as well as avirulent race-1 and avirulent race-32 on the same leaf tissues.

Maintain the infiltrated plants at 28°C under greenhouse condition to observe the initiation of susceptible water-soaking reaction and generation of transformants.

Isolation and determination of transformants: Use the infiltrated areas with total DNA of the virulent strain and viable cells of avirulent strains for isolation of transformants. Isolate the possible transformants (the infiltrated areas are aseptically macerated with sterilized water and the serial dilution was plated on nutrient sucrose-agar medium (in g): Peptone, 5; beef extract, 3; sucrose, 5; agar, 20; and distilled water, 1 L; pH 7.0) after 12 hours of infiltration (of DNA of virulent strain and bacterial culture of avirulent strain) at 6 hours interval up to 2 days on nutrient agar medium.

Observation

Count the colonies, and subcultured individual colonies and infiltrate into cotton leaves to test their virulence (ability to induce the water-soaking reaction). The avirulent bacterium which acquire DNA of virulent strain through transformation become virulent. Determine the number of transformants formed due to recombination.

35.2 TRANSFER OF GENETIC MATERIAL THROUGH CONJUGATION

Joshua Lederberg and Edward L. Tatum in 1946 first of all demonstrated conjugation in bacteria using triple auxotrophs of *E. coli* and shared the Nobel Prize in 1958 for their work on bacterial genetics. Bacterial conjugation, a mode of sexual mating in which a plasmid or other genetic material is transferred by a donor (male) to a recipient (female) cell via a specialized appendage. Wollman et al. in 1952 demonstrated that this gene transfer during conjugation was polar, that is, there were definite donor (F^+), and recipient (F^-) strains, and gene transfer was nonreciprocal. This differentiation or existence of different mating types in bacteria, was established by Francois Jacob and Elie L. Wollman, and is determined by the presence of a plasmid (fertility or F factor) within the donor cell, which allows it to synthesize a sex pilus or recipient pilus. A plasmid is an extrachromosomal closed circular double-stranded DNA that is smaller than and replicates independently of the cell chromosome. These cells having plasmids are referred to as F^+. The recipient cell is a related species or genus that has a recognition site on its surface but is without a plasmid and is denoted by F. Bacterial conjugation exhibits three patterns of genetic transfer: (1) F^+ factor transfer, (2) Hfr transfer, and (3) F' conjugation or seduction. In F^+ factor transfer only a copy of F factor (i.e., plasmid or episome) is transferred during conjugation and the donor genes are not usually transferred. In Hfr (high-frequency) recombination the F factor becomes incorporated into the bacterial chromosome and transfer of bacterial genes to recipient takes place and the cells are designated as high frequency recombinants. Seduction, a specialized form of conjugation, in which the F factor at the time of its separation from an Hfr chromosome picks up some bacterial genes to become an F' plasmid which readily transfers these genes to other bacteria during conjugation.

Demonstration of conjugation in bacteria can be performed in the laboratory by using two auxotrophic strains of *E. coli* (one strain able to synthesize methionline, but not threonine, that is, $thr^- met^+$ and other strain able to synthesize threonine but not methionine, that is, $thr^+ met^-$, and both unable to grow in a minimal medium) which are mixed in a nutrient broth and incubated for several hours and then plated on a minimal medium. If colonies will appear on the minimal agar medium after incubation due to the genetic recombination.

Material Required

Twelve-hour nutrient broth auxotrophic strain of *E. coli* ($thr^- met^+$) male (M)-strain 1; 12 hour nutrient broth auxotrophic strain of *E. coli* ($thr^+ met^-$) female (F)-strain 2; trypticase soy agar (TSA) plates (complete medium), minimal agar plate, sterile 1 ml pipettes, sterile 13 × 100 mm test tube. Beaker with 70 percent ethyl alcohol, bent glass rod, mechanical pipetting device, glass marker pencil, and so on.

Protocol

With separate sterile 1 ml pipettes, aseptically transfer 1 ml from each culture (i.e., auxotrophic strains I and II) into a sterile 13 × 100 mm test tube (labeled conjugation tube). Mix the two cultures by gently rotating the tube between the palms of your hands. Incubate the mixed culture at 28°C for 30 minutes. Following incubation, vigorously agitate the mixed culture to terminate the genetic transfer. With a sterile 1 ml pipette, aseptically transfer 0.1 ml from the conjugation tube to an appropriately labeled minimal agar plate. Using a sterile glass rod spread the inoculums over the entire plate.

Prepare control plates to establish that the two auxotrophic strains of *E. coli* grew well on a complete medium (TSA) but not on minimal agar (MA) using the spread plate technique as described below: Aseptically add 0.1 ml of each *E. coli* strain (i.e., female and male) to its appropriately labeled agar plates: TSA-Male, TSA-Female, MA-Male, and MA-Female. With a sterilized bent glass rod spread the medium over the entire surface of the agar plate. Incubate all the inoculated plates (2 TSA plates, 3 MA plates; i.e., control and experimental) in an inverted position for 48 hours at 28°C. Observe the two TSA and two minimal agar control plates inoculated with the individual *E. coli* strain as well as minimal plate inoculated with the mixed culture. Count the number of colonies on all the five plates and record the results.

Results

Trypticase soy agar (TSA) control plates will exhibit the growth of the two auxotrophic strains of *E. coli* (i.e., thr$^+$ met$^-$ and thr$^-$ met$^+$) on a complete medium but no growth of these strains on the minimal agar (control) plates. Recombinant phototrophic colonies of *E. coli* will be observed on the experimental minimal agar plate as a result of conjugation of two auxotrophic strains of *E. coli*.

During conjugation the chromosomes of two auxotrophs associate and undergo recombination, thus producing a recombinant strain with genetic composition of thr$^+$ met$^+$. The number of colonies counted on the experimental minimal agar plate be compared to control to estimate the percentage of recombinants produced.

The same type of experimentation can be performed for plant pathogenic bacterial strain having different biochemical requirements.

35.3   TRANSFER OF GENETIC MATERIAL THROUGH TRANSDUCTION

Transduction require a bacteriophage as a vector for the transfer of genetic material (DNA) from one bacteria to another. The bacteriophage acquires a portion of genetic material of the infected bacterium. When it is liberated and infect another bacterial cell, the genetic material of the first cell becomes integrated with the DNA of the second bacterium. Thus, different characters may be transferred from one bacterial cell to another, which shows new or different characters. The mechanism may be operative in plant pathogenic bacteria to transfer the genomic character and can be demonstrated.

Material Required

Pathogenic but lytic culture of *Xanthomonas*, nonpathogenic lysogenic culture of same bacterium, phage of the bacterium, nutrient broth, nutrient agar plates, bacteriological filter, test host plant, and so on.

Procedure

Obtain a pathogenic but lytic culture of *Xanthomonas* (prove the pathogenicity of the culture on host plant and its lytic nature by phage reaction). Obtain a nonpathogenic but lysogenic culture of the same bacterium (prove avirulence of the culture on host plant and its lysogenic nature by phage reaction). Grow the pathogenic test *Xanthomonas* culture in nutrient broth for 24 hours at 27 ± 2°C and add 0.5 ml of lytic phage solution into it. Incubate the inoculated broth at $27 \pm 2°C$ for 12 hours and filter through bacteriological G-5 filter to obtain the phages. Grow the nonpathogenic *Xanthomonas* culture in nutrient broth for 24 hours at $27 \pm 2°C$ temp and add 0.5 ml of phage solution obtained by lysis of pathogenic culture. Incubate the inoculated broth at $27 \pm 2°C$ for 12 hours and streak on the nutrient agar plate. On the basis of colony character and experience select the colonies, obtain their slant and test for pathogenicity on host plant.

Observation

If the avirulent culture turn pathogenic during the experimentation, it denotes the transfer of genetic material through transduction process.

Note: The pathogenic character on *Xanthomonas* is borne on chromosomal DNA rather than plasmid DNA (Borkar, 1982) and for transfer of the genetic material through phages the integration of bacterial chromosome occurs with phage chromosome.

36 Population Studies of Plant Pathogenic Bacteria on/in Host Plant

The bacterial population plays an important role in the induction and appearance of disease symptoms. Even a single bacterium can initiate the infection (Klement and Goodman, 1967) but the time taken is relatively longer. More bacterial population means quicker symptoms development.

36.1 ESTIMATION OF BACTERIAL POPULATION BY LEAF WASH METHOD

Material Required

Nutrient agar plates, distilled sterile water, magnetic stirrer, fresh leaves of plant, pipettes, and so on.

Protocol

Collect the symptomless fresh leaves of plant and weigh them. In case the leaves are of large size, cut them into small pieces of 3 cm^2. Suspend these leaf material (approximately 1 g) in 25 ml of distilled sterile water in small conical flask and agitate on rotary shaker for 10 minutes. Pipette out 0.2 ml of aliquot of this agitated suspension and make a serial dilution up to 10^{-6}.

Plate 0.1 ml of this suspension on nutrient agar plates and spread with glass rod and incubate at $27 \pm 2°C$ in incubator for 3 days.

Observation

Observe and count the bacterial colonies in plates of each dilution and determine the total bacterial population per leaf or per gram of leaf.

Note: The saprophytic bacterial colonies appear on the nutrient agar plates with 24 hours while the plant pathogenic bacterial colonies appears after 48–72 hours only. Make sure to count these colonies to determine the epiphytic population of plant pathogenic bacteria.

36.2 ESTIMATION OF EPIPHYTIC POPULATION BY LEAF IMPRESSION METHOD

Material Required

Fresh leaves of plant, nutrient agar plates, antibiotic aureofungin.

Protocol

Prepare nutrient agar plates with 100 ppm aureofungin. Put the dorsal side of the leaf on the nutrient agar medium and press the leaf with slight pressure so as to make the leaf impression on the medium. Take out the leaf after making impression.

On another nutrient agar plate, put the ventral side of the same leaf and press the leaf with slight pressure so as to make the leaf impression on the medium. Take out the leaf after making impression. Incubate the plates at $27 \pm 2°C$ in incubator for 48 hours and note the bacterial population on the leaf imprint on media.

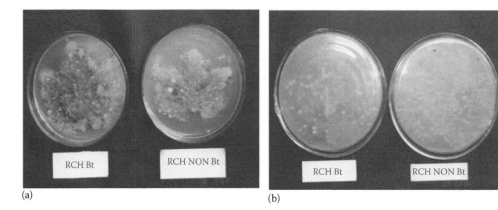

(a)

(b)

FIGURE 36.1   (a) Fungal epiphytes on cotton leaf. (b) Bacterial epiphytes on cotton leaf. (Courtesy of Dr. S. G. Borkar and Miss. Ashwini Bhosale, Department of Plant Pathology, Mahatma Phule Krishi Vidyapeeth, Rahuri.)

Observation

Observe the bacterial colony, types, and their locations on the leaf imprint. (See Figure 36.1.)

36.3   ESTIMATION OF ENDOPHYTIC POPULATION BY SERIAL DILUTION METHOD

Material Required

Infected disease sample, nutrient agar plates, water blank, pestle and mortar, pipettes, and so on.

Protocol

Measure the infected area or portion on the leaves/twig and cut them into small pieces. Sterilize these with 0.1 percent mercuric chloride solution with three washings of distilled sterile water. Macerate the disease sample in 5 ml sterilized water in pestle and mortar and allow to settle for 5 minutes. Pipette out 0.2 ml of aliquot of this suspension and make serial dilution up to 10^{-6}. Pipette out 0.1 ml of each dilution and plate on nutrient agar media and spread with glass rod. Incubate the plates at $27 \pm 2°C$ in incubator for 48–72 hours. Observe the plate and count the bacterial colonies.

Observation

Count the bacterial colonies of plant pathogenic bacteria and other endophytic bacteria in each dilution plate and determine the original population available in the sample.

36.4   DYNAMICS OF BACTERIAL POPULATION IN HOST PLANT

This exercise denote the multiplication of bacterial pathovar/race in the different cultivars of the same host plant so as to note their population in host cultivar of different genetic background.

Material Required

Inoculated leaves with bacterial pathogen, cork borer, nutrient agar plates, pestle and mortar, mercuric chloride, distilled sterile water, sterilized pipette, glass rod, and so on.

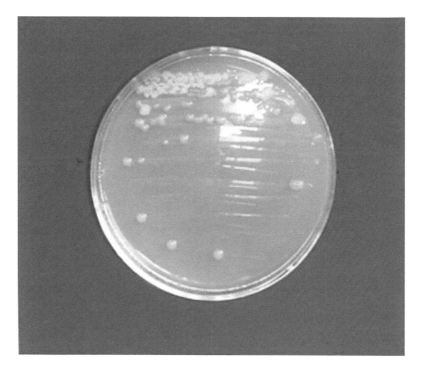

FIGURE 36.2 Status of Xam and endophytes in HR induced area in a cotton leaf. (Courtesy of Dr. S. G. Borkar and Ashwini Bhosale, Department of Plant Pathology, Mahatma Phule Krishi Vidyapeeth, Rahuri.)

Protocol

Infiltrate the suspension of young growth of Xcm-race-32 and Xcm-race-8 (at 0.1 extinction at 620 nm) in leaves of Acala-44 (susceptible to Xcm-32 and Xcm-8), 1-10B (susceptible to Xcm-32 but resistant to Xcm-8), and 101-102B (resistant to Xcm-32 and Xcm-8). At definite time intervals (4 H) the infiltrated areas are cut with the help of cork borer (8 mm in diameter) and taken out in the form of leaf discs. Wash three such leaf discs in distilled water and sterilize the leaf disc surfaces by dipping them in 0.1 percent mercuric chloride for 1 minute, followed by washing with distilled sterilized water. Macerate these leaf discs in 2 ml sterilized water and volume is made to 10 ml. From this suspension, make a serial dilution and plate 0.1 ml of each dilution on nutrient-sucrose-agar medium and incubate the plates in triplicate at 28°C for 4 days. Take the colony counts and calculated the bacterial population per cm^2 area (Borkar and Verma, 1984b). (See Figure 36.2.)

36.5 ISOLATION OF INTERCELLULAR FLUID TO STUDY BACTERIAL GROWTH

Intercellular fluid provide a perfect nutrition for the growth and multiplication of bacterial plant pathogen. Similarly, it can be used for isolation of bacterial metabolites like enzymes and toxins of bacterial plant pathogens.

Material Required

Fresh leaves of host plant, vacuum filtration apparatus, conical flask with beak, distilled sterile water, table top centrifuge with centrifuge tube, G-5 filters.

Procedure

The method to obtain intercellular fluid (IF) from leaves is the same as was used by Klement (1965). Pluck the leaves of the host plant and washed thoroughly with distilled water. These leaves are water infiltrated by dipping them under water in vacuum. Take out the leaves and remove excess water on their surface by pressing them in blotting paper. 65 g of infiltrated leaves are cut lengthwise, rolled, and placed in a centrifuge tube over a perforated support to prevent the leaf pieces from moving to the bottom. The leaves are centrifuged at 4000 rpm for 15 minutes. The intercellular fluid is driven out from the intercellular spaces and collected in the bottom of the centrifuge tube. 25 ml of IF is sterilized by filtration through bacteriological sintered glass filter (G-5 ultrathin) for further use to study the bacterial growth, enzyme production, or toxin production by the bacterial plant pathogen.

37 Preservation of Bacterial Disease Samples

Preservation of disease samples is very necessary to maintain the record of the disease and the types of symptoms produced or described at a given time. It is authenticated proof of the disease reported at a particular period.

37.1 DRY PRESERVATION OF SAMPLES

The method is used for preservation of leaf, twig, or root samples.

Material Required

Fresh disease samples, scissors, blotter paper, mercuric chloride solution, news paper, paper envelopes.

Protocol

Collect the infected leaf, twig, or root from the diseased plant. Wash in running tap water to remove the dirt and soak the excess water with blotter paper. Put the sample in mercuric chloride (0.1 percent) solution to get rid of unwanted saprophytic fungal and bacterial contaminants. Soak in the blotter paper and put the sample in its natural position in the newspaper sheet overnight under heavy pressure or clothing press. Put the sample in the paper envelop and properly label with name of disease, locality of collection, date of collection, name of collector, and so on. Check the sample at regular interval up to one month. The sample without degradation or contaminants/overgrowth should be finally used for preservation for long term.

37.2 MICROWAVE DRYING METHOD

Where the climatic conditions are not suitable for dry preservation method, the microwave drying method can be applied. In this method the color retention is better for the sample and it can be stored for a long time without fading.

Keep the disease sample leaves in microwave and run the microwave for 1 minute. Take out the leaves, keep in newspaper folds for 3 days under press conditions, and then store in paper bags.

37.3 PRESERVATION OF SAMPLES IN PRESERVATION SOLUTIONS

According to the color of disease leaves or fruits, the impregnation of the preservation solution has to be done to retain the same color of the host material for its preservation.

Impregnation of Green Color of the Leaf Sample

To save green color, copper acetate impregnating solution is used. Gradually increase the crystallization of copper acetate in 50 percent acetic acid solution (till it no longer dissolves). Dilute this with water 3 to 4 times as per the color of sample to be retain. Boil this mixture with the sample into it till the green color of the sample fade. Cool the solution. After 3–4 minutes the green color restores. Remove the sample, rinse with distilled water, and store in 5 percent formalin solution.

Note: The fruit samples, such as light green apples, pears, grapes, and so on, cannot be boiled and therefore are directly soaked in a mixture of 1 percent sulfite (100 ml), 95 percent alcohol (100 ml), and 800 ml water for 10–20 days. Then remove, rinse, and store in 5 percent formalin solution.

Impregnation of Red Color

For red fruit disease specimen like strawberry, pepper, potato, and other red plant samples, the impregnation solution used are cobalt nitrate (15 g), formalin (25 ml), chloride (10 g), and water (1 L).

Soak the specimen in this mixture for 2 weeks and then clean out by rinse with water. Then preserve in the mixture of sulfurous acid (30–50 ml), formalin (10 ml), ethanol 95 percent (10 ml), and water (1 L).

Impregnation of Yellow or Orange Color

Carotene-containing fruits such as lutein and apricot, pear, persimmon, citrus, and so on, are soaked in subsulfuric acid (containing SO_2, 5 percent) of 4–10 percent aqueous solution.

The impregnated specimen can be preserved in mixture of formalin (50 ml), ethanol 95 percent (300 ml), and water (2 L).

37.4   PRESERVATION OF LEAF SPOT/BLIGHT SAMPLES BY LAMINATION

Material Required

Infected leaf sample, mercuric chloride solution, blotter paper.

Procedure

Collect the infected (leaf spot/blight) leaves. Rinse them in tap water to remove the dirt. Soak the excess water in blotter paper. Put the leaf sample in mercuric chloride (0.1 percent) solution for 1 minute to get rid of unwanted saprophytic fungal and bacterial contaminants. Soak in blotter paper. Laminate the leaf in its natural shape.

37.5   PRESERVATION OF FRUIT SAMPLES BY WAX TREATMENT

Material Required

Infected fruit sample, liquid paraffin wax.

Procedure

Collect the infected fruit sample. Remove the dirt by swabbing with wet cotton followed by mercuric chloride solution (0.1 percent). Dip the fruit sample in liquid paraffin wax and take out to have wax coating. Preserve the sample in sampling containers or display boxes.

38 Measurement of Plant Bacterial Disease

38.1 GLOSSARY TO DESCRIBE SYMPTOMS

Black arm: Black arm refers to a bacterial infection when the main, secondary, and tertiary stem of the plant become black due to bacterial infection, for example, a black arm of cotton.

Blisters: A smooth raised growth on the leaves or fruit skin. These raised growth are black to brown with or without distinguished margins or halo. (See Figure 38.5.)

Blossom blight: The flower buds, flowers, or blossom get water-soaked, blackened, and wither; gives a blighted appearance. (See Figure 38.17.)

Cankers: A corky raised growth on the leaf surface, twig surface, or fruit skin; corky growths are generally yellowish in appearance. (See Figure 38.6.)

Crown galls: The infection led to formation of different sizes of galls at the crown level of the plant trunk. (See Figure 38.12.)

Defoliation: Shedding of leaves from the plant due to infection. (See Figure 38.18.)

Die-back: The infection led to death of young shoots and stems from top to bottom of the plant. (See Figure 38.11.)

Fire blight: When blight symptoms appear overnight and spread very quickly to the entire plantation or crop. (See Figure 38.4.)

Fruit spot: The bacterial infection led to different sizes and colors of spots on the fruit. (See Figure 38.14.)

Galls: Formation of galls of various sizes and shapes on the twigs, stem, and so on. (See Figure 38.9.)

Hypersensitive reaction: The bacterial infection led to quick browning or dull greenish reaction of infected leaf area. (See Figure 38.16.)

Leaf blast: Yellowing, browning, and drying of large areas of leaves, which gives blasted appearance to the foliage. (See Figure 38.3.)

Leaf blight: The blackening and withering of large areas of leaf due to infection, which gives blighted appearance to the foliage. (See Figure 38.2.)

Leaf spot: Formation of spots of various size, shape, and color on the dorsal or ventral surface of the leaf. These spots may be circular, ovoid, irregular, with or without halo, yellowish, brown, black, dark, faint, water-soaked, and so on. (See Figure 38.1.)

Necrosis of Root Bark: Brownish black necrotic area appear on bark of root system of infected plant which shows yellowing and dropping of leaves with wilting of a few branches of plants. Subsequently the entire plant wilt and die. (See Figure 38.19.)

Root rot: Rotting of roots, their disintegration, and dropping of plant foliage due to nonsupply of nutrient to plant. (See Figure 38.8.)

Seedling blight: Young growing seedling get blighted due to infection.

Soft rot: The infection leads to the maceration of the soft tissues of the leaves and fruit with pulpy appearance and bad odor. (See Figure 38.10.)

FIGURE 38.1   Symptoms of leaf spot on strawberry leaves. (Courtesy of Don Ferrin, Louisiana State University Agricultural Center, http://www.bugwood.org.)

FIGURE 38.2   Symptoms (rapid yellowing, wilting, and dying) of blight disease in rice. (Photo courtesy of International Rice Research Institute [IRRI], http://www.flickr.com.)

FIGURE 38.3 Symptoms of blast on almond leaves. (Courtesy of D. A. Golino, University of California, Davis, California, http://www.atlasplantpathogenicbacteria.it.)

FIGURE 38.4 Symptoms of fire blight on *Pyrus calleryana*, Bradford pear. (Courtesy of Charlie, http://www.flickr.com.)

FIGURE 38.5 Symptoms of blister on apple fruit. (Courtesy of T. van der Zwet, USDA.)

FIGURE 38.6 Symptoms of cankers on citrus leaves, twigs, and fruits. (Courtesy of S. G. Borkar and R. A. Yumlembam, PPAM, MPKV, Rahuri.)

FIGURE 38.7 Symptoms of wilt on tomato. (Courtesy of Don Ferrin, Louisiana State University Agricultural Center, http://ww.bugwood.org.)

Twig blight: The infection led to blackening or blightening of individual twigs randomly on the plant.

Vein blight: The bacterial infection led to blightening or blackening of leaf veins, for example, a leaf vein blight of cotton. (See Figure 38.13.)

Water-soaking: The bacterial infection led to water soaked, translucent spots/areas on leaves, twigs, or fruits. (See Figure 38.15.)

Wilt: Drooping and death of entire plant during its active growth phase. (See Figure 38.7.)

38.2   MEASUREMENT OF LEAF SPOT/BLIGHT DISEASE

The bacterial infection in crop plants can be recorded in term of incidence and intensity and then computed.

Incidence: Count the number of plants infected (in case of fruit orchards) and area of crop infected in the field (in case of field crops).

Intensity: Count the area of foliage infected out of the total foliage area. Give various grades for the foliage area infected.

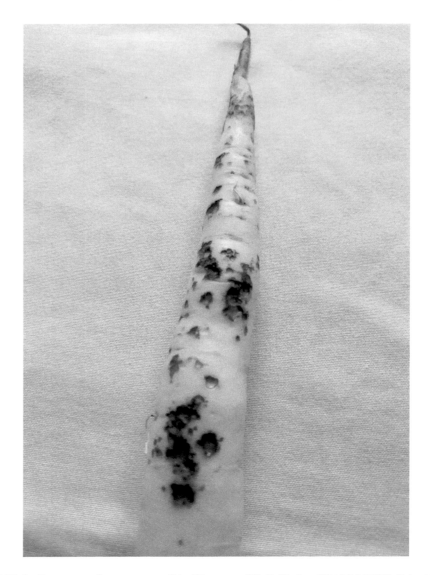

FIGURE 38.8 Symptoms of root rot on radish. (Courtesy of S. G. Borkar, PPAM, MPKV, Rahuri.)

Grade	Infection Area
5	More than 80 percent of leaf area or foliage is infected.
4	More than 60 percent and up to 80 percent leaf area or foliage is infected.
3	More than 40 percent and up to 60 percent leaf area or foliage is infected.
2	More than 20 percent and up to 40 percent leaf area or foliage is infected.
1	More than 1 percent and up to 20 percent leaf area or foliage is infected.
0	Less than 1 percent leaf area or foliage is infected.

FIGURE 38.9 Symptoms of gall formation on rose twigs. (Courtesy of Jennifer Olson, Oklahoma State University, http://www.bugwood.org.)

$$\text{Percent disease intensity} = \frac{\text{Sum of all the grade of observation}}{\text{Number of observation}} \times 100$$

$$\text{Disease pressure} = \text{Incidence} \times \text{Percent disease intensity}$$

Disease pressure of more than 10 values in a favorable epiphytotic condition can cause significant losses due to disease.

FIGURE 38.10   Symptoms of soft rot on onion foliage and bulb. (Courtesy of Howard F. Schwartz, Colorado State University, and Gerald Holmes, California Polytechnic State University at San Luis Obispo, http://www .bugwood.org.)

38.3   MEASUREMENT OF WILT DISEASE

The wilt disease is recorded in term of incidence.

Count the number of plants showing wilt symptoms.

Count total number of plants in the field of that crop and compute the percent wilt.

$$\text{Wilt percent} = \frac{\text{Number of wilted plants}}{\text{Total number of plants}} \times 100$$

More than 5 percent of wilt in the field can significantly reduce the yield.

FIGURE 38.11 Symptoms of die-back on pear. (Courtesy of Part of OSU Extension Plant Pathology Slide Set.)

FIGURE 38.12 Symptoms of crown gall on grapevine. (Courtesy of S. G. Borkar and Neeraja Agarwal, Department of Plant Pathology, Mahatma Phule Krishi Vidyapeeth, Rahuri.)

FIGURE 38.13 Symptoms of vein blight on cotton leaf. (Courtesy of Dr. S. G. Borkar and Ashwini Bhosale, PPAM, MPKV, Rahuri.)

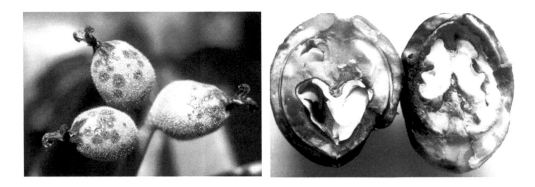

FIGURE 38.14 Symptoms of fruit spot on walnut fruits. (Courtesy of Gardan, L., French National Institute for Agricultural Research [INRA].)

FIGURE 38.15 Symptoms of water-soaking on cotton leaf. (Courtesy of S. G. Borkar and J. P. Verma, Division of Plant Pathology, IARI, New Delhi.)

FIGURE 38.16 Symptoms of hypersensitive on tobacco leaf. (Courtesy of S. G. Borkar and J. P. Verma, Division of Plant Pathology, IARI, New Delhi.)

FIGURE 38.17 Symptoms of blossom blight on apple tree. (Courtesy of Dr. Dave Rosenberger, retired plant pathologist, Cornell's Hudson Valley Lab.)

FIGURE 38.18 Symptoms of defoliation on citrus tree. (Courtesy of Gottwald, T. R., Graham, J. H., and Schubert, T. S., *Plant Health Progress*, 2002. doi: 10.1094/PHP-2002-0812-01-RV.)

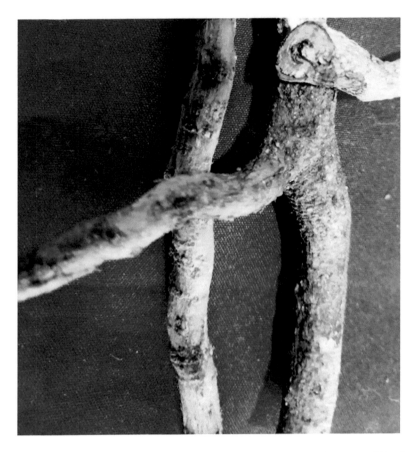

FIGURE 38.19 Symptoms of necrosis of root bark on pomegranate roots. (Courtesy of Dr. S. G. Borkar, Rupert Yulembam, and T. S. Ajayasree, Department of Plant Pathology, Mahatma Phule Krishi Vidyapeeth, Rahuri.)

39 Isolation of Antibacterial Compound

To control the bacterial infection, antibacterial compounds are required. The antibacterial compound can be present in certain plants or can be isolated from certain microorganisms. Testing of these antibacterial compounds against plant pathogenic bacteria and their identification at molecular level is important.

39.1 PRELIMINARY ASSESSMENT TECHNIQUE TO DETERMINE PRESENCE OF ANTIBACTERIAL COMPOUND IN PLANT BY COMPUTING BACTERIAL POPULATION

The method given by Yumlembam and Borkar (2014) is as follows.

I. Numeration of Bacterial Epiphytes on Leaf Surface

Material Required

Test plant, nutrient agar plates, and so on.

Protocol

The leaf imprint method is used to numerate bacterial population on the leaf surface of test plants. For these, leaves of test plant are collected, air blown for removal of dirt/dust, and then imprinted on the nutrient agar plates (both dorsal and ventral side of leaves on the respective plates). Incubate the leaf imprinted plates for 48 hours in BOD incubator at $28 \pm 1°C$ and note the growth of bacterial colonies with their population density.

II. Numeration of Bacterial Endophytes in Leaf Tissue

Protocol

For numeration of bacterial endophytes in the leaves of test plants, wash the leaves of the respective plants in tap water to remove the dirt. These leaves are surface sterilized in mercury chloride (0.1 percent) for 2 minutes and then wash thrice with distilled sterilized water. The sterilized 5 gm leaves are macerated in the sterile mortar and pestle in 10 ml of water and allowed to stand for 10 minutes. The supernatant is streaked on the nutrient agar plates with the help of sterilize inoculating needle. Incubate the inoculated plates at $28 \pm 1°C$ and observed for the development of bacterial colonies up to 3 days and note the growth of different colonies with their population density.

Observation

The absence of bacterial endophytes is indicative of probable presence of antibacterial compound in the test plant and the absence of either Gram-negative or Gram-positive bacteria or both is indicative that the antibacterial compound is active against what type of bacteria.

39.2   ASSESSMENT OF ANTIBACTERIAL PROPERTIES OF PLANT

Material Required

Test plant, test bacterial culture, water blank, nutrient agar plates, grinder, Whatman filter paper, and G-4 filters.

Protocol 1

Crush the plant sample (either roof, leaves, fruit, seed, bark, etc., as per test) in 5 ml of distilled water with a pestle and mortar and allow to settle for 10 minutes. Streak the supernatant on plate and incubate the plates at $27 \pm 2°C$ for 3 days. Observe the growth of bacterial colonies on streak portions.

Observation

Formation of bacterial growth on streaks indicate that the sample does not possess antibacterial activity. Development of no bacterial colonies on streaks indicate the sample possesses antibacterial compound.

Protocol II

Now wash the same sample (5 g) in running tap water to remove dirt, dry and crush in 10 ml of distilled sterile water, and allow to setter for 10 minutes. Filter the supernatant through a double layer of muslin cloth, followed by Whatman filter no. 42 and G-5 filter. Prepare the plates of test bacterium in nutrient agar medium by mixing the bacterial culture in luck warm nutrient agar media and

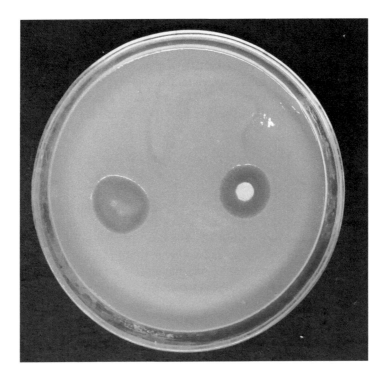

FIGURE 39.1   The plant extract containing antibacterial molecule form the zone of inhibition against test bacterium. (Courtesy of Dr. S. G. Borkar and Maria D'Souza, Department of Plant Pathology, Mahatma Phule Krishi Vidyapeeth, Rahuri.)

plating in sterilized Petri plates. When the medium with bacterial culture is solidified, place 0.1 ml of test plant extract at the center of the plate and incubate at $27 \pm 2°C$ for 48 hours in an incubator. Observe the plate for formation of inhibition zone by plant extract.

Observation

Formation of inhibition zone indicate the presence of antibacterial molecules in the plant extract and that the plant part possesses antibacterial properties. (See Figure 39.1.)

39.3 EFFICACY OF AQUEOUS EXTRACT OF MEDICINAL PLANT AGAINST PLANT PATHOGENIC BACTERIA (*XANTHOMONAS* SPP.)

Material Required

Test plant, test bacterium, nutrient agar plates, distilled sterile water, and so on.

Protocol

Suspend bacterial cultures in sterile distilled water, adjust to 0.1 OD at 620 nm (10^7 cfu/ml) and use as inoculums. From this add 0.5 ml each of bacterial inoculums respectively to 10 ml of aqueous sterilized leaf extract, 5 ml of sterilized aqueous leaf extract mixed with 5 ml of sterilized nutrient broth, and 5 ml nutrient broth as control. Incubate the tubes at $28 \pm 1°C$ for 48 hours to allow *Xanthomonas* to grow in these leaf extracts and nutrient broth. After 48 hours of incubation, a loop from the incubated suspension is streaked on NAS medium to detect *Xanthomonas* colonies.

Observation

No bacterial growth from aqueous leaf extract/leaf extract with nutrient broth, indicate the presence of antibacterial molecule/compound in leaf extract of test plant.

39.4 PREPARATION OF PLANTS EXTRACT IN SOLVENTS

For preparation of 5 percent leaf extract in solvent (5 gm of leaves in 100 ml solvents), 10 percent leaf extract (10 gm leaves in 100 ml solvents), and 25 percent extract (25 gm leaves in 100 ml solvents), wash the leaf samples thoroughly with tap water and macerate separately with respective solvents, viz., sterilized distilled water, ethanol, and methanol. The solvent extracts is then sieved through double-layer muslin cloth and then centrifuge at $4000 \times g$ for 30 minutes. The supernatant is filtered through Whatman no. 1 filter paper and sterilized through G4 bacterial filters to obtain sterile extracts. These solvent extracts are vacuum-dried and can be used for further studies of antibacterial properties.

39.5 ASSESSMENT AND QUANTIFICATION OF EQUIVALENT ANTIBACTERIAL COMPOUND IN SOLVENT PLANT EXTRACT

Material Required

Bacterial lawn of test bacterium, plant extract in solvents, piper disc, known concentration of antibiotics, and so on.

Protocol

For assessment of antibacterial properties of solvent extract, bacterial lawn is prepared by adding bacterial suspension (2 ml) to nutrient agar medium (20 ml) and poured in Petri plates. A sterilized

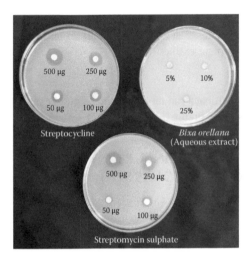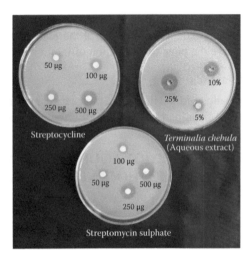

FIGURE 39.2   Zone of inhibition produced by aqueous extracted antibacterial compound of *Bixa orellana* and *Terminalia chebula* on XCC lawn. (Courtesy of Dr. S. G. Borkar and Rupert Yulembam, Department of Plant Pathology, Mahatma Phule Krishi Vidyapeeth, Rahuri.)

paper disc (Whatman no. 1, 5 mm) is dipped in the sterilized solvent extract, that is, aqueous, ethanolic, and methanolic extract, respectively, and is placed on the previously poured bacterial plates.

The disc of selective synthetic antibiotics streptomycin sulfate at 50, 100, 250, and 500 ppm concentration is used as control. Incubate the plates at $28 \pm 1°C$ for 24 hours to observe the inhibition zone around the disc.

Observation

Presence of clear zone indicate the presence of antibacterial activity in the respective extract. The inhibition zone of known concentration of antibiotics is used for comparison and for quantification of equivalent antibacterial compound in the solvent plant extract. (See Figure 39.2.)

39.6 ISOLATION OF ANTIBACTERIAL COMPOUND FROM PLANT SOURCE

Material Required

Acqueous extract of plant sample, different solvents, separating funnels, vacuum evaporator, vials, and so on.

Protocol

Prepare the aqueous extract of plant sample which have shown the antibacterial activity against the test bacterium. Take 25 ml aqueous extract (filtered through Whatman filter paper) in separating funnel and add equal volume of appropriate solvent. Shake the solvent extract vigorously and allow to settle for 10 minutes. Remove and collect the solvent phage, dry it in a vacuum evaporator. Dissolve the dried substance in 2 ml distilled sterile water and store in sterile vial.

Note: Different solvent must be used for the extraction of bactericidal compound. Prepare the plates of test pathogen on nutrient agar medium as described earlier and plate 0.1 ml of solvent extracted substance at the center of the plate. Incubate the plate at $27 \pm 2°C$ in an incubator for 48 hours to observe the zone of inhibition.

Note: The same procedure has to be followed to determine the solubility of the bactericidal compound in different solvents for further proper extraction and solvent selection.

Observation

Note the zone of inhibition in plates of different solvent extracted material and determine the suitable solvent to be used for isolation of antibacterial compound.

39.7 ISOLATION OF ANTIBACTERIAL COMPOUND FROM MICROBIAL SOURCE

Material Required

Cell-free extract of microbial growth, different solvent, separating funnel, vacuum evaporator, vials, and so on.

Protocol

Grow the antibacterial microbe in respective broth (use potato dextrose broth for fungal and nutrient broth for bacterial growth) for 7 days on rotary shaker at $27 \pm 2^{\circ}$C temp. Filter the broth through Whatman filter paper no. 42 to collect the cell-free extract for fungal growth or through G-5 filter for bacterial growth through vacuum filtration. Take 25 ml of cell-free extract in separating funnel and add equal volume of appropriate solvent. Shake the solvent extract vigorously and allow to settle for 10 minutes. Remove and collect the solvent phase and dry it in vacuum evaporator. Dissolve the dried substance in 2 ml distilled sterile water and store in sterile vial.

Note: A different solvent must be used for the extraction of the bactericidal compound. Prepare the plates of test pathogen on nutrient agar medium as described earlier and place 0.1 ml of solvent extracted substance at the center of the plate. Incubate the plate at $27 \pm 2^{\circ}$C in incubator for 48 hours to observe the zone of inhibition.

Note: The same procedure has to be followed to determine the solubility of the bactericidal compound in different solvent for further proper extraction and solvent selection.

Observation

Note the zone of inhibition in plates of different solvent extracted material and determine the suitable solvent to be used for isolation of antibacterial compound.

39.8 TESTING EFFICACY OF ANTIBACTERIAL MOLECULE

Material Required

Extracted molecule, test pathogen, nutrient agar medium, paper disc, known antibiotics, water blanks, and so on.

Protocol

Prepare 1 ml suspension of extracted molecule. Prepare 25, 50, 100, 250, and 500 ppm solutions of known antibiotic (streptomycin sulfate, streptocycline, bacterinol, etc). Prepare the nutrient agar plates with test bacterium as described earlier and allow the medium to solidify. Dip separate sterilized Whatman paper disc in suspension of extracted molecule and in various concentrations of antibiotic solution. Place these discs on the plate of test bacterium when the medium is solidified. Mark the place of the disc accurately. Incubate the plates at $27 \pm 2^{\circ}$C for 48 hours and record the zone of inhibition.

Observation

Record and measure the zone of inhibition of various concentrations of known antibiotic and the zone of inhibition of test molecule. Compare the zone of inhibition to calculate the concentration of antibacterial substance present in test molecule. Calculate the quantity of antibacterial molecular extracted per liter of extract.

39.9 IDENTIFICATION OF ANTIBACTERIAL MOLECULE

Identification of antibacterial molecules can be done by using gas chromatography mass spectroscopy (GC.MS) by using Quattro micro GC-MS unit. The column type used is DB-SMS with a column length of 30 m, fused silica capillary column (0.25 mm I.D; 0.25 μm film thickness), and using carrier gas as helium. The inlet pressure is held at 5 psi for 1 minute then raised at 20 psi/min and held for 4.5 minutes. The flow rate is maintained at 1 ml/minute with an initial column temp of 50°C and final temperature of 280°C. The rate of temperature change in the column is maintained as 25°C/minute for 6 minutes till the final temperature reaches 150°C, then it is maintained at 5°C/minute for 36 minutes until it reaches 280°C, with a hold time of 4 minutes.

Follow the following procedure for GC-MS analysis of test compound.

I. Preparation of Sample

Dilute 10 percent of ethanol and methanol plant extract with their respective solvent viz. Ethanol extract in ethyl acetate and methanol extract in methanol to make up the final concentration as 20 ppm (This is accomplished by adding 10 μl of test sample in 2 ml of solvent from where 40 μl is further diluted in 1 ml solvent making the final concentration 20 ppm).

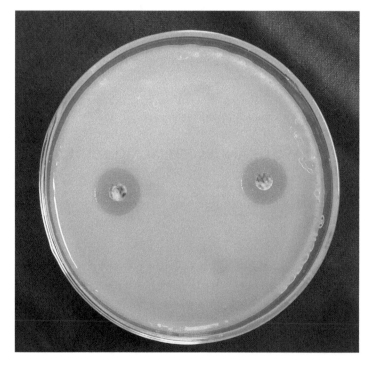

FIGURE 39.3 Inhibition zone formed by antibacterial molecule isolated from medicinal plant *Simarouba glauca* against *Xanthomonas*. (Courtesy of Dr. S. G. Borkar and Rupert Yulembam, Department of Plant Pathology, Mahatma Phule Krishi Vidyapeeth, Rahuri.)

II. Purification of Sample

Centrifuge the sample at 10 000 rpm for 4 minutes at 18°C for purification. Transfer the supernatant to vials for GC-MS analysis.

III. Injection of Sample

Get the initial oven temperature to 50°C. Rinse the syringe thrice with respective solvent viz. ethylacetate for ethanol extract and in methanol for methanol extract. It is again rinsed three times with the sample before injecting the sample. Inject 2 μl volume of sample in GC-MS column.

IV. Detection of Compound

Run the full scan of the sample with a temperature programming of 280°C, that is, 36 minutes. Carry out the qualitative analysis of the compound present in the sample by relative retention time (RRT) and comparison of the size (height or area) of the peak. Using library search of the GC-MS unit (NIST MS search 2.0) the probable compound can be identified. (See Figure 39.3.)

40 Isolation of Antibacterial Microbes

Some of the microbes like fungi, actinomycetes, yeast, and certain bacterial spp. show antagonism against bacterial plant pathogens and are therefore important sources of antibacterial microbial agents.

40.1 ISOLATION OF ANTIBACTERIAL FUNGI

Material Required

Diseased plant sample, soil sample with decomposed leaves in plant basin, nutrient agar plates, water blank, pestle and mortar, potato dextrose agar plates.

Protocol

Step 1

Crush the disease sample or decomposed diseased leaves in soil (1 g) in 5 ml of sterile water. Pipette out 0.2 ml macerate on the nutrient agar and potato dextrose agar plates separately and spread with bend glass rod. Incubate the plates at $27 \pm 2°C$ in incubator for 3 days.

Observation

Record the development of fungal and yeast colonies along with the bacterial plant pathogenic colonies. Please note the associated fungal or yeast colonies with plant pathogenic bacterial colonies. Such colonies needs to be separated.

Separate the fungal and yeast colonies associated with bacteria, pick them and transfer on potato dextrose medium for their growth.

Step 2

Prepare the lawn of plant pathogenic bacterial culture by plating nutrient agar medium with bacterial culture. After solidification of the medium put a fungal disc in the centre of the plate in invested position with slight pressure. Repeat this for each of the fungal isolate. Incubate the plates for 3 days.

Observation

Observe the zone of inhibition formed by the fungus in the bacterial lawn. Those fungal isolates which form zone of inhibition in the bacterial lawn are the antibacterial fungal bioagents.

40.2 ISOLATION OF ANTIBACTERIAL BACTERIA

Material Required

Diseased plant sample, soil sample with decomposed leaves in plant basin, nutrient agar plates, water blank, pestle and mortar.

Protocol

Step 1

Crush the disease sample or decomposed diseased leaves in soil (1 gm) in 5 ml of sterile water. Pipette out 0.2 ml macerate on the nutrient agar plates and spread with bend glass rod. Incubate the plates at 27 ± 2°C in an incubator for 3 days.

Observation

Record the development of all bacterial colonies along with the bacterial plant pathogenic colonies. Please note the associated bacterial colonies with plant pathogenic bacterial colonies. Such colonies needs to be separated.

Separate the saprophytic bacterial colonies associated with plant pathogenic based on their characteristic, pick them and transfer on nutrient agar medium for their growth.

Step 2

Prepare the lawn of plant pathogenic bacterial culture by plating nutrient agar medium with bacterial culture. After solidification of the medium put a paper disc charged with saprophytic bacterial culture in the center of the plate and apply slight pressure on the disc. Repeat this for each of the saprophytic bacterial isolate. Incubate the plates for 3 days.

Observation

Observe the zone of inhibition formed by the saprophytic bacterial cultures in the bacterial lawn of plant pathogenic bacteria. Those saprophytic bacterial isolates which form zone of inhibition in the bacterial lawn are the antibacterial bacterial bioagents.

40.3 TESTING ANTIBACTERIAL PROPERTY OF MICROBIAL CULTURE

Material Required

Test fungal/bacterial culture, test plant pathogenic bacterial culture, nutrient agar plate, sterile flask, water blank, nutrient broth, and so on.

Protocol

Prepare and sterilized 50 ml nutrient broth in conical flask. Prepare the suspension of test fungal/ bacterial culture as well as plant pathogenic bacterial culture (0.1 OD) under sterile conditions. Add 0.5 ml of test microbial culture and 0.5 ml of test plant pathogenic bacterial culture in nutrient broth, agitate in a shaker for 10 minutes, and incubate at 27 ± 2°C in incubator for 5 days. Take a loopful suspension from this and streak on nutrient agar plates. Incubate these plates at 27 ± 2°C in incubator for 48 hours to observe the development and growth of test plant pathogenic bacteria.

Observation

Observe the plates for the presence and growth of plant pathogenic test bacterium. Absence of plant pathogenic bacterial colonies in the plate indicate the antibacterial property of the test fungal/ bacterial bioagent.

40.4 MASS PRODUCTION OF ANTIBACTERIAL CULTURE

40.4.1 CARRIER-BASED MASS PRODUCTION

Material Required

Nutrient broth, antibacterial culture, rotary shakers, Incubators, talc powder.

Protocol

Grow antibacterial culture in nutrient broth in required quantities for 4–5 days by incubating alternatively on rotary shaker and incubators at $27 \pm 2°C$. Mix the grown broth culture in talc powder at appropriate quantities so that the talc powder is completely wet while mixing the solution. Dry the mixture under a shed to get dried talc powder. Test the population of antibacterial culture in talc powder by serial dilution plate technique. It should be 10^8 cfu/gram of powder. Fill the packets and sealed properly.

Preparation of carrier-based formulation of *Pseudomonas* fluorescence: The efficient strain of *Pseudomonas fluorescens* obtained from antagonistic test is multiplied on King B broth at $28 \pm 2°C$ for 3 days in an incubator. Sterilized the talc powder in an autoclave at $121°C$ at 15 pounds of pressure for 20 minutes. Mix 400 ml broth culture of *Pseudomonas* fluorescence with 1 kg talc powder. Mix it uniformly and keep for drying in aseptic condition for removing excess moisture. Take the initial count and pack the formulation in sterile polythene bags. Test the viability of biocontrol agent pf in carrier regularly.

40.4.2 LIQUID-BASED MASS PRODUCTION

Standard liquid medium for mass production of *Pseudomonas fluorescens*: Protease peptone, dipotassium hydrogen ortho phosphate (K_2HPO_4), glycerol, $MgSO_4.7H2O$, trehalose, arabinose, phosphovinyl pyruvate.

Preparation of standard liquid formulation: Sterilize the standard liquid medium and inoculate with an efficient strain of *Pseudomonas fluorescens* under sterile condition in conical flasks. Incubate these flask at $28 + 2°C$ on a rotary shaker for 8 hours per day for 4–5 days. Remove the flasks after vigorous growth of bacterium and incubate at $28°C$ for 1–2 days. Take the initial count and store the formulation in sterilized plastic bottles. Test the viability of biocontrol agent pf in liquid formulation regularly.

41 Preparation of Bactericidal Formulations

To control the bacterial infection, the formulations of available bactericide or bactericidal compounds have to be prepared in the form of sprays or paste to apply against the disease.

41.1 PREPARATION OF BACTERICIDAL SPRAY

41.1.1 PREPARATION OF BORDEAUX MIXTURE

Material Required

Copper sulfate, lime, water, plastic tub or tank.

Protocol

Dissolve 1 kg of copper sulfate in 20 L of water in a plastic container, mix well, and keep overnight. Dissolve 1 kg of hydrated lime in 20 L of water in another plastic container, mix well, and keep overnight. Sieve the supernatant from both the container with double layer muslin cloth in another big container and make the volume to 100 L with water. This gives 1 percent Bordeaux mixture. Adjust the pH of the Bordeaux mixture to approximately 7.0 with the help of the lime water.

To make the lower concentrations of the Bordeaux mixture, add appropriate quantities of water into 1 percent Bordeaux mixture. To make the higher concentration increase the quantity of copper sulfate and lime appropriately. The mixture thus prepared is ready for the spraying to control the bacterial infections on plant.

41.1.2 PREPARATION OF BACTERICIDAL MIXTURE

Material Required

Chemical to be used as bactericide, water, plastic tub, and so on.

Procedure

Select the bactericide or antibiotics to be used for spray. Generally 250 ppm, 500 ppm, 1000 ppm, or 0.2 percent solution of bactericide or antibiotics are used.

Chemical (g)	Water (L)	Concentration to be Obtained
25	10	250 ppm
50	10	500 ppm
100	10	1000 ppm
20	10	0.2 percent
10	10	0.1 percent

Dissolve the requisite quantity of antibiotic or bactericide in water as shown in the table to obtain the required concentration of bactericide; stir well. Sieve through a double layer muslin cloth and use for spraying to control the bacterial infection on plant.

41.2 PREPARATION OF BACTERICIDAL PASTES

Bactericidal pastes are necessarily applied and used to control the bacterial infection of twigs, stem, or main trunk of the plant.

41.2.1 PREPARATION OF BORDEAUX PASTE

Material Required

Copper sulfate, lime, water, plastic tub.

Procedure

Add 450 g of copper sulfate and 450 g of lime in 3.5 L water and mix well to prepare paste.

41.2.2 PREPARATION OF RAHURI PASTE

This is an oil-based paste formulation prepared by Professor Borkar and his student Rubina Pokharel. The paste is patented in India (CBR no. 4708 of 27.3.2015) for control of bacterial infection on twigs and stems of pomegranate. The paste also rejuvenates the infected plants by sprouting of new shoots on infected stems.

Bibliography

Alarcon, B., Gorris, M. T., Cambra, M., and Lopez, M. M. 1995. Serological characterization of potato isolates of *Erwinia carotovora* subsp. *atroseptica* and subsp. *carotovora* using polyclonal and monoclonal antibodies. *J. Appl. Microbiol.* 79:592–602.

Alcorn, S. M., and Ark, P. A. 1953. Softening paraffin embedded plant tissues. *Stain Technol.* 28:55–56.

Alegria, M., Souza, D., Andrade, M., Docena, C., and Khater, L. 2005. Identification of new protein–protein interactions involving the products of the chromosome and plasmid-encoded type IV secretion loci of the phytopathogen Xanthomonas axonopodis pv. citri. *J. Bacteriol.* 187:2315–2325.

Alvarez, A. 2004. Integrated approach for detection of plant pathogenic bacteria and diagnosis of bacterial diseases. *Annu. Rev. Phytopath.* 42:339–366.

Amann, R. and Fuchs, B. M. 2008. Single cell identification in microbial communities by improved *fluorescence* in situ hybridization techniques. *Nature* 6:339–348.

Amann, R. I., Binder, B. J., Olson, R. J., Chisholm, S. W., Devereux, R., and Stahl, D. A. 1990a. Combination of 16S rRNA-targeted oligonucleotide probes with flow cytometry for analyzing mixed microbial population. *Appl. Environ. Microbiol.* 56:1919–1925.

Amann, R. I., Krumholz, L., and Stahl, D. A. 1990b. Fluorescent-oligonucleotide probing of whole cells for determinative, phylogenetic and environmental studies in microbiology. *J. Bacteriol.* 172:762–770.

Arnow, L. E. 1937. Colorimetric determination of the components of 3, 4-dihydroxyphenylalanine tyrosine mixtures. *J. Biol. Chem.* 118:531.

Azegami, K., Nishiyama, K., Watanabe, Y., Kadota, I., Ohuchi, A., and Fukazawa, C. 1987. *Pseudomonas plantarii* sp. nov. the causal agent of rice seedling blight. *Int. J. Syst. Bacteriol.* 37:144–152.

Bar-Joseph, M., and Garnsey, S.M. 1981. Enzyme Linked immunosorbent assay (ELISA): Principles and application for diagonosis of plant viruses, in *Plant Diseases and Vectors*, ed. K. Maramorosch, and K. F. Harris, pp. 35–58. New York, London: Academic Press.

Bdliya, B. S., and E. Langerfeld. 2005. A semi-selective medium for detection, isolation and enumeration of *Erwinia carotovora* ssp. *carotovora* from plant material and soil. *Trop. Sci.* 45(2):90–96.

Bergervoet, J. H. W., Peters, J., Van Beckhoven, J. R. C. M., van den Bovenkamp, G. W., Jacobson, J. W., and Van der Wolf, J. M. 2007. Multiplex microsphere immune detection of potato virus Y, X, and PLRV. *J. Virol.* 117:97–107.

Bhosale, A. N. 2010. *Studies on bacterial blight of Bt-cotton*. MSc thesis. Rahuri, India: Mahatma Phule Agriculture University. p. 88.

Birajdar, K. S. 2007. *Effect of edible and non-edible oils on survival and virulence of citrus canker bacterium Xanthomonas axonopodis pv. vitri*. MSc thesis. Rahuri, India: Mahatma Phule Agriculture University. p. 93.

Borkar, S. G. 1982. *Races of Xanthomonas campestris pv. malvcearum and their interaction with compatible/ non compatible host*. PhD thesis. New Delhi, India: Indian Agricultural Research Institute. p. 171.

Borkar, S. G. 1989a. Glycoprotein secretion factor and glycoprotein secretion by bacterial clone *Xanthomonans corylina*. *Indian J. Plant Pathol.* 7(1):36–37.

Borkar, S. G. 1989b. Immunoglobin content in the antiserum of *Xanthomonans corylina* and its affinity towards the same antigen as influenced by immunization period. *Indian J. Experimental Biol.* 27:464–466.

Borkar, S. G. 1989c. Use of specific immunoglobulin under immunofluroscence to detect latent infection of *Xanthomonans corylina* in field samples of noisetier. *Indian J. Plant Pathol.* 7(1):23–26.

Borkar, S. G. 1990. One race horizontal resistance model. *J. Cotton Res. Develop.* (4):220–227.

Borkar, S. G. 2002. Scheme for classification of races of *Xanthomonas campestris pv. Viticola*, a leaf blight pathogen of grape vine. *Indian J. Plant Pathol.* 20(1&2):67–69.

Borkar, S. G. and Verma, J. P. 1984a. Effect of phenolic substances on pigmentation and pathogenicity of *Xanthomonans campestris pv. malvacarum*. *Appl. Microbiol. Biotechno.* 19:435–436.

Borkar, S. G., and Verma, J. P. 1984b. Population dynamics of *X.c. pv. malvacearum* in compatible/incompatible reaction of cotton cvs. *Acta phytopathologica Academiae Scienatiarum Hungaricae* 20(1–2):31–34.

Borkar, S. G., and Verma, J. P. 1988. DNA content of *X.c.pv. malvacearum* and its relation to virulence and drug resistance. *Indian J. Plant Pathol.* 6(1):52–55.

Borkar, S. G., and Verma, J. P. 1989a. Exopolysaccharides, a water soaking inducing factor by bacterial blight of cotton pathogen *X.c.pv. malvacarum*. *Coton et fibre Tropicals Facs* 2:149–153.

Borkar, S. G., and Verma, J. P. 1989b. Water-soaking reaction inhibitory metabolite excreted by *X.c.pv. malvacearum. Folia Microbiologica* 34:515–520.

Borkar, S. G., and Verma, J. P. 1991a. Inhibition of susceptible/hypersensitive reaction by exopolysaccharide of avirulant strain *of X.c.pv. malvacearum. Folia Microbiologica* 36:173–176.

Borkar, S. G. and Verma, J. P. 1991b. Transformation of virulence factor of *X.c.pv. malvacearum* in plants. *Folia Microbiologica* 36:169–172.

Borkar, S. G., Verma, J. P., and Singh, R. P. 1980. Transmission of *Xanthomonas malvacearum* of cotton through spotted bollworm. *Indian J. Entomol.* 42(4):390–397.

Boyer, G. L., Kane, S. A., Alexander, J. A., and Aronson, D. B. 1998. Siderophore formation in iron-limited cultures of frankia sp. *Canadian J. Bot.* 77:1316–1320.

Bradford, M. M. 1976. A rapid and sensitive method for the quantification of microgram quantities of protein utilizing the principle of protein-dye binding. *Anal. Biochem.* 72(1–2):248–254.

Breed, R. S., Murray, E. G. D., and Smith, N. R. 1957. Sources and characteristic of additional Pseudomonas species. *Bergey's Manual of Determinative Bacteriology*, 7th ed., Baltimore: Williams & Wilkins. p. 136.

Brenner, D. J., Krieg, N. R., and Staley, J. T. 2007. The Proteobacter, Part B: The Gammaproteobacter, Xanthomonas. *Bregey's Manual of Systematic Bacteriology* Baltimore: Williams & Wilkins. vol. 2, p. 323.

Carattoli, A., Bertini, A., Villa, L., Falbo, V., Hopkins, K. L., and Threlball, E. J. 2005. Identification of plasmids by PCR-based replicon typing. *J. Microbiol. Methods* 63:219–228.

Carroll, R. B., and Lukezic, F. L. 1971. Preservation of corynebacterium insidiosum in a sterile soil mix without loss of virulence. *Phytopathology* 61:688–690.

Couturier, M., Bex, F., Berquist, P. L., and Maas, W. K. 1988. Identification and classification of bacterial plasmids. *Microbiol. Rev.* 52:375–395.

Curuso, P., Bertolini, E., Cambra, M., and Lopez, M. M. 2003. A new and cooperational polymerase chain reaction (co-PCR) for rapid detection of *Ralstonia solanacearum* in water. *J. Microbiol. Methods* 55:257–272.

Datta, N., and Hedges, R. W. 1971. Compatibility groups among fi-R factors. *Nature* 234:222–223.

Datta, N., and Hughes, V. M. 1983. Plasmids of the same Inc group in enterobacteria before and after the medical use of antibiotics. *Nature* 306:616–617.

De Silva, D. P. P., Jones, P., and Shaw, M. W. 2002. Identification and transmission of piper yellow mottle virus and cucumber mosaic virus infecting black pepper in Sri Lanka. *Plant Pathol.* 51:537–545.

Delong, E. F., Wickham, G. S., and Pace, N. R. 1989. Phylogenetic stains: Ribosomal RNA-based probes for the identification of single cells. *Science* 243:1360–1363.

Desmarais, S. M., Cava, F., De Pedro, M. A., and Huang, H. C. 2014. Isolation and preparation of bacterial cell wall for compositional analysis by Ultra performance liquid chromatography. *JOVE. e.* 51183.

Diano, M., Le Bivic, A., and Hirn, M. 1987. A method for the production of highly specific antibodies. *Anal. Biochem.* 166:224–229.

Dickey, R. S. 1967. Techniques for histological studies of plant material. *Proceedings of the First Workshop on Phytobacteriology*, ed. R. N. Goodman, pp. 23–26.

Dubois, M., Gilles, K., Hamilton, J. K., Rebers, P. A., and Smith, F. 1951. A colorimetric method for the determination of sugars. *Nature* 168:167.

Dye, D. W. 1962. The inadequacy of the usual determinative tests for the identification of *Xanthomonas* spp. *N. Z. J. Sci.* 5:393–416.

Fargier, E., and Manceau, C. 2007. Pathogenicity assay restrict the species *Xanthomonas campestris* into three pathovars and reveal nine races within *Xanthomonas campestris* pv. *campestris. J. Plant Pathology* 56:805–818.

Friedman, B. 1964. Carbon source and tetrazolium agar to distinguish virulence in colonies of *Erwinia carotovora. Phytopathology* 54:494–495.

Gabriel, D. W., Kingsley, M. T., Hunter, J. E., and Gottwald, T. 1989. Reinstatement of *Xanthomonas citri* (ex Hasse) and *X. phaseoli* (ex Smith) to species and reclassification of all *X. campestris* pv. *citri* strains. *Int. J. Syst. Bacteriol.* 39:14–22.

Gavini, F., Mergaert, J., Beji, A., Mielcarek, C., Izard, D., Kersters, K., and Ley, J. De. 1989. Transfer of *Enterobacter Agglomerans* to *Pantoea* gen. nov. as. *Pantoea Agglomerans* comb. nov. and Description of *Pantoea* dispersa sp. nov. *Int. J. Syst. Bacteriol.* 39(3):337–345.

Ghosh, R., Nagarardhini, A., Sengupta, A., and Sharma, M. 2015. Development of loop-mediated isothermal amplification (LAMP) assay for rapid detection of fusarium oxysporium f.sp.ciceri wilt pathogen of Chickpea. *BMC Res Notes* 8:40.

Gillan, A. H., Lewis, A. G., and Anderson, R. S. 1981. Quantitative determination of hydroxamic acids. *Anal. Chem.* 53:841–844.

Giovannoni, S. J., DeLong, E. F., Olsen, G. J., and Pace, N. R. 1988. Phylogenetic group specific oligodeoxynucleotide probes for identification of single microbial cells. *J. Bacteriol.* 170:720–726.

Goto, M. 1983. *Pseudomonas ficuserectae* sp. nov. the causal agent of bacterial leaf spot of ficus. *Int. J. Syst. Bacteriol.* 33:546–550.

Götz, A., Pukall, R., Tietze, E., Prager, R., Tschäpe, H., van Elsas, J. D., Smalla, K. 1996. Detection and characterization of broad-host-range plasmids in environmental bacteria by PCR. *Appl. Environ. Microbiol.* 62:2621–2628.

Graber, P., and William, C. A. 1953. Methode permettant L'etude conjugee des proprietes electrophoretique et immunochiminique d'un. mélange de proteins. Application au serum snaguin. *Biochem. Biophys. Acta.* 10(1):193–194.

Gross, D. C., and Devay, J. E. 1977. Production and purification of syringomycin, a phytotoxin produced by *Pseudomonas syringae*. *Physiol. Plant Pathol.* 11(1):13–28.

Gross, D. C., Cody, Y. S., Proebsting, E. L., Radamaker, G. K., and Spotts, R. A. 1983. Distribution, population dynamics, and characteristics of ice-nucleation active bacteria in deciduous fruit tree orchards. Applied and environmental. *Microbiology* 46(6):1370–1379.

Guckert, J. B., Cooksey, K. E., and Jackson, L. L. 1988. Lipid solvent system are not equivalent for analysis of lipid classes in the microeukaryotic green algae, chlorella. *J. Microbiol. Methods* 8:139–149.

Harper, S. J. 1980, 2012. Human humoral or circulating antibodies belong to five classes Ig A, IgG, IgM, IgE, in *Radioiodination: Theory, Practice, and Biomedical Application*, ed. M. K. Dewanjee. Hagerstown, Maryland.

Hiramatsu, K., Lgarashi, M., Morimoto, Y., Baba, T., Umekita, M., and Akamatsu, Y. 2012. Curing bacteria of antibiotic resistance: Reverse antibiotics, a novel class of antibiotics in nature. *J. Antimicrobial Agents* 39 (6):478–485.

Hoang, D. D., Pham, V. D., and Loan, L. C. 2010. Study on the use of combination resistance genes in rice lines against *Xanthomonas aryzae* pv. *oryzae* in Cuulong river delta. *Omonrice* 17:147–151.

Huang, N., Angeles, E. R., Domingo, J., Magpanty, G., Singh, S., Zhan, G., Kumarvadivel, N., Bennett, J., and Khush, G. S. 1997. Pyramiding of bacterial leaf blight resistance genes in rice marker assisted selection using RFLP and PCR. *Theor. Appl. Genetics* 95:313–320.

Jensen, W. A. 1962. *Botanical Histochemistry*. San Fransisco: W. H. Freedman.

Johansen, D. A. 1940, 1999. Selected staining methods for plant microtechnique Johansen's, Safranin, and Fast Green, in *Plant Microtechnique and Microscopy*, ed. S. E. Ruzin. Oxford: Oxford University Press.

John, H. A., Birnstil, M. L., and Jones, J. K. 1967. Formation and detection of RNA–DNA hybrids at the cytological level. *Nature* 223:582–587.

Kadam, S. S. 2010. *Studies on bacterial leaf spot of Grapevine*, MSc degree. Rahuri, India: Mahatma Phule Agriculture University. p. 113.

Kale, P. B., Chimote, V. P., Raghuwanshi, K. S., Kale, A. A., Jadhav, A. S., and Borkar, S. G. 2012. Microbial, biochemical, pathogenicity and molecular characterization of *Xanthomonas axonopodis* pv. Punicate form pomegranate. *J Pure Appl Microbiol* 6(4):1699–1706.

Kamoun, S., Kamdar, H. V., Tola, E., and Kado, C. I. 1992. Incompatible interaction between crucifers and *Xanthomonas campestris* involve a vascular hypersensitive response: Role of the hrp x locus. *Molecular Plant Microbe Interaction*. http://aps.net.org.

Kauffman, H. E., Reddy, A. P. K., Hsieh, S. P. Y., and Merca, S. D. 1973. An improved technique for evaluating resistance of rice varieties to Xanthomonas oryzae. *Plant Disease Rep.* 57:537–541.

Kelman, A. 1954. The relationship of pathogenicity in *Pseudomonas solanacearum* to colony appearance on a tetrazolium medium. *Phytopathol.* 44:693–694.

Klement, Z. 1965. Method of obtaining fluid from the intercellular spaces of foliage and the fluid's merit as a substrate for phytobacterial pathogens. *Phytopathol.* 55:1033–1034.

Klement, Z., and R. N. Goodman. 1967. The hypersensitive reaction to infection by bacterial plant pathogens. *Annual Rev. Phytopathol.* 5:17–44.

Krieg, N. R., and Halt, J. C. (eds.). 1984. *Bergey's Manual of Systematic Bacteriology*, 1st ed., Baltimore: Williams & Wilkins. vol. 1, p. 301.

Laemmli, U. K. 1970. Cleavage of structural protein during assembly of the Head of Bacteriophage T$_4$. *Nature* 227(5259):680–685.

Lau, A., Sorrell, T. C., Lee, O., Stanley, K., and Halliday, C. 2008. Colony multiplex-tandem PCR for rapid, accurate identification of fungal cultures. *J. Clin. Microbiol.* 46:4058–4060

Lavermicocca, P., Lonigro, S. L., Evidente, A., and Andolfi, A. 1999. Bacteriocin production by *Pseudomonas syringae* pv. *ciccaronei* isolation and partial characterization of the antimicrobial compound. *J. Appl. Microbiol.* 86:257–265.

Lederberg, J. and Tatum, E. L. 1946. Gene recombination in *Escherichia coli. Nature* 158:558.

Lee, K., Yum, J. H., Yong, D., Lee, H. M., Kim, H. D., and Docquier, J. D. 2005. Novel acquired metallo-B-lactamase gene, bla SIM-1, in a class 1 integron from acinetobacter baumannii clinical isolates from Korea. *Antimicrob. Agnets Chemother.* 49:4485–4491.

Lewis, P. J., Thaker, S. D., and Errington, J. 2000. Compartmentalization of transcription and translation in bacillus subtilis. *EMBO. J.* 19:710–718.

Lievens, B., Claes, L., Vanachter, A. C. R. C., Cammue, B. P. A., and Thomma, B. P. H. J. 2006. Detecting single nucleotide polymorphism using DNA arrays for plant pathogen diagnosis. *FEMS Microbiol Lett* 225:129–139.

Liu, W. T., Marsh, T. L., Cheng, H., and Forney, L. J. 1997. Characterization of microbial diversity by determining terminal restriction fragment length polymorphisms of genes encoding 16S rRNA. *Appl. Environ. Microbiol.* 63:4516–4522.

Lopez, M. M., Bertolini, E., Olmos, A., Caruso, P., Gorris, M. T., Llop, P., Penyalver, R., and Cambra, M. 2003. Innovative tools for detection of plant pathogenic bacteria and viruses. *Int. Microbiol.* 6:23–243.

Mary, L. S., and Vidaver, A. K. 1986. Isolation and characterization of syringacin W-1, a bacteriocin produced by *Pseudomonas syringae* pv. *syringae. Canadian J. of Microbiology.* 32(3):231–236.

Mastan, A., Sreedevi, B., and Kumari, J. 2014. Evaluation of the in vitro antioxidant and antibacterial activities of secondary metabolites produced from lichen. *Asian J. Pharma. Clin. Res.* 7:193–198.

Matharia, L., and Joseph, L. 2007. Antigen antibody reaction, in *Methods for General and Molecular Microbiology*, 3rd ed., eds. C. Reddy, T. Beveridge, J. Breznak, G. Marzluf, T. Schmidt, L. Snyder, pp. 138–167. Washington, DC: ASM Press.

Meyer, J. M., and Abdallah, M. A. 1978. The *fluorescent* pigment of *Pseudomonas flourescens*, Biosynthesis, purification and physical–chemical properties. *J. Gen. Microbiol.* 107:319–328.

Miyajimat, K., Tanil, A., and Akita, T. 1983. *Pseudomonas fuscovaginae* sp. nov. nom. rev. *Int. J. Syst. Bacteriol.* 33:656–657.

Moter, A., and Gobel, U. B. 2000. Fluorescence in situ hybridization (FISH) for direct visualization of microorganisms. *J. Microbial Methods* 41(2):85–112.

Nagrale, D. T. 2011. *Studies on bacterial collar rot of banana.* PhD thesis. Rahuri, India: Mahatma Phule Agriculture University. p. 131.

New, P. B., and Kerr, A. 1971. A selective medium for *Agrobacterium radiobacter* biotype 2. *J. Appl. Bacteriol.* 34:233–236.

Novick, R. P. 1987. Plasmid incompatibility. *Microbial Review* 51:381–395.

Olmos, A., Bertolini, E., and Cambra, M. 2002. Simultaneous and co-operational amplification (Co-PCR): A new concept for detection of plant viruses. *J. Virol. Methods* 106:51–59.

Ouchterlony, O. 1948. Antigen antibody reaction in gel. *Acta pathologica et microbiologica scandinarica.* 26(4):507–515.

Oudin, J. 1946. Method of immunochemical analysis by specific precipitation in gelled medium. *C.R. Acad. Sci. Paris* 222:115–116.

Pardue, M. L., and Gall, J. G. 1969. Molecular hybridization of radioactive DNA to the DNA of cytological preparations. *Proc. Natl. Acad. Sci. USA* 64(2):600–604.

Pierce, L., and Mccain, A. H. 1992. Selective medium for isolation of pectolytic *Erwinia* sp. *Plant Disease* 76:382–384.

Pruvost, O., Hartung, J. S., Civerolo, E. L., Dubois, C., and Perrier, X. 1992. Plasmid DNA fingerprints distinguish pathotypes of *Xanthomonas campestris* pv. *citri. Phytopathol.* 82:485–490.

Pugsley, A. P., and Oudega, B. 1987. Methods for studying colicins and their plasmids, in *Plasmids, a Practical Approach*, ed. K. G. Hardy, pp. 105–161. Oxford: IRL Press.

Rademaker, J. L. W., and De Bruijn, F. J. 1997. Characterization and classification of microbes by rep- PCR genomic fingerprinting and computer assisted pattern analysis, in *DNA Markers: Protocols, Applications, and Overviews*, eds. G. Caetano-Anolles and P. M. Gresshoff, pp. 151–171. New York: John Wiley & Sons.

Rawlins, T. E., and Takahashi, W. N. 1952. *Plant Histochemistry and Virology.* Millbrae, CA: The National Press.

Ristaino, J. B., Madritch, M., Trout, C. L., and Parra, G. 1998. PCR amplification of ribosomal DNA species identification in the plant pathogen genus phytopathora. *Appl. Envir. Micro.* 64:948–954.

Robinson, K., Bellaby, T., and Wakelin, D. 1996. The mycobacterial component of complete Freund's adjuvant induces expulsion of the intestinal nematode trichinella spiralis in mice. *Appl. Parasitol.* 37:23–31.

Rohlf, F. J. 1994. *Numerical taxonomy and multivariate analysis system version 1.80 manual.* New York: Exeter Software.

Sanchez, A. C., Brar, D. S., Huang, N., Li, Z., and Khush, G. S. 2000. Sequence tagged site marker assisted selection for three bacterial blight resistance genes in rice. *Crop Science* 40:792–797.

Sathyanarayana, N., and Verma, J. P. 1993. Possible role of plasmids in the virulence of Xanthomonas campestris pv. malvacearum. *Indian Phytopath.* 46(2):165–166.

Schaad, N. W., and Frederick, R. D. 2002. Real time PCR and its application for rapid plant disease diagnostics. *Canadian J. of Plant Pathology* 24:250–258.

Schleifer, K. H., and Kandler, O. 1972. Peptidoglycan types of bacterial cell walls and their taxonomic implications. *Bacteriol. Rev.* 36:407–477.

Schonhuber, W. B., Fuchs, S., Juretschko, S., and Amann, R. 1997. Improved sensitivity of whole cell hybridization by the combination of horseradish peroxidase-labelled oligonucleotide and tyramide signal amplification. *Appl. Environ. Microbiol.* 63:3268–3273.

Schroth, M. N., Thompson, J. P., and Hildebrand, D. C. 1965. Isolation of *Agrobacterium tumefaciens*—A radiobacter group from soil. *Phytopath.* 55:645–647.

Schwartz, D. C., and Cantor, C. R. 1984. Separation of yeast chromosome sized DNAs by pulsed field gradient gel electrophoresis. *Cell* 37(1):67–75.

Schwyn, B., and Neiland, J. B. 1987. Universal chemical assay for the detection and determination of siderphores. *Anal. Biochem.* 160:47.

Sellner, L. N., Coelen, R. J., and Mackenzie, J. S. 1992. A one tube one manipulation RT-PCR reaction for detection of Ross River virus. *J. Virol. Methods.* 40:255–264.

Silverman, A. P., and Kool, E. T. 2005. Quenched probes for highly specific detection of cellular RNAS. *Trends Biotechnol.* 23:225–230.

Smale, B. C., and Worley, J. F. 1956. Evaluation of 2,3,5 triphenyl tetrazolium chloride for obtaining pathogenic types from stock cultures of halo blight and common blight organisms. *Plant Dis. Reptr.* 40:628.

Stanley, K. K., and Szewczuk, E. 2005. Multiplexed tandem PCR: Gene profiling from small amounts of RNA using SYBR Green detection. *Nucleic Acids Res.* 33:e180.

Suggs, S. V., Hirose, T., Miyake, T., Kawashima, E. H., Johnson, M. J., Itakura, K., and Wallace, R. B. 1981. Use of synthetic oligonucleotide for the isolation of specific cloned DNA sequences, in *Developmental Biology Using Purified Gases*, eds. D. Braun and C. F. Fox, pp. 683–693. New York: Academic Press.

Sundin, G. W. 2007. Genomic insights into the contribution of phytopathogenic bacterial plasmids to the evolutionary history of their host. *Annu. Rev. Phytopathol.* 45:129–151.

Van Ooyen, A. 2001. Theoretical aspects of pattern analysis, in *New Approaches for the Generation and Analysis of Microbial Typing Data*, eds. L. Dijkshoorn, K. J. Towner, and M. Struelens, pp. 31–45. Amsterdam: Elsevier.

Verma, J. P., and Borkar, S. G. 1981. Transmission of *X.c.pv malvacearum* through red cotton bug. *Annals of Agril. Research* 2(1):57–63.

Verma, J. P., and Singh, R. P. 1975. Studies on the distribution of races of *Xanthomonas malvacearum* in India. *Indian Phytopath.* 28:459–463.

Versalovic, J., Schneider, M., De Bruijn, F. J., and Lupski, J. R. 1994. Genomic fingerprinting of bacteria using repetitive sequence based polymerase chain reaction. *Method Mol. Cell Biol.* 5:25–40.

Vicente, J. G., Conway, J., Roberts, S. J., and Taylor, J. D. 2001. Identification and origin of *Xanthomonas campestris* pv. *campestris* races and related pathovars. *Phytopathology* 91:492–499.

Vicente, J. G., Ignatov, A., Conway, J., Roberts, S. J., and Taylor, J. B. 1998. *Development of an improved brassica differential series for the identification of races of Xanthomonas campestris. pv. campestris.* 7th International Conference on Plant Pathology, pp. 2. 2. 71.

Vidavar, A. K. 1967. Synthetic and complex media for the rapid detection of *fluorescence* of phytopathogenic *Pseudomonads*: Effect of the carbon source. *Appl. Microbial.* 15:1523–1524.

Vidavar, A. K., Mathys, M. L., Thomonas, M. E., and Schuster, M. L. 1972. Bacteriocins of the phytopathogens *Pseudomonas syringae*, *P. glycinea* and *P. Phaseolicola*. *Canadian J. of Microbiology* 18(6):705–713.

Wagner, M., Roger, A., Flax, J., Brusseau, G., and Stahl, D. 1998. Phylogeny of dissimilatory reductases supports an early origin of sulfate respiration. *J. Bacteriol.* 180:2975–2982.

Ward, J. H., Jr. 1963. Hierarchical grouping to optimize an objective function. *Journal of American Statistical Association* 58:236–244.

Wells, J. M., Boligala, R. C., Hung, H.-Y., Weisburg, W. G., Mandelco-Paul, L., and Brenner, D. J. 1987. *Xyella fastidiosa* gen. nov., sp. Nov: Gram negative, *Xylem* limited *fastidious* plant bacteria related to *Xanthomonas* spp. *International Journal of Systemic Bacteriology* 37(2):136–143.

White, T. J., Bruns, T., Lee, S., and Taylor, J. 1990. Amplification and direct sequencing of fungal ribosomal RNA genes for phylogenetics, in *PCR Protocols: A Guide to Methods and Applications*, eds. M. A. Innis, D. H. Geland, J. J. Sninsky, and T. J. White, pp. 315–322. San Diego: Academic Press.

Willems, A., Gillis, M., Kersters, K., Van den Broecke, L., and Deley, J. 1987. Transfer of *Xanthomonas ampelina (Panagopoulos* 1969) to a New Genus, *Xylophilus* gen. nov., as *Xylophilus ampelinus (Panagopoulos* 1969) comb. nov. *International Journal of Systemic Bacteriology* 37(4):422–430.

Williams, J. G. K., Kubelik, A. R., Livak, K. J., Rafalski, J. A., and Tingey, S. V. 1990. DNA polymorphisms amplified by arbitrary primers are useful as genetic markers. *Nucleic Acids Res.* 18:6531–6535.

Wollman, E. L., Jacob, F., and Hayes, W. 1956. Conjugation and genetic recombination in *E.coli* K 12. Cold spring Harbor Symp. *Quant. Biol.* 21:141–162.

Wood, S. A., Allen, N. D., Rossant, J., Auerbach, A., and Nagy, A. 1993. Non injection method for the production of embryonic stem cell embryo chimeras. *Nature* 365:87–89.

Yoo, S. M., Jong, H. C., Lee, S. Y., and Yoo, N. C. 2009. Applications of DNA microarray in disease diagnostics. *J. Microbiol. Biotechnol.* 19:51–54. doi: 10.4014/jmb.0803.226.

Yumlembam, R. A. 2011. *Evaluation of antibacterial property of medicinal and aromatic plants against plant pathogenic Xanthomonads*. MSc thesis. Rahuri, India: Mahatma Phule Agriculture University. p. 109.

Yumlembam, R. A., and Borkar, S. G. 2014. Assessment of antibacterial properties of medicinal plants having bacterial leaf endophytes against plant pathogenic *Xanthomonads*. *Indian Phytopath.* 67(4):353–357.

Appendix

A.1 CLEANING SOLUTION FOR GLASSWARE

SULFURIC ACID DICHROMATE

Sodium dichromate	25.0 g
Sulfuric acid (conc.)	1000.0 ml
Distilled water	50.0 ml

Dissolve the dichromate crystals in 50 ml of warm water. Allow to cool to room temperature and add acid slowly to the preparation.

This preparation can be used for the removal of residual organic matter from laboratory glassware. Glassware should be soaked in this solution for a number of days, then rinsed in running tap water at least 10 times, then rinsed twice in single distilled water and finally rinsed, once in double glass distilled water.

A.2 PREPARATION OF BIOCHEMICAL TEST REAGENTS

DIPHENYLAMINE REAGENT (FOR DETECTION OF NITRATES)

Diphenylamine	0.7 g
Conc. sulfuric acid	60.0 ml
Conc. hydrochloric acid	11.3 ml
Distilled water	28.8 ml

Dissolve the diphenylamine in a mixture of sulfuric acid and distilled water. Allow it to cool. Add hydrochloric acid gently. Allow to stand for 12 hours. Sedimentation indicates that the reagent is saturated.

KOVAC'S REAGENT (FOR DETECTION OF INDOLE)

P-Dimethylaminobenzaldehyde	5.0 g
Amyl alcohol	75.0 ml
Hydrochloric acid (conc.)	25.0 ml

Dissolve the dimethylaminobenzaldehyde in the amyl alcohol. Then add the hydrochloric acid to the above preparation. Store the reagent in a glass-stoppered bottle in refrigerator.

Motility Studies Reagent

Carboxy-methyl-cellulose	2.0 g
Sucrose (0.2 M)	98.0 ml
Distilled water	1000.0 ml

Prepare 0.2 M sucrose solution by adding 68.4 g of sucrose in 1000 ml of distilled water.
Dissolve the carboxy-methyl-cellulose in 100 ml of warm distilled water in a waring blender. Add 98 ml of sucrose (0.2 m) to the above. Make the total volume by the addition of more distilled water.

Nessler's Reagent (for Detection of Ammonia)

Potassium iodide	50.0 g
Distilled water (ammonia free)	25.0 ml
Mercuric chloride (saturated)	35.0 ml
Potassium hydroxide (50 percent aqueous)	400.0 ml

Dissolve the potassium iodide in 35 ml of distilled water and add saturated solution of mercuric chloride until a slight precipitate persists, add potassium hydroxide. Dilute to 1000 ml by addition of distilled water. Allow to settle for one week, decant supernatant liquid, and store in tightly stoppered brown bottles.

Nitrate Test Solutions (for Detection of Nitrites)

Solution A	
Sulfanilic acid	8.0 g
Acetic acid, 5 N*	1000.0 ml
Solution B	
Alpha-naphthylamine	5.0 g
Acetic acid, 5 N*	1000.0 ml

* To prepare 5 N, take 287 ml of concentrated acetic acid and adjust with distilled water to 1000 ml.

Tetramethyl-para-phenylenediamine dihydrochloride (for Oxidase Activity)

Tetramethyl-para-phenylendediamine dihydrochloride	5.0 g
Distilled water	50.0 ml

Prepare the solution of dihydrochloride in the distilled water immediately before use. If the preparation becomes darkened, discard it.

Trichloroacetic Acid (Gelatin Hydrolysis Test Reagent)

Trichloroacetic acid (c.p.)	5.0 g
Distilled water	100.0 ml

Dissolve the acid in the water with constant stirring.

Trommsdorf's Reagent (for Detection of Nitrite)

Zinc chloride solution (20 percent)	100.0 ml
Starch	4.0 g
Potassium iodide	2.0 g
Distilled water	100.0 ml

Prepare 100 ml of 20 percent aqueous $ZnCl_2$ solution and add slowly with constant stirring to a mixture of 4.0 g of starch in water. Heat until the starch is dissolved and solution becomes clear. Dilute with water and add potassium iodide. Make the volume to 1 L by addition of more water, filter and store in a brown-stoppered bottle.

VDRL Buffered Saline Solution (pH 6.0) (for Antigen Preparation)

Formaldehyde (neutral, c.p.)	0.5 ml
Sodium phosphate ($Na_2HPO_4.12H_2O$)	0.093 g
Potassium phosphate (KH_2PO_4)	0.170 g
Sodium chloride	10.0 g
Distilled water	1000.0

Dissolve all the constituents in distilled water. Check the pH of the resulting solution which should be 6.0. Store the reagent in screw-capped or glass-stoppered bottles.

MOTILITY AGAR

Peptone	10.0 g
Sodium chloride	5.0 g
Agar	3.5 g
Distilled water	1000.0 ml

A.3   PREPARATION OF STAINS

Acid Fast Stain (Carbol fuchsin [Ziehl's])

Solution A	
Basic fuchsin (90 percent dye content)	0.3 g
Ethyl alcohol (95 percent)	10.0 ml
Solution B	
Phenol crystals (c.p.)	5.0 g
Distilled water	95.0 ml

Dissolve the solutes of each solution in the respective solvents indicated. Then mix solutions A and B and allow to stand overnight. Filter the final preparation and store in a glass-stoppered bottle.

ACID-ALCOHOL

Ethyl alcohol (95 percent)	97.0 ml
Hydrochloric acid (37 percent)	3.0 ml

Add the acid to the alcohol slowly. Store the solution in a properly labeled, glass-stoppered bottle.

METHYLENE BLUE

Methylene blue (90 percent dye content)	0.3 g
Distilled water	100.0 ml

CAPSULE STAIN (CRYSTAL VIOLET [1 PERCENT])

Crystal violet (85 percent dye content)	1.0 g
Distilled water	99.0 ml
Copper sulfate ($CuSO_4.5H_2O$)	20.0 g
Distilled water	80.0 ml

CYTOPLASMIC MEMBRANE STAIN

Plasmolysis	
10 percent aqueous solution of potassium nitrate	
Bouin's Fixative Stain	
Saturated solution of picric acid	15 parts
Formalin	6 parts
Glacial acetic acid	1 part
Victoria Blue Stain	
1 percent aqueous solution of Victoria blue	1 part
Water	1.5 parts

FLAGELLA STAIN

Loeffler's Flagella Mordant	
20 percent aqueous tannic acid	100.0 ml
Ferrous sulfate	20.0 g
10 percent basic fuchsin in alcohol	10.0 ml
Distilled water	40.0 ml

Dissolve ferrous sulfate crystals in distilled water by warming and add the remaining ingredients.

LOEFFER'S FLAGELLA STAIN

1 percent basic fuchsia in alcohol	20.0 ml
3 percent aniline water	80.0 ml

FUNGAL STAINS

Water–iodine solution	
Gram's iodine (as in Gram's stain)	10.0 ml
Distilled water	30.0 ml

LACTO PHENOL-COTTON BLUE (STAIN)

Lactic acid	20.0 ml
Phenol crystals	20.0 g
Glycerol	40.0 ml
Distilled water	20.0 ml
Cotton blue (1 percent aqueous)	2.0 ml

Add lactic acid and glycerol/glycerin to the distilled water and mix thoroughly. Add phenol crystals and heat gently in hot water with frequent agitation until the crystals completely dissolve. Add the dye and mix thoroughly. Store the stain in a brown bottle.

GRAM STAIN (CRYSTAL VIOLET [HUCKER'S])

Solution A	
Crystal violet (90 percent dye content)	2.0 g
Ethyl alcohol (95 percent)	20.0 ml
Solution B	
Ammonium oxalate	0.8 g
Distilled water	80.0 ml

Dissolve crystal violet in ethyl alcohol and the ammonium oxalate in distilled water. Mix solutions A and B.

GRAM'S IODINE

Iodine	1.0 g
Potassium iodide	2.0 g
Distilled water	300.0 ml

Dissolve iodine and potassium iodide in distilled water.

ETHYL ALCOHOL (95 PERCENT)

Ethyl alcohol (100 percent)	95.0 ml
Distilled water	5.0 ml

SAFRANIN

Safranin (2.5 percent solution in 95 percent ethyl alcohol)	10.0 ml
Distilled water	100.0 ml

NEGATIVE STAIN (NIGROSIN)

Nigrosin (water soluble)	10.0 g
Distilled water	100.0 ml
Formalin	0.5 ml

Dissolve nigrosin in distilled water and immerse the mixture in boiling water bath for 30 minutes. Then add formalin (as preservative). Filter the solution twice through double filter paper. Store in small tubes.

NUCLEAR STAIN (GIEMSA STAIN)

Giemsa stain (powdered)	3.8 g
Glycerol	125.0 ml
Methyl alcohol (solvent)	100.0 ml

Dissolve the stain in methyl alcohol. Add glycerol and shake well in bottle with glass beads. Filter if necessary. Keep tightly closed and stoppered bottle at all times.

BACTERIAL SPORE STAIN

Malachite Green (5 percent)	
Malachite green	5.0 g
Distilled water	100.0 ml
Safranin	
Safranin 0 (2.5 percent solution in 95 percent ethyl alcohol)	10.0 ml
Distilled water	100.0 ml

Viability Stain

Loeffler's Methylene Blue	
Methylene blue	0.5 g
Potassium hydroxide solution (1 percent)	1.0 ml
Ethyl alcohol	30.0 ml
Distilled water	100.0 ml

Warm water to 50°C. Add methylene blue, stir well and then add other ingredients. Filter the contents.

Litmus Solution

Litmus granules	100.0 g
Ethyl alcohol (40 percent)	500.0 ml
Sodium hydroxide (1 N)	100.0 ml

Grind the litmus and transfer to a flask containing 250 ml of ethyl alcohol (40 percent). Boil the solution for 1 minute. Decant the liquid and add the rest of the ethyl alcohol (i.e., 250 ml) to the granules. Boil for 1 minute.

Decant the liquid and add it to the first extract. Make up the volume to 500 ml adding more of 40 percent ethyl alcohol. Add 1 N sodium hydroxide, drop-wise until the solution becomes purple.

To test the litmus solution, boil a test tube of distilled water and a test tube containing tap water. Add litmus solution, one drop to each tube. The tap water should turn blue and the distilled water, purple

Methyl Red Solution

Methyl red	0.04 g
Ethyl alcohol (absolute)	40.0 ml
Distilled water	60.0 ml

Dissolve methyl red in ethyl alcohol and add water.

A.4   CULTURE MEDIA

SOURCES OF READY-TO-USE CULTURE MEDIA

Dehydrated media are commercially available and the directions for the preparation of these from the dehydrated products are adequately described by the manufactures. To prepare them for laboratory use, glass distilled water is added and mixture is autoclaved according to the manufacturer's instructions. Suppliers of prepared (dehydrated) media include

Albimia Laboratories, Inc.
16 Clinton Street
Brooklyn, New York 11201

Difco Laboratories Inc.
920 Henry Street
Detroit, Michigan 48200

Baltimore Biological Laboratories
1640 Gorsuch Avenue
Baltimore, Maryland 21203

Sigma Chemical Company
P.O. Box 14508
St. Louis, MO 63178

Baired and Tatlock (London) Ltd.
Freshwater Road, Chadwell Heath
Essex, UK

Oxoid Ltd.
Southwark Bridge Road
London SE 1, UK

HiMedia Laboratories Pvt. Limited
23 Vadhai Industrial Estate
Mumbai 400 086, India

Index

T - #0889 - 101024 - C342 - 254/178/15 - PB - 9781032096001 - Gloss Lamination